LASERS AND ELECTRO-OPTICS RESEARCH AND TECHNOLOGY

NONLINEAR PERFORMANCE AND CHARACTERIZATION METHODS IN OPTICS

Lasers and Electro-Optics Research and Technology

Additional books in this series can be found on Nova's website under the Series tab.

Additional e-books in this series can be found on Nova's website under the e-book tab.

LASERS AND ELECTRO-OPTICS RESEARCH AND TECHNOLOGY

NONLINEAR PERFORMANCE AND CHARACTERIZATION METHODS IN OPTICS

JINGSONG WEI
EDITOR

nova publishers
New York

For permission to use material from this book please contact us:
Telephone 631-231-7269; Fax 631-231-8175
Web Site: http://www.novapublishers.com

Library of Congress Cataloging-in-Publication Data

ISBN: 978-1-62808-093-3

Library of Congress Control Number: 2013940380

Published by Nova Science Publishers, Inc. † New York

Contents

Preface		**vii**
Chapter 1	The Development and Application of Third-Order Optical Nonlinearities Measurement Techniques with Phase Object *Zhongguo Li and Yinglin Song*	**1**
Chapter 2	Theory and Applications of the Z-Scan Technique through Beam Dimension Measurements *George Tsigaridas*	**41**
Chapter 3	Z-Scan Technique Using Fresnel–Kirchhoff Diffraction Method *Ehsan Koushki*	**63**
Chapter 4	Measurements of Nonlinear Optical Parameters of Transparent and Nontransparent Materials Using Single-Shot Techniques *R. A. Ganeev*	**79**
Chapter 5	Pseudo-Nonlinearity and Elimination in Z-Scan Measurement *Rui Wang, Hui Yan and Jingsong Wei*	**95**
Chapter 6	Thermo-Optical Coefficient Measurement with Laser Pulse Width Tunable Z-Scan Method *Jingsong Wei*	**125**
Chapter 7	Z-Scan Technique Considering the Fresnel Reflection Loss *Jingsong Wei*	**147**
Chapter 8	Trapped Modes and Symmetry Breaking of Photonic and Atomic Waves in Rotating Double-Well Potentials *Yongyao Li, Wei Pang and Boris A. Malomed*	**171**
Chapter 9	Sharp Criterion of Global Existence for A Class of Nonlinear Schrödinger Equation *Chuang Xu and Runzhang Xu*	**195**
Chapter 10	Quenching Phenomena for Second and Fourth Order Nonlinear Parabolic Equations with Special Absorption Term *Junsheng Zhao, Xiaotong Qiu and Senyang Wu*	**213**
Index		**231**

PREFACE

Nonlinear optical phenomena, such as third harmonic generation, four-wave mixing, stimulated Brillouin scattering, self-phase modulation, multi-photon absorption, optical phase conjugation, etc., have been widely investigated and well understood since the invention of Laser. Some applications of the nonlinear optical phenomena have always been advanced in various fields, such as, optical information storage, all-light computing, optical switches, optical limiters, micro-nano-fabrication etc. The nonlinear optics has been developed into a subfield of optics, there is no indication that either the developments of nonlinear optics or its applications are slowing down. In nonlinear optics, the characterization and measurement techniques of the nonlinear index are critical for understanding the nonlinear physical process and developing new applications. In this book, different optical nonlinearity characterization and measurement methods are introduced and nonlinear problems are discussed theoretically.

Chapters 1 through 7 are about the optical nonlinearity characterization and measurement methods. The first chapter presents the third-order optical nonlinearity measurement techniques with phase object. The second chapter describes a modification of the well-known z–scan technique based on the direct measurement of the beam dimensions in the far field. The third chapter introduces a nonlinear index characterization method based on Fresnel–Kirchhoff diffraction integrals. The fourth chapter gives single–shot techniques for the measurement of the nonlinear optical refraction of the materials with limited and high transparency. The fifth chapter analyzes three kind of pseudo-nonlinearity and the corresponding elimination methods in the z-scan measurement. The sixth chapter is about the measurement of the thermal-optic coefficients with laser pulse width tunable z-scan method. The seventh chapter introduces an improved z-scan technique by considering Fresnel reflection loss.

Chapters 8 and 9 are some studies about the nonlinear Schrödinger equations. Chapter 8 introduces basic types of trapped modes and their spontaneous symmetry breaking phenomenology in nonlinear systems including the 1D rotating ring and both cases of the self-focusing and self-defocusing cubic nonlinearities. Chapter 9 deals with two problems: the first is to discuss the nonhomogeneous nonlinear Shrödinger equation with critical exponent, and the second is to discuss a class of nonlinear Shrödinger equations with combined power-type nonlinearities and harmonic potential.

Chapter 10 analyzes the quenching phenomena of the Cauchy problem for the second-order nonlinear parabolic equation on unbounded domain and fourth-order nonlinear parabolic equations with special absorption term.

I hope that this book can drive the nonlinear optics to continue to advance. This book is very useful for advanced undergraduates, graduate students and researchers working in related nonlinear optical fields.

Professor Dr. Jingsong Wei
Shanghai Institute of Optics and Fine Mechanics
Chinese Academy of Sciences
Shanghai 201800, China

In: Nonlinear Performance and Characterization Methods … ISBN: 978-1-62808-093-3
Editor: Jingsong Wei

Chapter 1

THE DEVELOPMENT AND APPLICATION OF THIRD-ORDER OPTICAL NONLINEARITIES MEASUREMENT TECHNIQUES WITH PHASE OBJECT

Zhongguo Li[1] and Yinglin Song*[1,2]

[1]Department of Physics, Harbin Institute of Technology, Harbin, PR China
[2]School of Physical Science and Technology, Soochow University, Suzhou, PR China

ABSTRACT

In this chapter, we review the recent progress in the third-order optical nonlinearities measurement methods, with a focus on results from our laboratory. The third-order optical nonlinearities measurement techniques with phase object are a kind of experimental tool for investigating nonlinear light-matter interaction on variable time scales. These techniques hold great promise for advances in the characterization of the structure-property relationship and fundamental physical processes in systems ranging from liquids to solid films, biological assemblies and novel nanomaterials. The history and theoretical underpinnings of the technique are briefly discussed. A few representative measurement methods including a 4f coherent imaging system with phase object (NIT-PO), single-shot transmittance measurement with phase object (T-PO) and time-resolved pump-probe with phase object (PO-Pump probe), are described and summarized in detail. This chapter is designed both for specialists in the field and general physics and chemistry readership.

Keywords: Nonlinear optics, measurement, phase object

* Corresponding author e-mail: ylsong@hit.edu.cn.

1. INTRODUCTION

Nonlinear optics, a subfield of optics, emerged after the invention of Laser. The demonstration of the second harmonic generation (SHG), made with a ruby laser in 1961 by Peter Franken at the University of Michigan [1], has led to a revolution in both science and technology. Since then, a large number of nonlinear optical phenomena such as the Third harmonic generation (THG), Four-wave mixing (FWM), Stimulated Brillouin scattering (SBS), Self-phase modulation (SPM), Multi-photon absorption, Optical phase conjugation, etc., had been proposed theoretically and discovered experimentally [2-13]. The development of nonlinear optics (NLO) has transformed the scientific field of optics and opens up a new perspective for previously undreamed of knowledge on the elementary light-matter interaction processes taking place in the world. Presently, nonlinear optics has become a highly advanced and intersectional science [14-16]. Also, the remarkable validity and versatility of nonlinear optical phenomena hold great promise for novel diverse applications in the fields of communication, biomedicine, energy resources, etc. Some of those well-established applications include novel light source, all-light computing, cancer diagnostics and solar cells, to name only a few [17-19]. Even after fifty years of advancement, there is no indication that either the developments of nonlinear optics or its applications are slowing down. Rather, the opposite is true.

The nonlinear optical measurement technique is an experimental tool for investigating and characterizing the nonlinear optical properties of materials, mostly in the condensed phase, on variable time scales. These techniques can provide a wealth of information on the photo-physical properties and dynamic processes of systems after strong light excitation, which helps us to understand the structural-properties relationship of different materials [20-22]. The power of nonlinear optical measurement techniques relies on their sensitivity and assessment to the material's properties in a way usually inaccessible by other linear optical techniques. With the rapid and enormous progress of laser technology during the last 50 years [23], modern state-of-the-art nonlinear optical measurement techniques enable us to visualize, trace and manipulate the microscopic light-matter interaction process in all forms of matter to the benefit of mankind [24-27].

The nonlinear optical effect is induced by macroscopic nonlinear polarization in the materials under strong light. The light-induced nonlinear polarization is related to a variety of microscopic parameters and can give way to a wide range of distinct nonlinear optical phenomena. In general, any nonlinear optical phenomena could be used to characterize the material's nonlinear properties per se. Therefore it is reasonable to accept the prospect of nonlinear optical measurement techniques. It would have been an impossible task to review even an insignificant part of it.

In this chapter, we provide a comprehensive review on fundamentals, advancements and potential applications of a novel type of third-order nonlinear optical measurement technique, using phase object. The discussion will be mainly descriptive, attempting to communicate the principles without getting mired in details, either theoretical or experimental. Admittedly, the subjects included reflect the research interests and the expertise of the author's group. We have made a conscious decision not to include such important and widely employed techniques as z-scan. There is extensive literature devoted to the z-scan method. This includes both original publications and book chapters - see, e.g., Refs. [28-42]. This field, in itself, is large enough to warrant its own dedicated review. Another important subject that we do not include in this review is the Optical kerr effect (OKE) method - there is extensive literature and excellent reviews on this subject - see, e.g., Refs. [43-46]. We only briefly touch upon these techniques in Section 3. Also, due to limitations of time and space we do not cover the Third harmonic generation (THG). Another important class of measurement technique that we leave mostly outside of this review article is the Four wave mixing (FWM). There exist a significant number of reviews on the fundamentals of four wave mixing techniques and its variants [47-51]. We only introduce the principles of FWM technique in section 3, in a brief discussion. We hope that our selection and description will be of benefit to readers of optics and material science.

This chapter is organized as follows: Section 2 gives a general introduction on nonlinear optics and a historical background of Zernike phase contrast microscopy. Then a brief description of the theoretical basis for the optical nonlinearities measurement technique is reported. In section 3 we address the configurations and principles for various nonlinear optical measurement methods using phase contrast including static single-shot measurement techniques and transient nonlinear dynamics measurement methods. The Experimental demonstration and procedures are outlined when necessary, and their unique features compared to other techniques are summarized in detail. Finally, section 4 describes the current results and suggests future potential applications of this kind of technique.

2. Background

2.1. Introduction of Nonlinear Optics

Nonlinear optics involves nonlinear interactions between the optical field and matter. At present, there are extensive books and excellent reviews on nonlinear optics research. Due to limitations of time and space we could not cover this subject systematically, but only give a brief introduction to the principle.

Nonlinear optics is based on nonlinear polarization induced by sufficiently intense light (usually laser), which are no longer scales linear to the electric amplitude of incident light wave. According to the electromagnetic theory, the motions of the light wave are described by the classical Maxwell equations:

$$\begin{cases} \nabla \times E = -\dfrac{\partial B}{\partial t} \\ \nabla \times H = J + \dfrac{\partial D}{\partial t} \\ \nabla * D = \rho \\ \nabla * H = 0 \end{cases} \tag{1}$$

where the electric displacement is $D = \varepsilon_0 E + P$ and P is the macroscopic polarization within the media induced by incident electric field E.

In linear optics, the macroscopic polarization depends linearly on the electric field:

$$P = \varepsilon_0 \chi^{(1)} E \tag{2}$$

where $\chi^{(1)}$ is the linear susceptibility. All of the optical properties of a material (absorption, refraction, reflection, etc.) are associated with linear polarization P_L.

However, when the incident light intensity exceeds a threshold value, the expression above is no longer valid. Now we can expand the macroscopic polarization in a power series of the incident electric field [52]:

$$P = P_L + P_{NL} = \varepsilon_0 \chi^{(1)} E + \varepsilon_0 \chi^{(2)} EE + \varepsilon_0 \chi^{(3)} EEE + ... \tag{3}$$

where $\chi^{(2)}$ and $\chi^{(3)}$ are the second-order nonlinear optical susceptibility and third-order nonlinear optical susceptibility, respectively. Because the electric field E and polarization P are vector, the nonlinear optical susceptibility $P_{NL}^{(n)}$ is a tensor. For a material with inversion symmetry, all even orders of $P_{NL}^{(n)}$ vanished. Thus the lowest order optical nonlinearity that occurs in all media is due to $P_{NL}^{(3)}$. Third-order nonlinear optical susceptibility $P_{NL}^{(3)}$ is a fourth rank tensor with 81 elements, but the number of non-zero elements could be greatly reduced by the symmetry of material based on the point group theory.

Furthermore, because the optical field could be characterized by a variety of parameters, such as wavelength, frequency, polarization and wave vector, there are a large number of nonlinear interactions involved in the same order of nonlinear polarization. Thus, the nonlinear optical susceptibility has many different forms. Let us consider the third-order nonlinear polarization in a material irradiated with a monochromatic plane wave [53]:

$$\begin{aligned} P_{NL}^{(3)} &= \frac{1}{4}\varepsilon_0 \chi_{ilkm}^{(3)} (E_{\omega_1}^3 e^{i3\omega_1 t} + c.c) + \frac{3}{4}\varepsilon_0 \chi_{ilkm}^{(3)} (|E_{\omega 1}|^2 E_1) + \\ & \frac{3}{4}\varepsilon_0 \chi_{ilkm}^{(3)} [\frac{1}{2}(E_{\omega 1}^2 E_{\omega 2} e^{i(2\omega 1 + \omega 2)t} + c.c)] + ... \end{aligned} \tag{4}$$

where the indices *i, l, k* and *m* denote the Cartesian components of the nonlinear polarization and the incoming light fields, respectively. The electric component of the monochromatic plane wave is expressed as

$$E(r,t) = \frac{1}{2}\sum_{k=1}^{3}(E_{\omega_k k}(r,\omega_k)e^{iw_k t} + c.c)$$

From Equation (4), it could be seen that the nonlinear polarization has many expressions depending on the frequency of the input light wave, and for each expression there are many different non-zero elements denoted by its Cartesian indices.

In many cases, we do not only care about the value of third-order nonlinear polarization but also its time response. In some literature, the nonlinear polarization is written as a convolution of electric fields with the nonlinear response function of media [54]:

$$P_{NL}^{(n)}(t) = \int_0^\infty dt_n \int_0^\infty dt_{n-1} \ldots \int_0^\infty dt_1 * E(t-t_n)E(t-t_n-t_{n-1}) \ldots E(t-t_n-\ldots-t_1) * S(t_n, t_{n-1}, \ldots, t_1) \quad (5)$$

where $S(t_n, t_{n-1}, \cdots t_1)$ is the nonlinear response function of the sample material. The nonlinear response function is defined for positive times only according to the causality. The physical meaning of this expression is rather straightforward: the interactions of light fields with material at time 0, $t_1 \cdots$ and $t_{n-1} + \cdots + t_1$ generate a nonlinear polarization, which at time $t_n + t_{n-1} + \cdots t_1$ acts as a source of radiation to induce the nonlinear optical phenomenon.

The aim of the nonlinear optical measurement technique is to characterize the sign, magnitude, dispersion and evolution of nonlinear polarization for investigating nonlinear light-matter interactions in various phases. These techniques can be performed in either the time domain or the frequency domain. In this review, we are mainly concerned about the time domain nonlinear optical measurement technique.

2.2. Zernike Phase Contrast Microscopy

The image and visualization of unstained objects that are transparent to the light wave, such as living cells, small organisms, tissue sections, fibers, and so on, is a challenging yet often vital requirement in interdisciplinary research such as biology and optics. These specimens are called *phase objects* because they do not absorb light but slightly retard the phase of the light propagated. However, conventional optical microscopy such as bright field microscopy is unable to detect these phase differences, because our eyes as well as the CCD camera are only sensitive to light intensity. So the acquired image of bright field microscopy is lacking in contrast and the structural details of the sample are almost invisible. Therefore, a technique that can analyze optical phase disturbance is required.

Frits Zernike, a Dutch physicist and mathematician, developed the phase contrast microscopy that can produce high-contrast images of transparent objects in the 1930's. Because the phase contrast technique could convert minute phase modulations of the sample into corresponding intensity modulations in the image, structural details could be viewed with high resolution without staining or killing the sample. Because of this huge advantage, this technique gained widespread applications in biological and material research soon after its discovery and promoted Zernike's award of the prestigious Nobel Prize in physics in 1953 [55-57].

The Zernike phase contrast microscopy exists in many different forms such as the point diffraction, dark central ground filtering, and field absorption methods. However, one unique component exists in all of forms of Zernike phase contrast microscopy: the phase plate. This device distinguishes the phase contrast microscopy to the conventional microscope. A schematic representation of the operation of a full-field Zernike phase contrast microscopy is illustrated in Figure 1. The phase ring is placed at the objective's back focal plane (Fourier spectrum plane).

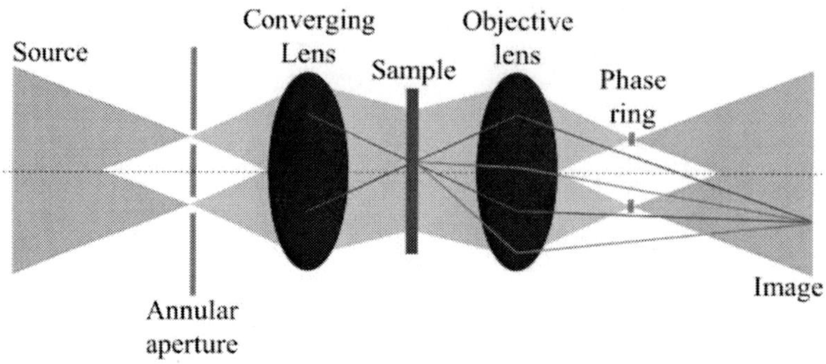

Figure 1. Re-plotting of schematic of the Zernike phase contrast microscopy (Ref. [58])

The beam, illuminated by a light source (usually a tungsten-halogen lamp), is directed through a specialized ring annulus, which is placed at the front focal plane of the condenser. Wavefronts passing through the annulus illuminate the specimen and are collected by an objective lens placed after it. The light illuminating the specimens could be divided into two parts: one part is the zeroth-order undiffracted wavefront, which passes around the specimen undeviated (referred to as the surround (S) wave, red line), and the other is the deviated part that passes through the specimen and is diffracted (diffracted (D) wave, green line). Because different regions of the specimen have different refractive indexes, phase differences will be introduced in D wave passing through the specimen. Therefore D waves carry the information about the specimen's internal fine structure. Under the conditions of the Köhler illumination, S waves that do not interact with the specimen are focused as a bright ring in the rear focal plane of the objective, which is conjugated to the ring annulus placed at the condenser's front focal plane. The D wave diffracted by the specimen passes through the objective and spreads over the entire rear focal plane of the objective. Because these two waves arrive at different locations on the rear focal plane, selective altering of these different wave's phase and/or amplitude is possible. A specially designed phase plate is positioned in the objective rear focal plane which interacts with the S wave. The ring area of the phase plate where the S wave passes through is optically thinner or thicker than the rest of the plate, so that the S wave, after the phase plate, is phase-shifted. At the far field image plane, the phase-contrast image is formed by the interference of the phase-shifted S wave with the undisturbed diffracted D wave, which is carrying the structure information of the specimen. Based on the theory of interference of light, the D wave and the S wave can either interfere destructively or constructively based on their phase differences. The phase of the S wave agrees with that of the D wave, the image brightness will be enhanced; or the brightness will be decreased. So the phase modulations caused by the specimen's structure could be translated into intensity modulations in the image. This method offers a means, within the small phase variations limit, of relating the image brightness to the magnitude of the optical path variation (change due to the refractive index difference of transparent specimen).

In Zernike phase contrast microscopy, the intensity of an image is dependent on a variety of factors including the degree of phase advancement or retardation at the phase plate, the relative sign of this phase shift, and so on. The design and manufacture of the phase plate is of paramount importance for the sensitivity of phase contrast microscopy. The relative image intensity differences could be further enhanced if the amplitude of the S wave is decreased at the same time when its phase is altered (this can be done by suitable absorbing material). The latest phase contrast microscopes are equipped with a highly sophisticated phase plate so it can enable specimens such as protein molecules to be detected.

The idea of Zernike phase contrast microscopy can be symbolized in the vector diagram using a system of polar coordinates [59]. The schematic representation of the light field in conventional bright-field microscopy and phase contrast microscopy are shown in Figure 2(a) and (b), respectively.

In this system, the electric field component of illuminating light field is expressed as a vector in the complex plane, and the horizontal axis and vertical axis represent the amplitude and phase of the input light wave, respectively. The radius of the vector represents the norm of a light wave, which is related to the light intensity. Because the sample in phase contrast microscopy has no absorption, the vector radius remains unchanged, only the angle of the vector is rotated, relative to the phase alteration caused by the specimen or phase plate.

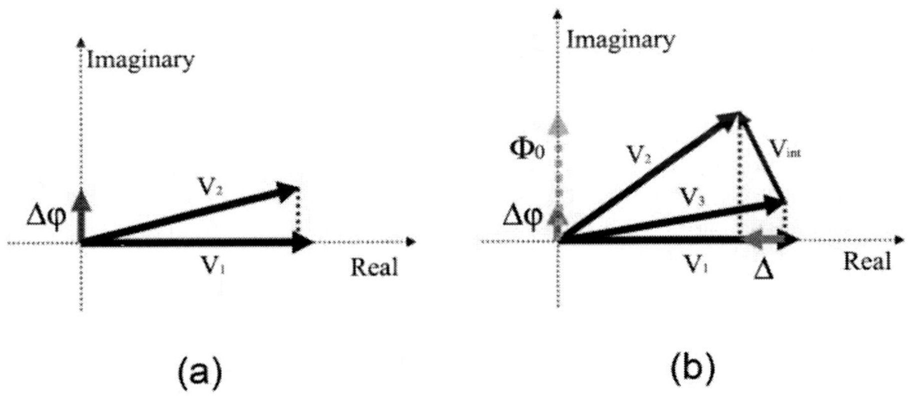

Figure 2. Schematic vector diagrams for the light wave in conventional microscope (a) and Zernike phase contrast microscope (b).

The initial light wave illuminating the transparent specimen is shown as vector 1 (V_1), which is lying in the horizontal axis. Because the thickness or refractive index of the specimen has slight variations due to the internal fine structure, there is a relative phase shift ($\Delta\varphi$: red arrow in Figure 2) between the light wave passing through the specimen and the wave that is not, due to the optical path difference. Usually, the phase shift caused by the specimen is very small, typically $\pi/3$ for individual cells in tissue culture. Thus the light wave at the image plane (V_2) will rotate counterclockwise or clockwise relative to V_1 due to the phase variation. Therefore the projection length of V_2 on the real axis changes synchronously. Detection of the specimen image depends on the relative differences between the projection value of V_2 and V_1 on the real axis. If the projection value of V_2 is significantly different from that of V_1, then the specimen acquires a large amount of contrast and becomes visible in the microscope eyepieces. Otherwise, the specimen remains transparent and invisible to our eyes or the CCD camera.

In conventional microscopy, as shown in Figure 2(a), there is no phase plate to modulate the phase of light wave. Due to the small phase alteration caused by the specimen, the vector V_2 rotates very slightly relative to vector V_1. Therefore the amplitudes of the projection length of V_2 and V_1 are nearly the same, so the acquired image is completely low in contrast and the transparent specimen is almost invisible.

In the phase contrast microscope, the light wave is seen after the specimen is distinguished into an S wave (V_1) and the D wave (V_3). The phase of the D wave is altered by the variation of thickness or refractive index of the specimen. After passing through the phase plate with fixed phase delay ϕ_0, the S wave (V_1) is transformed into $V_{2,}$ which rotates counterclockwise relative to the real axis. At the image plane, the phase modulated S wave (V_2) will interfere with the D wave (V_3), creating an interference vector (V_{int}). The output image's contrast will depend critically on the variation of projection length of V_{int} on the real axis (Δ, violet arrow in Figure 2(b)). Because the phase plate to the S wave V1 induces the additional phase delay, the corresponding amplitude variation is much larger than that in a conventional microscope. Hence the generated image acquires lots of contrast, and the specimen becomes visible relative to the background.

2.3. Nonlinear Optical Measurement Using Phase Object

The Zernike phase contrast technique is operated by transforming the invisible phase alteration caused by the specimen into the measurable amplitude variation of the output image. In conventional Zernike phase contrast microscopy, the phase alteration is usually modulated by the variation of the thickness or differences in the linear refractive index between the specimen and its surroundings. However, there are other mechanisms that can give a phase disturbance to the input wave front. One example is the nonlinear refraction which is caused by the nonlinear polarization under strong laser irradiation. The generalization of the Zernike technique is also valid with this kind of phase alteration mechanism. Thus it is workable, in theory, to characterize the nonlinear refraction properties of sample materials using the Zernike phase contrast technique.

The idea of measuring optical nonlinearities using the phase contrast principle was first proposed and developed by the research group of Georges Boudebs in Universite d'Angers, France. The schematic setup of a typical configuration for an optical nonlinearities measurement system using the phase contrast principle is shown in Figure 3. The main optical system consists of two convex lenses arranged in a confocal 4*f* system. This system can be classed as common path interferometry where the signal and reference beam travel along the same optical axis and interfere at the output plane of the system. Similar to the conventional Zernike phase contrast microscopy, there is a phase plate in the system. However, contrary to the original Zernike phase contrast technique, the phase plate is arranged at the object plane of the 4*f* system, which will perturb a portion of the light wave passing through the phase plate with a fixed phase delay. Therefore there are two waves leaving the phase plate: an unperturbed wave and a phase altered wave (reference). Both waves are focused by a convex lens to the nonlinear sample located at the rear focal plane (Fourier plane). Within the sample, the nonlinear refraction induced by a concentrated laser beam will modulate the phase of the incident light wave, where the sample could be treated as a phase filter. This phase modulation will give rise to interference between the unperturbed wave and the coaxial reference wave, leading to an amplitude variation of the reference beam. Thus the intensity profile of the image at the output plane changes relatively, allowing visualization of the nonlinear refraction property of the sample materials.

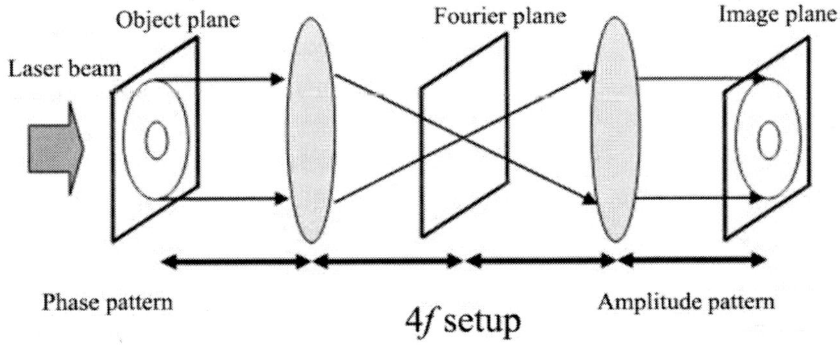

Figure 3. A schematic representation of the nonlinear imaging measurement system based on a confocal $4f$ system.

As in the Zernike phase contrast microscopy, the design of the phase plate is of paramount importance for the maximization of measurement sensitivity of the technique. The development and maturity of the optical nonlinearities measurement method using the phase contrast principle is a process of improvement and optimization of phase plate. Early in 1996, soon after proposition of the idea of measuring optical nonlinearities using a phase contrast principle, two kinds of phase plates: two slits [60] and microscopic rectangular objects (including phase or/and amplitude rectangular objects) [61], were introduced and verified experimentally or numerically, respectively. Later, the circular aperture as the phase plate and the direct Gaussian laser beam illumination were proposed and evaluated experimentally [62]. The experiments with these kinds of phase plates could verify the magnitude of a third-order nonlinear refractive index but could not distinguish the sign of it. Therefore these phase plates are far from ideal and a better phase plate design was required. Finally, in 2004, the bi-layer phase plate in which a transparent dielectric film was coated on a glass plate inside a circular aperture was developed and validated experimentally, which can measure the sign and magnitude of the third-order nonlinear refractive index simultaneously [63]. This discovery marks the maturity of this kind of nonlinear measurement technique.

Since then, the research on the nonlinear optical nonlinearities measurement technique using phase contrast has been progressing rapidly. A variety of modified measurement configurations were proposed and verified experimentally. An approximate analytical formula for the intensity profile at image plane was given by J.-L. Godet, and the measurements of phase plate design for optimizing the experimental sensitivity was given. In 2006, Li *et al* extended this measurement technique to the characterization of samples with nonlinear absorption and nonlinear refraction synchronously. Later, the optimal parameters of phase plate for both nonlinear absorption and refraction measurements were calculated numerically. In 2010, a detailed theoretical description for calculating the nonlinear refractive coefficient analytically was proposed by shui *et al* [64-71].

Due to space limitations, we do not present extensive discussions on details of the related theory and experiments. To compensate for the lack of information, the reader who is interested in this may go to references for a deeper understanding.

3. Nonlinear Optical Measurement Techniques with Phase Object

Materials with large optical nonlinearities and fast response time have the potential to meet the increasing demands in fields such as information processing, data storage and laser protection. A variety of materials such as organic polymers and semiconductors are found to be well-suited to such applications [72-79]. Although there have been considerable research achievements in finding the candidate material for future photonic devices, there are still several problems that must be overcome to enable full exploitation of nonlinear optical materials: (1) characterizing the overall nonlinear optical properties of a large number of materials, including nonlinear absorption and nonlinear refraction; (2) understanding the microscopic physical process contributing to the nonlinear optical response, including the light-matter interaction mechanism and their response time. Therefore the development of nonlinear optical measurement techniques is critical in nonlinear optics research. Nonlinear optical measurement techniques are typically classified as static measurement techniques or transient measurement techniques. The former is concentrated in evaluating the nonlinear optical coefficient of materials, such as two-photon-absorption (TPA) coefficient or nonlinear refractive index. The latter concerns the dynamic response that takes place in the materials after laser irradiation.

3.1. Static Optical Nonlinearities Measurement Techniques

The static optical nonlinearity measurement techniques are simple but effective tools for recording the third-order nonlinear optical properties of materials with high accuracy and signal-to-noise ratio. The aim of these techniques is expanding the database of candidate materials for future photonic and optic-electronic devices such as all-optical switches and optical limiters. The materials that have been studied include simple liquids, polymer solutions, organic semiconductor thin-films, bulk and nanostructure inorganic semi-conductors, etc.

In this section, the principles and variety of modifications of the optical nonlinearity measurement methods using the phase contrast principle are outlined and summarized. We begin with an in-detail description on theory of the 4*f* coherent imaging measurement technique with phase object. This is followed by a brief introduction on other nonlinear measurement techniques, such as the single shot transmittance measurement technique with phase object, and a modified z-scan method with phase object. Finally, the unique features of these measurement methods compared to other techniques are highlighted and summarized.

3.1.1. 4f Coherent Imaging Measurement Technique with Phase Object

In this section, we introduce the 4*f* coherent imaging measurement technique with phase object (4*f* NIT-PO or 4*f* method for short) in the first place. There are two reasons for this. First, the 4*f* method is the first nonlinear optical measurement method using phase contrast. The theoretical principle of the 4*f* method is also the backbone of other static optical nonlinearity measurement methods. Second, the feature of the 4*f* method also shows the representative characteristics of static optical nonlinearity measurement methods comparing them to other multiple-shot measurement techniques, such as z-scan.

The schematic diagram of the phase object (PO) used in the $4f$ imaging method is shown in Figure 4. It is prepared by depositing a layer of transparent dielectric in the center of an aperture on a flat glass plate. The central transparent dielectric layer will induce phase retardation to the input laser beam. Because the size of the aperture is small compared with that of the input laser beam, the output beam could be considered a top-hat beam.

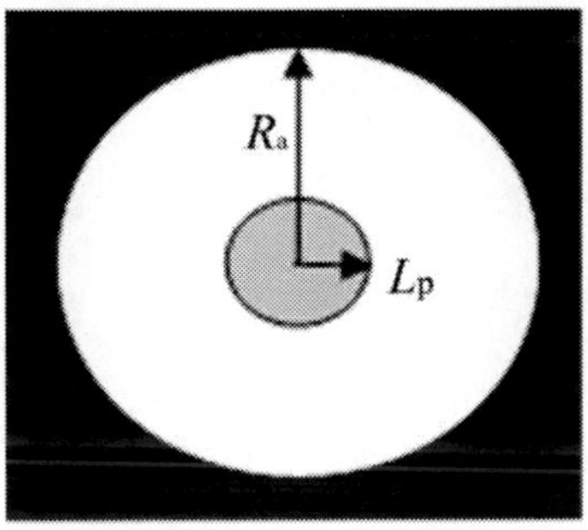

Figure 4. Schematic configuration of the phase object with circular aperture. The outside black area is opaque. L_p and R are the radius of phase object and aperture, respectively.

The experimental setup of the $4f$ imaging system with phase object (PO) is shown in Figure 5. The input laser illuminates on an aperture with a phase object (PO). Then the beam is divided into two beams: an intense beam and a weak one. The intense arm of the optical setup is a $4f$ nonlinear imaging system for measuring the optical nonlinearities. And the other arm is the reference, which is used to monitor the energy fluctuation of the laser pulse. The sample is arranged at the Fourier plane of the $4f$ system, and a CCD camera is located at the output plane of the $4f$ system to record the image.

A linearly polarized plane wave field illuminates the phase object: $E = E_0(t)\exp[-i(\omega t - kz)] + \text{c.c.}$, where ω is the angular frequency, k is the wave vector in free space, and $E_0(t)$ is the amplitude of electric field of light wave. We define the transmittance function of the PO with an aperture as $t_p(x,y)$, so the amplitude of the electric field after the PO is $O(x,y,t) = E * t_p(x,y)$. The field at the rear focal plane of the convex lens L_1 is the spatial Fourier transform of $O(x,y,t)$:

$$S(u,v) = \frac{1}{\lambda f_1} FT[O(x,y)] = \frac{1}{\lambda f_1} \iint O(x,y) \exp[-2\pi i(ux + vy)] dxdy \tag{6}$$

where FT denotes the spatial Fourier transform operation, $u = x/\lambda f_1$ and $v = y/\lambda f_1$ are the spatial frequencies in the focal plane, λ is the wavelength of the laser pulse, and f_1 is the focal length of lens L_1.

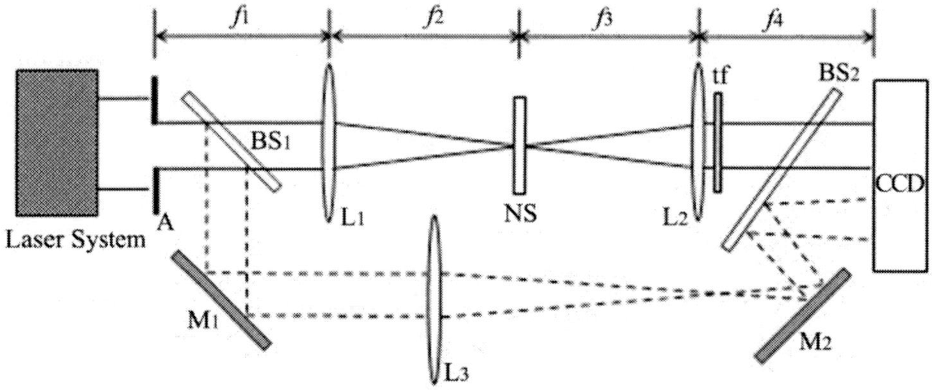

Figure 5. Schematic setup of the 4*f* coherent imaging system with a phase object (NIT-PO). A is the phase object with aperture. NS is the sample.

The input electric field before the sample is $S(u,v)$, and the peak intensity I of the laser beam is proportional to $|S(u,v)|^2$. If the sample thickness is less than the Rayleigh length of the beam, then the sample could be regarded as "thin". Then the beam propagation equations for the phase and the irradiance within the sample could be described using the slowly varying envelope approximation (SVEA) as follows:

$$\begin{cases} \dfrac{dI}{dz'} = -(\alpha_0 + \beta I) \\ \dfrac{d\Delta\phi}{dz'} = kn_2 I \end{cases} \tag{7}$$

where $\Delta\phi$ represents the nonlinear phase change and z′ is the propagation length in the sample. α_0, β, and n_2 are the linear absorption coefficient, non-linear absorption coefficient and the effective nonlinear refractive index of the sample, respectively. The field at the exit surface of the sample can be written as:

$$S_L(u,v,t) = S(u,v,t)\exp(-\alpha_0 L/2)[1+q(u,v,t)]^{(ikn_2/\beta-1/2)} \tag{8}$$

where $q(u,v,t) = \beta I(u,v,t)L_{eff}$, $L_{eff} = [1-exp(-\alpha_0 L)]/\alpha_0$, L is the thickness of the sample.

At the exit plane of 4*f* system, the light intensity recorded by the CCD camera is expressed as:

$$I_{im}(x,y,t) = |U(x,y,t)|^2 = |FT^{-1}[S_L(u,v,t)H(u,v)]|^2 \tag{9}$$

where FT^{-1} denotes the inverse spatial Fourier transform, and $H(u,v) = circ[(u^2 + v^2)^{1/2}\lambda G/NA]$ is the coherent optical transfer function of the 4*f* system. The function $circ(\rho)$ is defined to be 1 if the radius is $\rho(u,v)\leq 1$, and 0 otherwise. N_A and G are the numerical aperture and magnifications of the 4*f* nonlinear imaging system, respectively.

As the image recorded by the CCD camera is the fluence distribution of the laser beam, this image can be expressed by the integration of the intensity:

$$F(x,y)=\int_{-\infty}^{+\infty} I_{im}(x,y,t)dt \tag{10}$$

Considering a top-hat beam experimental setup used in our experiment, the transmittance of the aperture is defined as:

$$t_a(x,y)=circ[(x^2+y^2)^{1/2}/R_a] \tag{11}$$

where R_a is the radius of the aperture. So the overall transmittance function of the phase object can be given by:

$$t_p(x,y)=t_a(x,y)\exp\{i\phi_L circ[(x^2+y^2)^{1/2}/L_P]\} \tag{12}$$

Using the equations above, the optical nonlinearity parameters of the sample can be evaluated by simulating the experimental images recorded by CCD camera numerically.

In the 4*f* experiment, three images need to be recorded for evaluating the sample's optical nonlinearities, including the linear image, nonlinear image and image without sample. (i) Linear image: this image is obtained at low excitation intensity by placing high-density neutral attenuation filters (tf) before the sample material. (ii) Nonlinear image: this image is acquired at high excitation intensity by placing the same neutral filters after the sample. (iii) Image without sample: this is acquired by removing the sample while the position of the neutral filter remains unchanged. By integrating all pixels of the linear image, linear transmitted energy of the laser pulse E_l is obtained. In the same way, the nonlinear transmitted energy of the laser pulse E_{nl} and the input laser pulse energy E_i can be acquired. The linear transmittance of the sample material is determined by $T_l = E_l/E_i$, and the nonlinear transmittance of the material is given by $T_{nl} = E_{nl}/E_i$. If the nonlinear transmittance is determined, the nonlinear absorption coefficient β can be deduced unambiguously. Then, the effective nonlinear refractive index n_2, the only parameter not known in Equation (8), can be evaluated by fitting the nonlinear image.

The fluence distribution of the image at the output plane of the 4*f* system, simulated numerically, is shown in Figure 6(a). Figure 6(b) gives the beam profile of the image at polar coordinates. Here we define the difference between the central PO area and annular ring area as the value of normalized transmittance ΔT. Based on the phase contrast principle, the normalized transmittance is positive when the nonlinear phase shift caused by the sample material is positive (positive phase contrast), or else normalized transmittance would be negative. Within the small phase shift limit, the amplitude of the normalized transmittance will scale linearly with the variation of the nonlinear refractive index. Therefore the sign and magnitude of the nonlinear refractive index of the sample can be determined simultaneously. The nonlinear absorption coefficient of the sample can be evaluated by comparing the fluence integration of the total beam area under nonlinear and linear conditions.

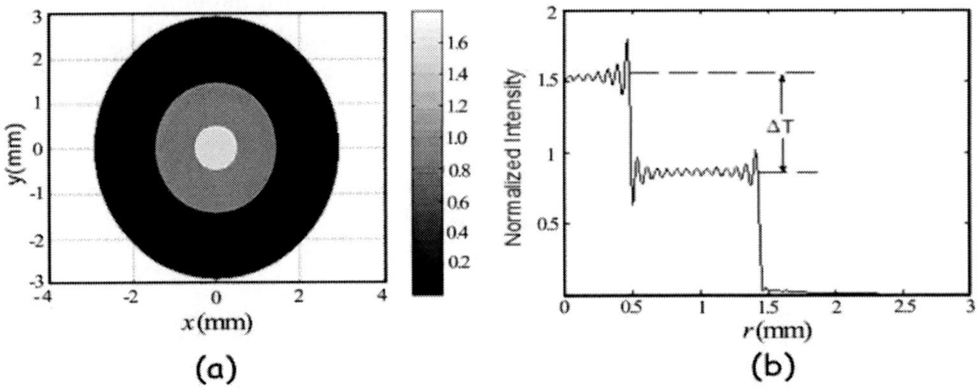

Figure 6. (a) Numerical simulation of a nonlinear image recorded by a CCD camera when the nonlinear refraction of the sample is positive. (b) Beam profile of the nonlinear image. The normalized transmittance is defined as the difference between the intensity value of the PO area and outside ring.

The $4f$ method achieves optical nonlinearities by distinguishing the beam profile difference on different regions of the output image. Thus this measurement technique can be classified as a beam distortion measurement method and the profile of the input laser beam is an essential parameter to the method's availability. A conventional $4f$ method assumes a Top-hat beam illumination after propagating through the PO with aperture. However, this assumption is not always true. For a laser emitted form laser source with alterable wavelength (such as optical parametric oscillator), the beam profile of the output laser is very complex and far from the ideal Gaussian distribution. It is difficult to get a top-hat beam after the phase object. In this case, the measured image has a lot of unwanted noise and the determination of nonlinear optical coefficients of the sample is rather difficult or even impossible.

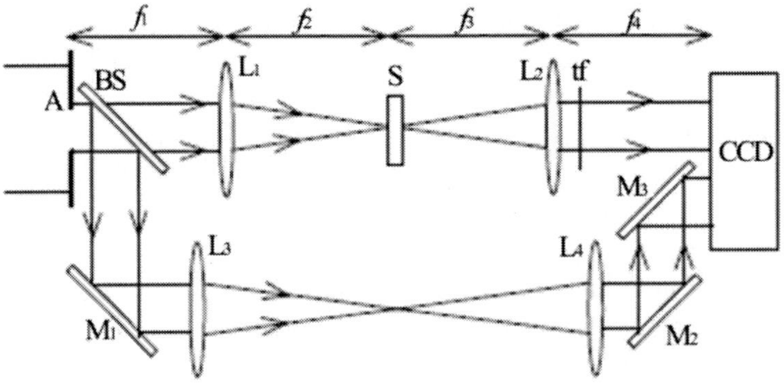

Figure 7. Schematic configurations of double $4f$ coherent imaging system with phase object. The reference path is also set up into a confocal $4f$ system.

Our research group had developed a modified $4f$ technique that can evaluate the nonlinear optical properties of material under a non-ideal Top-hat beam illumination, which greatly expands the practicality of this method. This is achieved by setting the reference path in a conventional $4f$ measurement system into another $4f$ system, just like the nonlinearity path. In this case, the reference path could record the energy fluctuation and beam profile simultaneously. By setting the reference image recorded by the CCD camera as the initial value of the numerical simulation, the beam profile disturbance can be eliminated. A schematic setup of the modified double $4f$ measurement technique is shown in Figure 7 [80], and the experiment demonstration is shown in Figure 8.

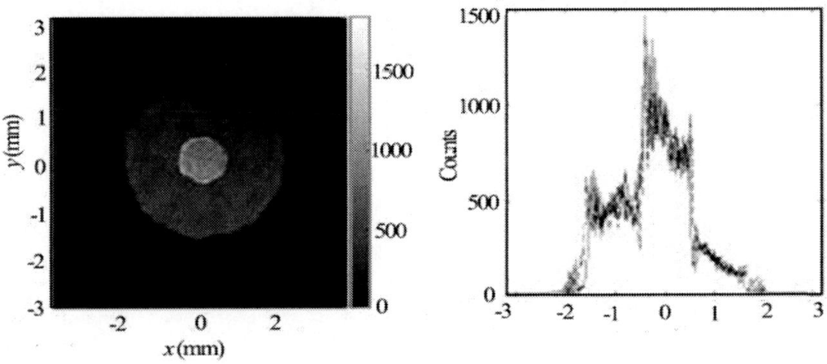

Figure 8. Experimental demonstration of the double $4f$ nonlinear imaging system with phase object. The left is the nonlinear image acquired experimentally, the right is the beam profile of the image (solid line) and theoretical fit (dash line).

3.1.2. Single-Shot Optical Nonlinearities Measurement with a Phase Object

In the above section, we introduced the theory and configuration of the $4f$ coherent imaging measurement technique with phase object. In the $4f$ method, the main optical system is a con-focal $4f$ system, and a CCD camera records the output image. Although this arrangement is proved to be effective and the fitting process is straightforward, there are certain limitations of this measurement method. First, the requirement for a CCD camera significantly increases the complexity and cost of the technique. Second, because the data fitting process is accomplished by integrating all image pixels recorded by the CCD camera, the recording speed is rather slow and it takes a lot of computer time to fit a result. This seems to be nothing serious for a single-shot measurement, but it seriously limits the practicality of the $4f$ method when hundreds of samples need to be measured or multiple shots should be taken on a sample alternately (which is the case in a time-resolved experiment). Last but not least, because the signal-to-noise ratio of the $4f$ measurement depends critically on the sensitivity of the CCD camera, this method is not suitable to measure those samples having a weak nonlinear optical response. From the perspective of practical applications, a much simpler, cheaper and faster optical nonlinearity measurement technique than the $4f$ method is needed.

The single shot transmittance measurement technique with phase object (T-PO) is an improved single-shot optical nonlinearity measurement technique that is based on the $4f$ nonlinear imaging method. The configuration of the transmittance measurement technique with a phase object is shown in Figure 9 [81].

Similar to the $4f$ method, a phase object (PO) is arranged in front of a convex lens, and the sample is located at the rear focal plane of the lens. One part of the input beam is divided by the beam splitter to a detector in order to monitor the energy fluctuation of the laser pulse. After the sample, the transmitted beam propagates to the far field where an aperture is arranged in front of the detector. The radius of the aperture is equal to the size of the diffraction pattern irradiated by the beam transmitting through the PO's central dielectric layer. The ratio of D_2/D_1 is measured under two different conditions: the linear condition (low laser irradiance) when the nonlinear optical effects are absent, and the nonlinear condition (high laser irradiance) when the nonlinear optical response of the material is activated. The normalized transmittance of the aperture is the ratio measured under a nonlinear condition divided by that under the linear condition, which is related to the nonlinear optical response of the sample.

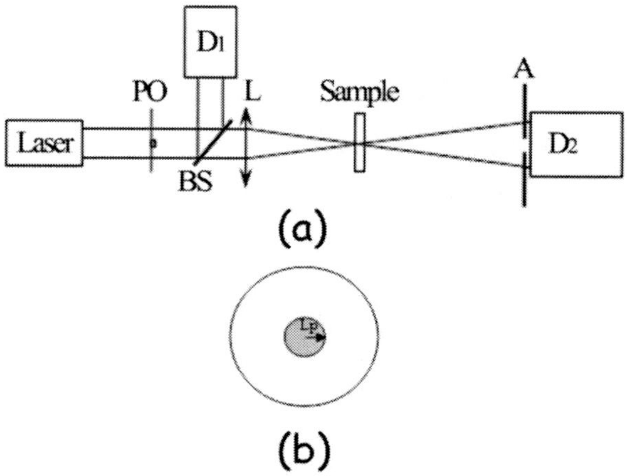

Figure 9. (a) Schematic setup of a single shot transmittance measurement technique with a phase object (T-PO). (b) Configuration of the phase object. L_p is the radius of the phase object. Note there is no aperture in the phase object. Therefore the laser beam after the phase object is still a Gaussian distribution.

The mathematical description of the beam propagation in the T-PO technique is rather straightforward. It should be noted that the PO can be arranged at an arbitrary position in front of lens L. Here, to simplify the process of the numerical simulation, we choose to place the PO at the front focal plane of lens L. Then the beam propagation process from the PO at the front focal plane to the sample at the rear focal plane can be described by a Fourier transform. The beam propagating equations for the phase and the irradiance within the sample are the same as those in the $4f$ method. The beam propagation process after the sample could be expressed by the Fresnel diffraction integral. The normalized nonlinear transmittance can be calculated by:

$$T_{nor} = \frac{\int_{-\infty}^{+\infty} \int_{0}^{r_a} 2\pi r_2 \left| E_a \right|^2 dr_2 dt}{\int_{-\infty}^{+\infty} \int_{0}^{r_a} 2\pi r_2 \left| E_a^{'} \right|^2 dr_2 dt} \tag{13}$$

where r_a is the radius of the aperture, E_a and $E_a^{'}$ are the amplitude of the electric component of the light field at the aperture plane under a nonlinear condition and linear condition, respectively. When r_a trends to infinity (open aperture condition), the normalized transmittance T_{nor} is only sensitive to the nonlinear absorption response of the sample materials. The nonlinear absorption coefficient and the nonlinear refractive index can be deduced by fitting the nonlinear transmittance to the experimentally measured one. If T_{nor} is larger than 1, the sign of nonlinear absorption and nonlinear refraction is positive; otherwise it is negative.

The principles of the T-PO method and the 4*f* method for optical refraction nonlinearities are the same. However, there are three major advantages of the T-PO method compared to the 4*f* method. First, the optical configuration of T-PO is significantly simplified, which is very beneficial for increasing the technique's availability. Second, the replacement of the CCD camera not only reduces the cost of the measurement method, but also greatly expands the measuring scope to the weak optical nonlinearities region by using a much more sensitive detector such as the Photomultiplier tube (PMT). Last but not least, the data processing speed of the T-PO method is greatly enhanced because there is no need for carrying on image integration in evaluating the normalized transmittance.

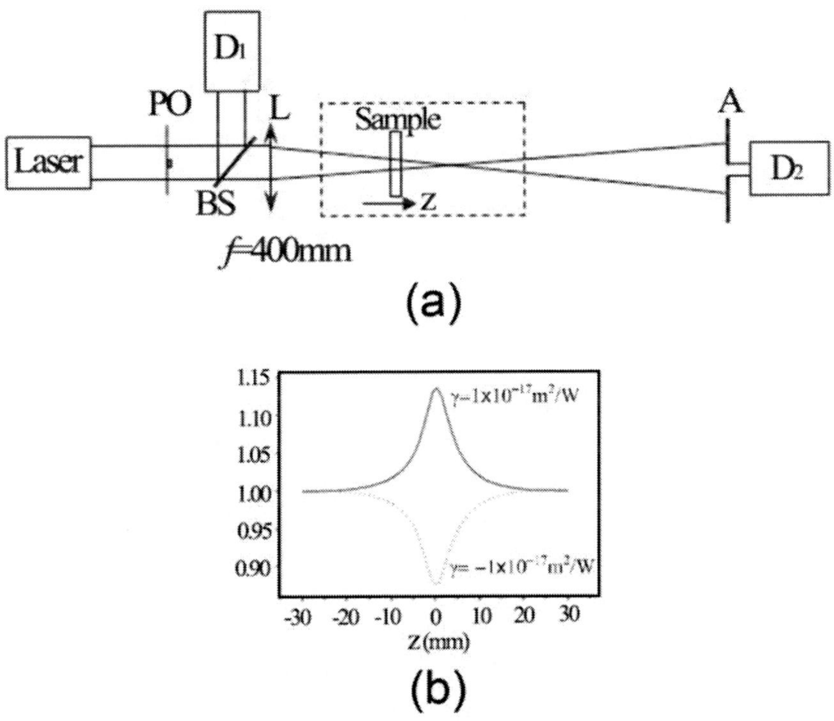

Figure 10. (a) Schematic configurations of z-scan measurement system with a phase object. The phase object is similar to the one in T-PO method. (b) Experimental curve of PO z-scan for pure third-order nonlinear refraction.

3.1.3. Z-Scan with Phase Object (PO z-Scan)

In the T-PO measuring technique described above, the sample material is arranged at the rear focal plane of the convex lens and remains still. In order to activate the nonlinear optical response of the sample, we need to either modify the light irradiance by a light limiter or arrange an additional density attenuator in the optical setup. However, the light intensity modulation could be carried out by moving the sample position as in the conventional z-scan experiments. This modified configuration is called the z-scan with phase object technique (PO z-scan). This method can be thought to be the combination of the conventional z-scan and T-PO method. In this method, the sample is placed on a stage that could move along the axis of the focused laser beam. A far field detector placed after an aperture records the fluence variation of the beam inside the PO. The main difference between the conventional z-scan and the PO z-scan is the shape of the normalized transmittance curve in the far field aperture position. In the conventional z-scan, the normalized nonlinear refraction curve is a peak-valley shape. In the PO z-scan, however, the measured normalized curve is a single peak or valley curve, as shown in Figure 10.

Because the conventional z-scan method is sensitive to multiple nonlinear optical mechanisms that could give rise to a change of the refractive index and the corresponding experimental curve is always a peak-valley shape, it is highly difficult or impossible to determine the underlying physical processes from a z-scan curve. However, in the PO z-scan measurement, different mechanisms will induce different shapes on the normalized transmittance curves. So it is possible to reveal the underlying nonlinear optical mechanisms by analyzing the shape of the experimental trace [82]. This unique feature is the key advantage of the PO z-scan measurement technique compared with the other z-scan methods, such as the Gaussian beam z-scan or Top-hat z-scan. A detailed description will be given in Section 4.

3.1.4. Summary and Discussion

At present, there are a variety of measurement techniques available to characterize the third-order nonlinear optical properties of materials. The representative methods are the z-scan and Frequency mixing. First in this section, we will briefly describe the fundamentals of these existing measurement techniques. Then the critical features of nonlinear measurement methods using phase objects will be highlighted and discussed.

Z-Scan

The z-scan method, which was proposed by Sheik-Bahae and co-worker, is the most widely used third-order optical nonlinearities measurement technique today. In the z-scan method, the sample is moved along the propagation direction of a focused laser beam (named z-axis), and passes through the focal region. Because the beam's spot size within the sample changes with the sample moving towards or away from the focus, the peak intensity is modulated accordingly. The strong illumination near the focal region will induce the nonlinear optical response such as a two-photon absorption (TPA) or a nonlinear refraction effect. Thus the sample experiences a phase and intensity modulation, which can be observed on its transmittance measured as a function of the sample position (z). The z-scan experiment can be performed in two geometries. One is the open-aperture (OA) geometry, where the detector collects all the transmitted light, and the nonlinear absorption effect of sample material is measured. The other is the closed-aperture (CA) geometry, the detector collects only the on-axis transmitted light and both the nonlinear refractive index and optical absorption can be measured. The normalized CA curve to OA curve is related to the pure nonlinear refraction effect of the sample.

The z-scan technique is simple, accurate and could measure the real and imaginary parts of the third-order optical susceptibility simultaneously. However, the z-scan method suffers from the necessity of multiple-shot incidents when the sample is moving relative to the beam waist, which may cause thermal damage to the sample. Moreover, the z-scan technique is sensitive to all nonlinear optical mechanisms that give rise to a change of nonlinear optical coefficients, and it is possible to distinguish the underlying physical mechanisms from z-scan experiments.

Frequency Mixing

In the frequency mixing method, one or two laser beams are incident on the sample material with the same or different frequencies. The nonlinear interaction in the sample gives rise to the signal at frequency ω_s, whose intensity is related to the magnitude of third-order nonlinear susceptibilities $\chi^{(3)}$. When the light wave at 3ω is produced by only one laser beam incident at frequency ω this is the Third harmonic generation (THG) method. Or when employing two incident beams at ω_a and ω_b, a light wave at ω_s can be obtained, where $\omega_s = 2\omega_a - \omega_b$ is any combination of three incident frequencies [76]. Because signal generation requires phase matching the frequency mixing technique is highly accurate and mostly sensitive to the bound electron nonlinearity. However, this method is not suitable for measurement in the resonance region.

Table 1. Features of representative static optical nonlinearity measurement techniques.

	$4f$	T-PO	Z-scan	PO Z-scan	FW
Sensitivity	Good	Good	Good	Good	Highest
Sample Movement	No	No	Yes	Yes	No
Distinguish the Real and Image Parts	Yes	Yes	Yes	Yes	No
System Simplicity	High	highest	Good	Good	Bad
Scope of Measurement	Both	Both	Both	Both	Parametric process only

There are many important advantages for measurement methods using phase objects compared to the frequency mixing technique. The first is that the frequency mixing technique can only detect the parametric (or coherent) process induced by bound-electron nonlinearity, and cannot measure those responses due to real excitations such as nonlinear absorption, free carrier effect, etc. This limits the measurement scope of the technique. Second, the configurations of frequency mixing are the most complicated of all of these measurement methods, which means the system is hard to build and maintain. The phase object measurement method is comparable to the z-scan technique in the field of optical nonlinearity measurements. In most cases, the measurement techniques using phase object and z-scan give the same information. In the field of nonlinear optics, phase object measurement techniques may never replace the standard z-scan technique, not in performance, but certainly in popularity. However, there are applications for which the z-scan technique is not suitable, especially in characterizing the optical nonlinearities of thin films. The phase object measurement technique is a promising candidate for such applications as they combine the fast, accurate measurement with a single-shot. The features of these static optical nonlinearity measurement techniques are summarized in Table 1.

3.2. Transient Nonlinear Dynamics Measurement Technique

The photo-physical dynamics after laser illumination plays a key role in lots of optoelectronic and photonic devices such as light-emitting diodes, all-light modulators, etc. Insights into these microscopic dynamics will help us to develop a better understanding of the light-matter interaction process. As the optical response of matter to incident light perturbation is always time-dependent, the time-resolved measurement on the optical nonlinearities is of paramount importance. The time-resolved pump-probe technique is a powerful tool for the characterization of light-matter interaction kinetics such as excited-state dynamics in various materials. The roles of free electrons, charged carriers, and other light-matter interactions can be partly separated by the macroscopic response of the materials at different timescales [83].

In this section, we provide a comprehensive review of time-resolved pump probe techniques using phase object, which could be used to investigate the nonlinear absorption and refraction kinetics simultaneously. The pump-probe system, based on a nonlinear imaging technique with phase object (4*f* pump-probe) is introduced first, and then the pump-probe system based on the single-shot transmittance measurement technique (PO pump-probe) is presented. The theorical and experimental tips of these methods are given in detail. Their advantages over other time-resolved measurement methods are highlighted and summarized. The remarkable features of these techniques have paved the way to other applications in the research fields of ultrafast liquids dynamics, free carrier decay and transfer dynamics in organic/inorganic semiconductors, etc.

3.2.1. Brief Introduction of Pump-Probe Technique

The idea of pump probe spectroscopy can be traced back before 1900. In 1864, Toepler used the spark photography technique to study sound-wave dynamics by using two short light sparks with a controlled time delay between them. This marks the creation of the Pump-probe technique. Later, Abraham and Lemoine improved the Pump-probe technique by deriving the pump and probe flash from the same spark. This update improves the resolution of pump-probe spectroscopy to the scale limit by the flash duration. The recent achievements in laser technology and other light sources have been greatly expanding the scope of the pump-probe technique, stimulating variable kinds of pump-probe combinations with different wave frequencies, pulse duration and bandwidth. The examples of modern pump probe techniques include one-color (degenerate) laser pump probes, two-color (non-degenerate) laser pump probes, optical pump-THz probes, IR laser pump-x-ray probes, to name only a few. The temporal resolution of the modern pump-probe technique is as short as several femtoseconds or even attoseconds time scale [84].

The schematic diagram of a typical pump-probe experiment is shown in Figure 11. First, a strong pump pulse irradiates the sample, generating a nonlinear polarization, which would modulate the macroscopic absorption, refraction or reflection of the sample. A time-delayed weak probe pulse is incident on the same spot to track the pump-induced changes in the optical response. It should be noted that the pump probe experiment is a third-order nonlinear optical process. Usually, the pump and probe pulses come from the same laser source and the temporal resolution of the experiment is mainly dependent on the probe pulse.

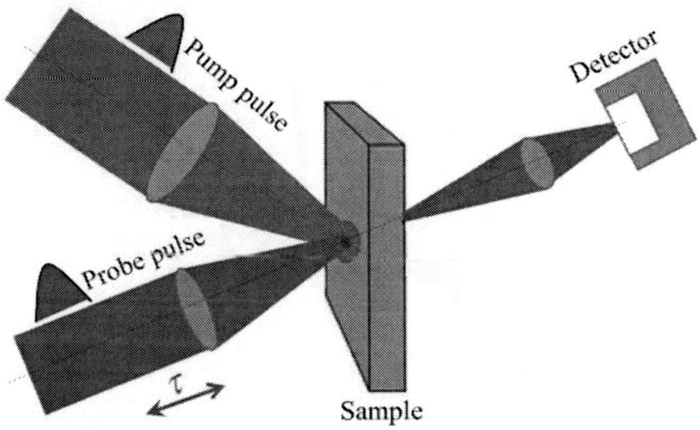

Figure 11. Re-plotting of schematic setup of a pump-probe experiment (Ref. [20]).

Most of the pump-probe experiments using laser pulses concern the absorption or reflection dynamics of the samples. This is understandable as follows: the amplitude variation of the transmitted or reflected light caused by the sample under strong laser irradiance is directly measurable by the detectors such as a silicon detector, photomultiplier tube or CCD. However, in many cases, the phase variation dynamics induced by the refractive index changes of the sample is highly crucial for understanding the light-matter interaction process in the non-resonance region. Therefore, a time-solved pump probe technique which could evaluate nonlinear refraction kinetics is desirable. The 4*f* pump-probe and the PO pump-probe techniques are the combination of static single shot optical nonlinearity measurement methods with time-resolved pump probe spectroscopy. These techniques are capable of measuring the nonlinear absorption and refraction dynamics of variable materials accurately and synchronously.

3.2.2. Pump-Probe System Based on the Nonlinear Imaging Technique with Phase Object

The arrangement of the time-resolved pump-probe system based on the 4*f* method (4*f* pump-probe) is shown schematically in Figure 12. The input laser beam is first divided into two arms: a strong pump beam and a weak probe beam. The probe arm is set into a standard 4*f* coherent imaging system with phase object while the pump beam passes through a time delay line. First, a beam expander and incident perpendicular to the phase object with an aperture to get a top-hat illumination spatially expands the probe beam. In the experiment, the sample is located at the Fourier plane of the 4*f* system and remains unmoved. The probe and pump beams are focused to the same region on the sample with an incident angle difference less than 10°. After the sample, the pump beam is blocked and the probe beam is recorded by a CCD camera arranged at the output plane of the 4*f* system. The introduction of the variable time-delay between the pump and probe beam enables us to detect the nonlinear response induced by the strong pump pulse. It should be noted that careful alignment of the 4*f* imaging system and the spatial/temporal superposition of the pump and probe are essential for producing the maximum contrast without introducing artifacts and noises.

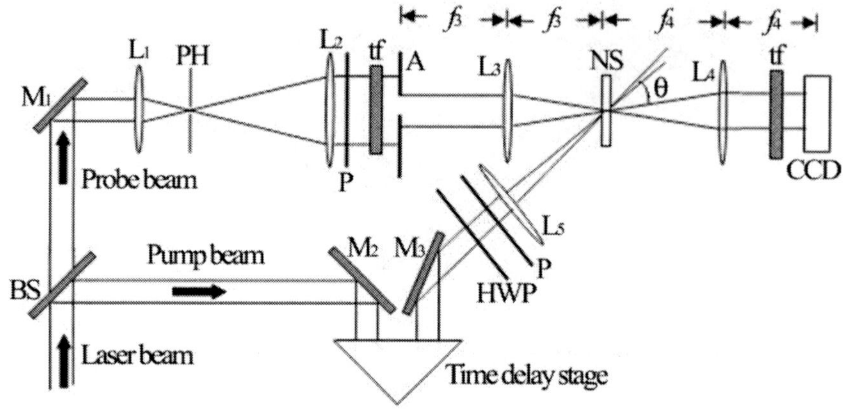

Figure 12. Schematic configuration of the time-resolved pump-probe system based on nonlinear imaging technique with phase object. The “tf” is marked as natural filter.

The theoretical description is similar to the standard $4f$ method. The probe beam propagation process from the phase object to the CCD camera is described by a Fourier transform and inverse Fourier transform. The normalized fluence recorded by the CCD camera can be expressed as:

$$NF(r)=\frac{\int_{-\infty}^{+\infty}\left|E_{pL}(r,t)\right|^{2}dt}{T_{0}\int_{-\infty}^{+\infty}\left|E_{p0}(r,t)\right|^{2}dt} \tag{14}$$

where E is the amplitude of the electric component of the incident probe beam and T_0 is the linear transmittance of the sample. The normalized transmittance relating to the nonlinear optical response is defined by the following calculation:

$$NT=\frac{E_{NL}}{E_{L}}=\frac{2\pi\int_{-\infty}^{+\infty}dt\int_{0}^{+\infty}rI_{PL}dr}{E_{0}*T_{0}} \tag{15}$$

The initial pump beam on the front surface of the sample can be expressed as:

$$I_{e}=I_{oe}\frac{\omega_{oe}^{2}}{\omega_{e}^{2}}\exp\left[-\frac{2r^{2}}{\omega_{e}^{2}}-\frac{(t-t_{d})^{2}}{\tau^{2}}\right] \tag{16}$$

where $\omega_e(z)=\omega_{0e}[1+(1+z/z_{0e})^2]^{1/2}$ is the pump-beam radius at z, z is the sample position away from the focus on the axis, $z_{0e}=(\pi\omega_{0e}^2)/\lambda$ is the Rayleigh diffraction length of the pump beam. ω_{0e}, I_{0e} and t_d are the pump beam radius at focus, on-axis peak irradiance at focus and time delay between the pump and probe beam, respectively.

In contrast with the single beam 4*f* coherent imaging or the T-PO method, the strong pump beam induces the nonlinear optical response of the sample in the pump-probe experiment. Thus the differential propagation equations for phase and irradiance of the probe beam within the sample are defined as:

$$\begin{cases} \dfrac{\partial I_p}{\partial z'} = -\alpha(I_e)I_p \\ \dfrac{d\phi}{dz'} = k\Delta n(I_e) \end{cases} \tag{17}$$

where $\alpha(I_e)$ and $\Delta n(I_e)$ denote the variations of the absorption coefficient and refractive index caused by the incident pump beam. The two characters have many different forms, which are related to the properties of the sample and the nonlinear mechanisms involved in the light-matter interaction.

An experimental demonstration of the time-resolved pump-probe with a 4*f* coherent imaging system is shown in Figure 13 [85]. The sample is a bulk semiconductor ZnSe and the experiments were performed with a 20 ps laser pulse at 532 nm. There are two mechanisms in the nonlinear optical response of the bulk semiconductor: bound electron nonlinearity and light-injected free carrier nonlinearity. The bound electron nonlinearity is an instantaneous process that follows the time duration of the pump pulse. The free carrier nonlinearity is a slow process, which has a long decay time. From the experimental curve these two mechanisms are clearly distinguished by their diverse temporal response. The magnitude of photo-physical parameters of semiconductor ZnSe can be determined by fitting the curve numerically.

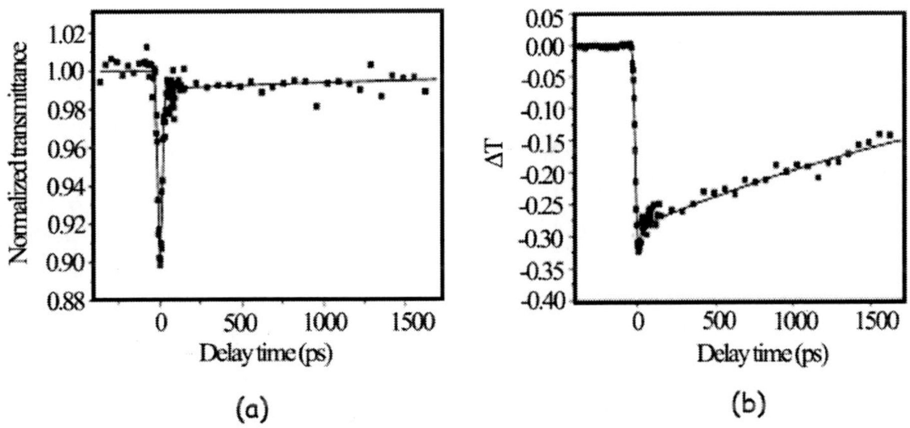

Figure 13. Measured normalized transmittance value as a function of temporal delay for ZnSe. Dots are experimental results and the solid line is a numerically simulated curve, (a) nonlinear absorption dynamics, (b) nonlinear refraction dynamics.

3.2.3. Pump-Probe System Based on the Single-Shot Transmittance Measurement with Phase Object

As reviewed in the single shot measurement techniques section above, the signal acquisition in the 4*f* pump probe method is accomplished by a CCD camera. The aftermath of image processing needs to integrate all pixels of an image recorded by the CCD camera for calculating the normalized fluence. Thus the experimental data processing of the 4*f* technique is relatively complicated and costs considerable computer time. Furthermore, in order to detect the nonlinear kinetics of the sample material, multiple shots need to be taken at variable delay times. The temporal resolution largely depends on the step between the image and image delays. In a typical 4*f* pump-probe experiment, at least hundreds of images need to be acquired for a variable delay time. These procedures significantly increase the complications and costs of the 4*f* pump-probe experiment.

The T-PO measurement technique is a simplification of the 4*f* coherent imaging system. The CCD camera is replaced by an energy detector and the data processing doesn't require intricate image integration. Both of these greatly simplify the experimental procedures. The full potential of the T-PO method is fully revealed by its combination with the pump probe technique. Because of its simplicity, high speed and accuracy, the practicality and availability of the pump-probe system based on the single-shot transmittance measurement with phase object (PO pump-probe) are superior to the pump-probe system based on the nonlinear imaging technique with phase object (4*f* pump-probe).

The configuration of the modified time-resolved pump-probe system with phase object is shown in Figure 14.

The transparent phase object is placed at the front focal plane of the convex lens L while the sample is arranged at the rear focal plane. After passing through the sample material, the probe beam propagates to the far field and is separated into two parts by a beam splitter and detected by two detectors simultaneously. The circular aperture in the front of the detector only transmits one small part of probe beam inside the PO in order to record the nonlinear refraction information.

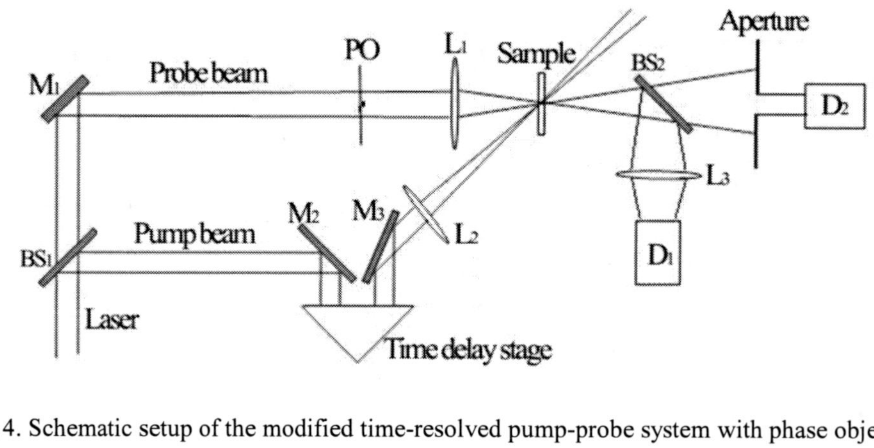

Figure 14. Schematic setup of the modified time-resolved pump-probe system with phase object.

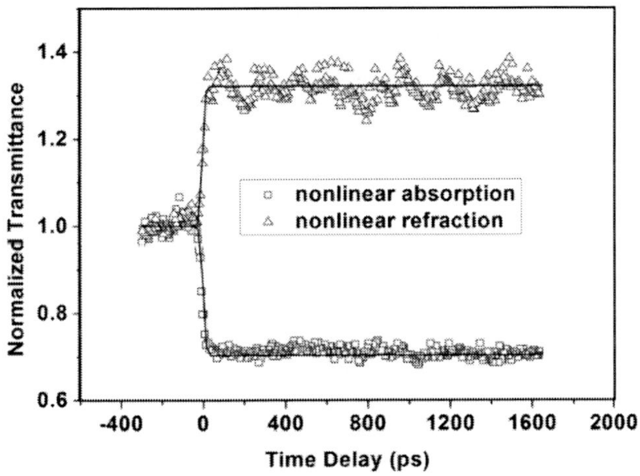

Figure 15. The measured normalized transmittance value as a function of the temporal delay of a phthalocyanine/DMSO solution. The hollow square and triangle represent the nonlinear absorption and refraction dynamics of the sample, respectively. The solid line is the numerically simulated curve.

The pump beam's expressions and the differential propagation equations of the probe beam are the same as those in the 4*f* pump-probe technique. However, the propagation procedure for the probe beam from the phase object to the far field aperture is achieved with three Fresnel diffraction integrals. The normalized transmittance of the aperture is defined as:

$$\Delta T(t_d) = \frac{2\pi \int_{-\infty}^{+\infty} \mathrm{d}t \int_0^{R_a} r_a \mathrm{d}r_a \left|E_a\right|^2}{2\pi \int_{-\infty}^{+\infty} \mathrm{d}t \int_0^{R_a} r_a \mathrm{d}r_a \left|E_a(I_e = 0)\right|^2} \tag{18}$$

where t_d, R_a and E are the delay time between the pump and probe, the radius of the aperture and the amplitude of the electric field of the probe beam, respectively. When R_a approaches infinity (open aperture condition), the experimentally recorded transmittance is related to nonlinear absorption dynamics; or else the aperture's normalized transmittance is related to the nonlinear refraction kinetics. An experimental example of the PO Pump-prove curve is shown in Figure 15 [86].

3.2.4. Summary and Discussion

There are several measurement techniques available to characterize the sample's nonlinear refraction dynamics. The representative measurement technique includes: nonlinear interferometer, four wave mixing (FWM), optical kerr effect (OKE) and time-resolved z-scan pump probe method. To start, the principles of these measurement techniques are briefly described. Then the primary advantages of the nonlinear measurement method using phase object will be highlighted and discussed.

Nonlinear Interferometer

In this method, the probe beam is divided into two paths to create two copropagating probe pulses with the same optical path (interferometer). The sample is arranged in one path of the probe beam and its optical response is modulated by the strong pump pulse. After passing through the sample, the altered probe beam and the reference probe beam are superimposed, and the detected beam signal is related to the pump-induced phase change [87, 88]. This method is accurate and polarization-resolved. However, in this kind of technique, the nonlinear absorption and refraction dynamics cannot be measured synchronously, and the system configuration is rather complicated.

Four Wave Mixing (FWM)

In this method, two or three pulses with different wave vectors, separated by a time-delay, are incident on the sample. They will induce a nonlinear polarization within the material which gives rise to a signal wave in the direction of phase matching. This method has the highest signal-to-noise ratio and is also polarization-resolved. It is sensitive to various microscopic processes such as inhomogeneous broadening, excited-state population and spatial diffusion. However, there is an inherent disadvantage of this technique. It cannot distinguish the real and imaginary parts of nonlinear susceptibilities, which is a serious problem especially when the sample exhibits nonlinear absorption. Although the introduction of an additional phase reference induced by a local oscillator may partly solve this problem, the optical configuration of the measurement system is highly sophisticated and hard to maintain.

Optical Kerr Effect (OKE)

In this method, the probe beam passes through two perpendicularly arranged polarizers located before and after the sample, and the polarization angle between the pump and probe beams is set to 45°. The sample's birefringence, induced by the pump beam, will allow some part of the probe beam to pass through the polarizer after the sample. This technique is straightforward, highly precise and applicable to various materials such as simple liquids, dilute solutions and other complex systems. But the major drawback of the OKE technique is that only the anisotropic nonlinear refraction response of the materials is observable.

Time-Resolved Z-Scan

In the time-resolved z-scan method, the probe path is set to be a standard z-scan configuration. The sample with nonlinear refraction response under strong laser irradiation can be a phenomenon logically treated as a nonlinear lens. This lens will cause the spatial distortion of the probe beam at the far field aperture plane, leading to a change of the aperture's transmittance. The original time-resolved z-scan experiment was performed by recording the difference of the transmittance value when the sample is placed at the valley and peak positions of the conventional z-scan curve [30]. This simplified time-resolved z-scan technique (sometimes called spatially masked or transient lens spectroscopy in literature [89]) detects the transmittance at the peak or valley position only. The optimal position to achieve the maximal signal is found to be 0.8 z_0 [90]. This method can measure the nonlinear absorption and refraction dynamics simultaneously and is polarization-resolved. However, the sensitivity of the time-resolved z-scan method is lower than those techniques introduced above. In all pump-probe systems, the relative spot sizes of the pump and probe beam within the sample and the spatial coincidence between them are very important parameters, which ultimately determine the sensitivity of the measurement. Because the sample material has to be arranged away from the probe beam's focus, the system's instability will reduce the measurement sensitivity prodigiously. Meanwhile, the corresponding measurement system is hard to align and maintain.

The signal generation in the four-wave-mixing technique requires phase matching, which causes a high signal-to-noise ratio. Although the sensitivity of phase object transient dynamics measurement techniques is lower than FWM or nonlinear interferometer, they offer the advantages of simple and robust configurations, easy maintenance, low cost and acceptable measurement accuracy. The phase object measurement techniques have exclusive advantages over the OKE method because their measurements are polarization-independent. This means that all tensor elements of nonlinear response function could be characterized by these techniques, which is of great benefit for the full understanding of the material's dynamics. Compared with the time-resolved z-scan technique, the combination of simple and straightforward structure, higher sensitivity and easier alignment makes those methods with phase object far superior. The features of these transient dynamics measurement techniques are summarized in Table 2.

Table 2. Features of representative transient dynamics measurement techniques

	Interferometer	FWM	OKE	TR Z-scan	$4f$ pump-probe	PO pump-probe
Measurement Sensitivity	High	High	High	Good	Good	Good
System Complicacy	High	High	Low	Medium	Medium	Low
Range of polarization	All	All	Anisotropic only	All	All	All

3.3. Reflection Nonlinear Optical Measurement Technique

So far, most of optical nonlinearity measurement techniques are employed to characterize transparent materials. In 1994, D. V. Petrov *et al* proposed a modified reflection z-scan (RZ-scan) method to investigate the optical nonlinearities of opaque materials [37, 38]. However, the RZ-scan method suffers from the drawback of multiple-shot illumination, which may cause thermal damage to the sample. Later, researchers raised an alternate RZ-scan technique using single-shot illumination [91] but the sensitivity of this technique is rather low. None of these measurement techniques are mature enough for extensive use because of the sensitivity limit and strict beam profile restrictions.

Analogous to the modification of the RZ-scan from the conventional z-scan, the standard $4f$ method could be transformed into the reflected $4f$ method to evaluate the nonlinear optical properties of opaque materials [92].

The arrangement of the RNIT-PO system is shown schematically in Figure 16. First, the input laser beam is spatially expanded and perpendicular incident on the phase object with the aperture to achieve the top-hat illumination. The sample is placed at the rear focal plane of the convex lens L_3. The reflected beam is collected by lens L_4 to the CCD camera. Convex L_3 and Convex L_4 are aligned in a confocal $4f$ system. The rest of the configuration is similar to that of the conventional $4f$ method.

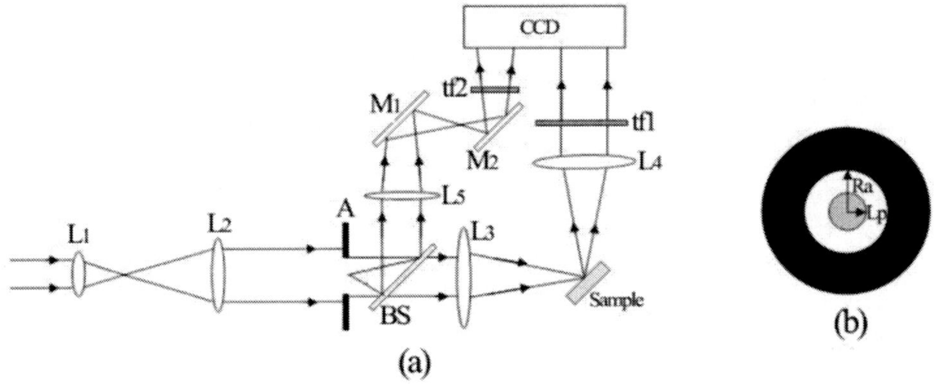

Figure 16. (a) Schematic setup of the RNIT-PO system. The sample is placed at the focal plane (Fourier plane). The phase object with aperture and CCD camera are at the input and output planes of the 4*f* system. (b) Schematic configuration of the phase object with circular aperture. The outside dark area is opaque. The inner circular area is the uniform phase shift area by depositing a transparent layer on glass plate with certain thickness.

The theory of the beam propagation process in the reflected 4*f* method is the same as that of the conventional 4*f* method. The first propagation process, from phase object to the sample, is a spatial Fourier transform and the second process, from sample to CCD camera, is an inverse Fourier transform. However, the modulation mechanism of the laser beam is completely different with the reflected 4*f* technique. There are two possibilities that may alter the phase or amplitude of the reflected beam. One is the variations of the sample's reflection coefficient caused by the surface effect. This is the case for most of the opaque materials (such as silicon in visible wavelengths). The other is the bulk nonlinear effect induced variation of the phase and/or the amplitude of the light wave reflected from the second surface. This is the case for a transparent film grown on an opaque substrate, as shown in Figure 17. To simplify the analysis, we assume only one reflection from the substrate.

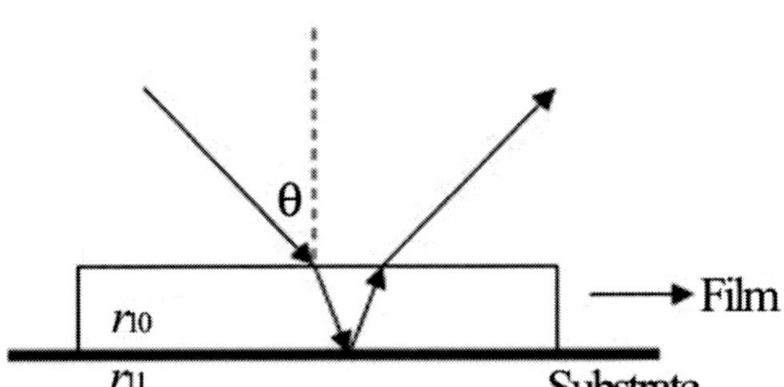

Figure 17. Schematic of beam reflection from a film coated on an opaque substrate.

For surface effect condition, the effective reflection coefficient is given by Martinelli *et al*:

$$r(u,v,t)=\frac{n^2\cos\theta-\sqrt{n^2-\sin^2\theta}}{n^2\cos\theta+\sqrt{n^2-\sin^2\theta}} \tag{19}$$

where $\hat{n} = n_0 + \gamma|S(u,v,t)|^2 + i\left(\frac{\alpha_0}{2k} + \frac{\beta}{2k}|S(u,v,t)|^2\right)$ is the complex refractive index, θ, n_0 and α_0 are the incidence angle, linear refractive index, and the linear absorption coefficient of the sample material, respectively. γ and β are the nonlinear refractive index and the nonlinear absorption coefficient of the sample, respectively.

For the condition of bulk effect, the effective reflection coefficient can be obtained as:

$$r(u,v,t) = \frac{n_1^2\sqrt{n_0^2 - \sin^2\theta} - n_0^2\sqrt{n_1^2 - \sin^2\theta}}{n_1^2\sqrt{n_0^2 - \sin^2\theta} + n_0^2\sqrt{n_1^2 - \sin^2\theta}} e^{-\alpha_0 L n_0/\sqrt{n_0^2 - \sin^2\theta}} \times [1 + q_1(u,v,t)]^{(jk\gamma/\beta - 1/2)}[1 + q_2(u,v,t)]^{(jkn_2/\beta - 1/2)} \tag{20}$$

with

$$q_1(u,v,t) = \beta L_{\text{eff}}|S(u,v,t)|^2,\ q_2(u,v,t) = \beta L_{\text{eff}}|S_l(u,v,t)|^2,$$

$$S_l(u,v,t) = S(u,v,t)\frac{n_1^2\sqrt{n_0^2 - \sin^2\theta} - n_0^2\sqrt{n_1^2 - \sin^2\theta}}{n_1^2\sqrt{n_0^2 - \sin^2\theta} + n_0^2\sqrt{n_1^2 - \sin^2\theta}} e^{-\alpha_0 L n_0/\sqrt{n_0^2 - \sin^2\theta}/2}[1 + q_1(u,v,t)]^{(jk\gamma/\beta - 1/2)}$$

$$L_{\text{eff}} = [1 - \exp(-\alpha_0 L n_0/\sqrt{n_0^2 - \sin^2\theta})]/\alpha_0,$$

L is the sample length, n_1 is the linear refraction coefficient of the substrate.

Using the Equation (20), the complex field of the reflected beam is obtained as:

$$S_L(u,v,t) = S(u,v,t)r(u,v,t) \tag{21}$$

Then the image at the output plane of the 4*f* system can be calculated numerically.

The case that the variation of the reflected beam is induced by the bulk contribution, owing to the nonlinear optical properties of the transparent film, is similar to the conventional 4*f* coherent imaging system, and the simulated numerical images of the reflected 4*f* method are shown in Figure 18.

One experimental demonstration of the reflected 4*f* technique is shown in Figure 19. The sample was a metal phthalocyanine film coated on the silicon. The experiments were performed with 20 ps laser pulse at 532 nm. As shown in Figure 19, the film exhibits positive nonlinear refraction. The total count of the nonlinear image is much smaller than that of the linear image, which shows that the film has a strong reverse saturation absorption (RSA). Based on the numerical simulation, the nonlinear absorption coefficient and the nonlinear refractive index can be determined.

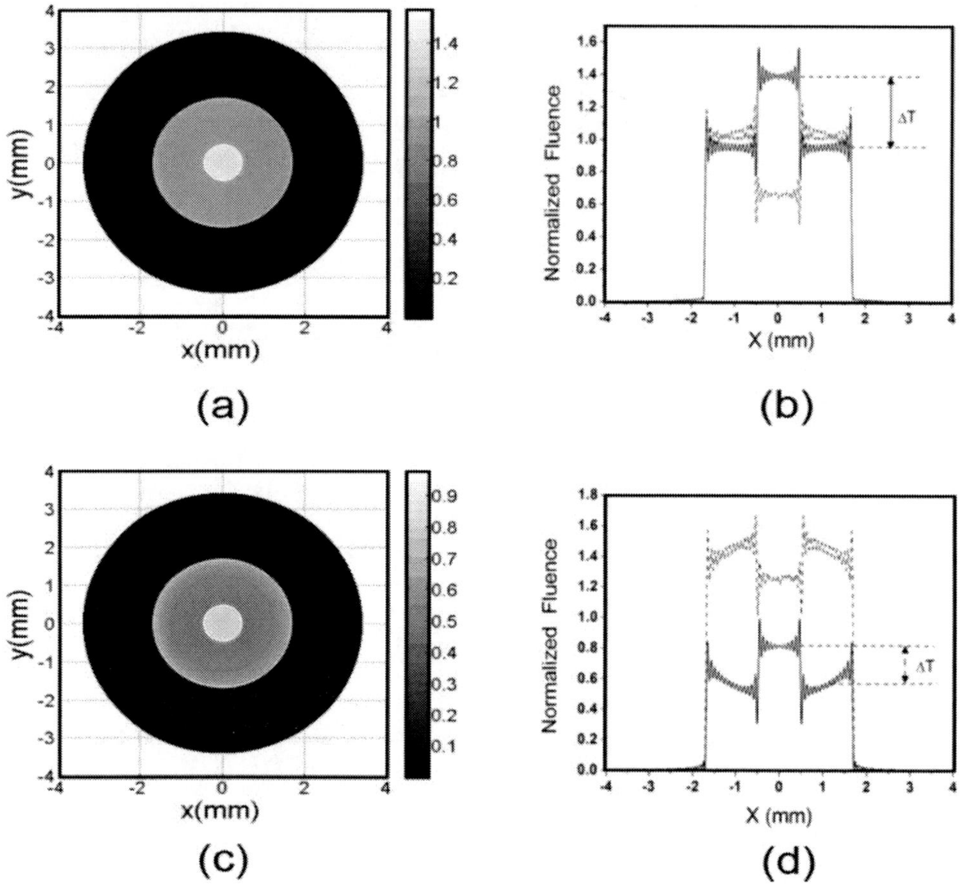

Figure 18. (a) Numerical simulation images reflected from nonlinear material ($\gamma > 0, I\_ = 0$) at the output of the 4*f* system. (b) The solid line is the profile of (a), the dash line is the result of a positive nonlinear refraction. (c) Numerical simulation of the top-hat beam images reflected from nonlinear material (($\gamma = 0, I\_ > 0$) at the output of the 4*f* system. (d) The solid line is the profile of (c), the dash line is the result of saturation absorption ($I\_ < 0$).

The reflected 4*f* method is superior to that of the reflected counterpart because there is no movement of the material, it is a single-shot measurement with precise theoretical analysis and high sensitivity.

4. Potential Applications

The optical nonlinearity measurement techniques with phase object are experimental tools for investigating nonlinear optical responses of materials on a variable time scale. The measurement techniques are used to disclose the fundamental light-matter interaction processes and microscopic mechanisms. This section focuses on the scientific and technical applications of the optical nonlinearity measurement techniques using the phase contrast principle.

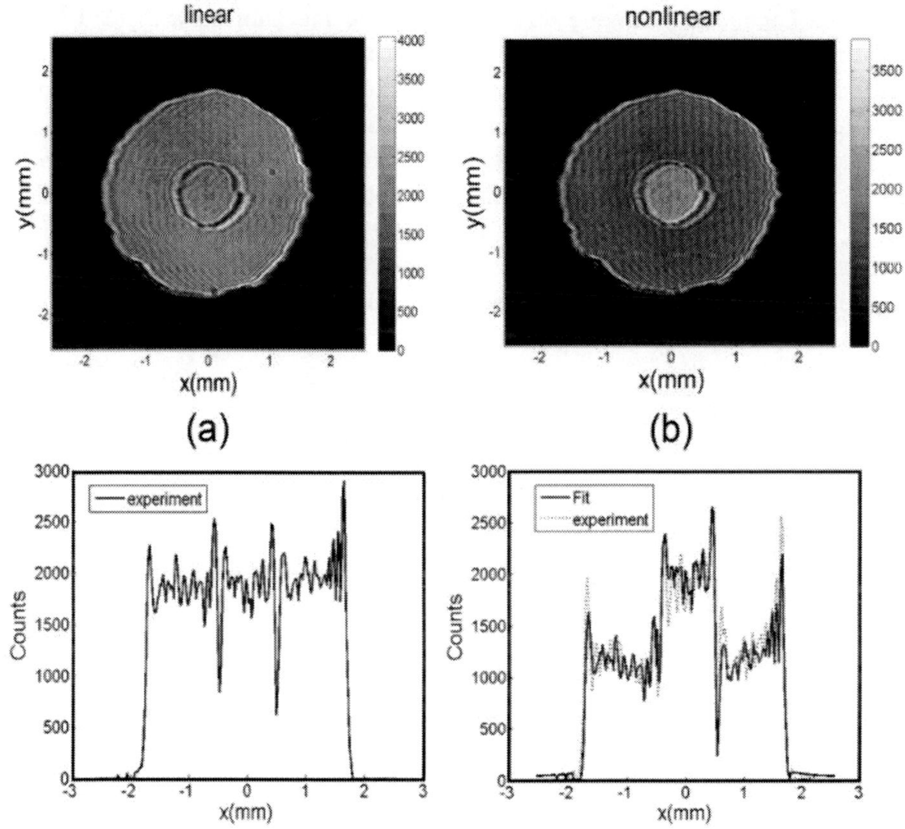

Figure 19. The experimental results of the metal phthalocyanine film measured using RNIT-PO technique. (a) The linear image corresponding to the excitation at low intensity. (b) The nonlinear image. The images in the lower row represent the corresponding beam profiles along the center of the image.

4.1. Distinguishing Different Nonlinear Mechanisms during the Z-Scan Experiment

As described above, the z-scan with phase object (PO z-scan) technique can be considered as the combination of a conventional z-scan configuration with a phase object. This method can measure the optical nonlinearities of materials based on the phase contrast principle. In the PO z-scan technique, the nonlinear phase modulation, due to the variation of the refractive index of the sample, leads to interference between the wave inside the PO and the wave around the PO, which changes the amplitude of the wave inside the PO. Thus the normalized transmittance of the far field aperture is altered by the amplitude modulation of the PO diffraction pattern. The experimental curve of the PO z-scan shows a single peak or valley for third-order optical nonlinearities, which is different from the peak-valley curve of the conventional z-scan experiment. The positions of the peaks map the strongest phase contrast, and the peak amplitude reflects the relative magnitudes of the phase contrast.

However, the prerequisite of a single peak or a valley PO z-scan curve is that only one nonlinear mechanism exists during the experiment. In the real experiment, this assumption is usually not met. Competition among different nonlinear mechanisms will lead to the shape distortion of the PO z-scan curve. Therefore we can use the shape of the PO z-scan trace to reveal and detect the nonlinear mechanism of the light-matter interaction during the experiment. One example is the thermal-lensing effect in the solution sample. When a part of the input laser energy is absorbed by the liquid sample, a temperature gradient will be introduced in the area of the beam spot. This temperature gradient will modulate the liquid's density along the gradient direction, which causes the variation of the refractive index of the liquid sample. This phenomenon is called the thermal-lensing effect. The total nonlinear refractive index of sample when both pure third-order optical nonlinearities and thermal-lensing effects are present could be expressed as:

$$\Delta n(r,z,t) = \gamma I + \Delta n_{th} \tag{22}$$

where the Δn_{th} is the variation of the refractive index caused by the thermal-lensing effect, and γ is the pure third-order nonlinear refraction index. The transient thermal index change Δn_{th} induced by the laser pulse propagation through the sample can be obtained by solving the photoacoustic wave equation non-statically:

$$\frac{\partial^2 \Delta n_{th}(r,z,t)}{\partial t^2} - v_s^2 \nabla_\perp^2 \left(\Delta n_{th}(r,z,t)\right) = \left({dn}/{dT}\right) v_s^2 \nabla_\perp^2 \left(\Delta T(r,z,t)\right) \tag{23}$$

where dn/dT represents the thermo-optic coefficient, v_s is the velocity of sound in the sample, ΔT is the temperature variation in the sample after the absorption of the laser energy. Using Equations (22) and (23), the normalized transmittance of the far field aperture of the PO z-scan can be determined.

An experimental demonstration for distinguishing the transient thermal-lensing effect from the pure third-order nonlinear refraction by using the PO z-scan technique is shown in Figure 20.

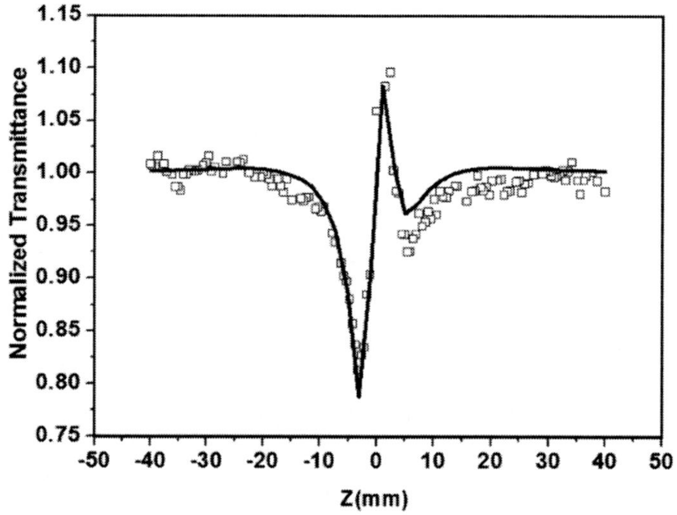

Figure 20. PO z-scan traces for the C_{70}/toluene solution. The solid curve is the theoretical fitting.

The sample is a C_{70}/toluene solution. The experiment was performed using 8 ns laser pulses at 532 nm light wavelength with a repetition rate of 1Hz. The parameters of the phase object are $L_p = 0.5\ mm$ and $\varphi_L = \pi/2$. The experiment was carried out at Gaussian beam illumination, and the beam waist at the focal plane was $\omega_0 = 24\ um$. It is clear that the shape of the PO z-scan trace under competition among multiple nonlinear mechanisms is very different from the one with a pure third-order nonlinear refraction. From the theory of a PO z-scan technique and the Equation (23), the photo-physical parameters of a transient thermal-lens effect can be evaluated consistently.

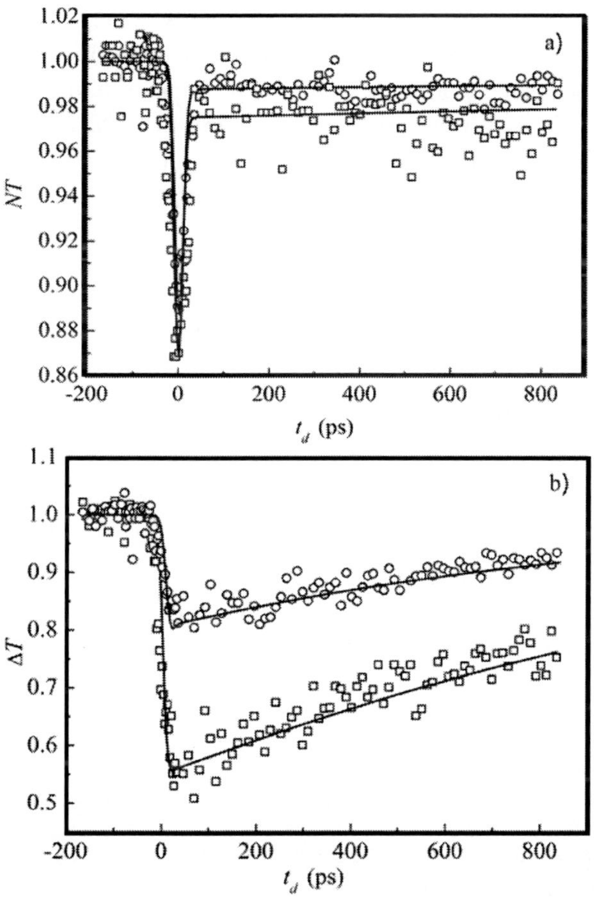

Figure 21. Experimental results of PO pump-probe for bulk semiconductor ZnS using picoseocnd laser pulse when pumped at 532 nm and probed at 680 nm. The peak pump intensities are $I_{e0} = 1.0\ GW/cm^2$ and $I_{e0} = 2.0\ GW/cm^2$. Solid curves are theoretical fittings to the experimental results.

4.2. Investigation on the Non-Degenerate Optical Nonlinearities in II-V Semiconductors

Identifying candidate materials with large and fast optical nonlinearities is an ongoing effort towards future photonic applications. Semiconductors are excellent materials for photonic devices, such as all-light switching, because of their large optical nonlinearities and flexible structures [93-95]. The characterization of nonlinear optical properties and response time is of paramount importance to pave the way for practical applications.

The pump-probe experiment is non-degenerate when the input pump and probe's wavelengths are different. In this case, the synchronous combination of photons with different energies will generate a nonlinear optical response. The non-degenerate optical nonlinearities are critical for understanding some fundamental physical processes, such as free-carrier refraction dispersion effects, and technical applications, such as mid-infrared detection. So far, data concerning the non-degenerate optical nonlinearities are rare, partly due to the lack of efficient and convenient measurement techniques.

The time-resolved pump probe with phase object (PO pump-probe) method is a simple and accurate technique which can extract the nonlinear absorption and refraction dynamics of materials simultaneously. This technique holds great promise in investigating degenerate and non-degenerate photo-physical processes in semiconductors. Our group has been employing this technique intensively to carry out a comprehensive investigation on the non-degenerate nonlinear dynamics of II-V semiconductors [96, 97]. A typical result is shown in Figure 21.

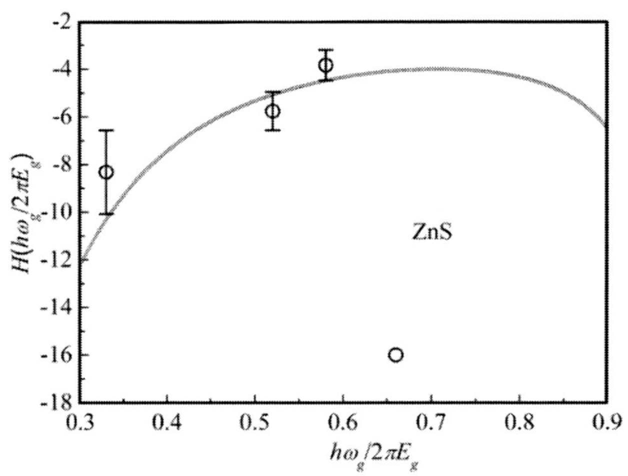

Figure 22. Free-carrier refraction dispersion function versus probe-beam photon energy for ZnS from Ref. [97]. From left to right is the probe wavelength at 1060 nm, 680 nm, 600 nm, 532 nm.

Figure 21(a) shows the nonlinear absorption dynamics measured under open aperture conditions. It is clear that the nonlinear absorption mechanism for ZnS at a picosecond timescale is a combination of non-degenerate two-photon-absorption (ND-2PA) and free-carrier absorption. The instantaneous non-degenerate two-photon absorption and the slow free-carrier absorption are clearly separable by their response time on the experimental curve. However, dynamics of the non-degenerate nonlinear refraction of ZnS (shown in Figure 21(b)) displays no obvious fast-response signal near zero-time-delay, indicating that the non-degenerate bound electronic Kerr effect is much smaller than the light-induced free-carrier refraction. These results demonstrate that the free-carrier refraction (FCR) is the dominant nonlinear refraction mechanism at a picosecond timescale. All of the photo-physical parameters of the semiconductors are based on a two-band model that can be evaluated by fitting the experimental curve numerically. Experiments performed at other wavelengths exhibit similar results, and the free-carrier refraction dispersion function versus the probe-beam photon energy could be determined, as is shown in Figure 22. Our ultimate goal is to survey and assess the transient carrier or the excited state refraction dispersion spectrum of the material. By analyzing the dispersion spectrum and the fine structure of the sample, a new insight into the mechanism of the light-matter interaction process will be evaluated.

CONCLUSION AND OUTLOOKS

In the last two decades considerable theoretical and experimental efforts have been devoted to the development of novel nonlinear optical measurement techniques. The recent development of optical nonlinearity measurement techniques, using phase object, provides a succinct and efficacious tool to investigate the material's photo-physical properties occurring upon photo-excitation. In this chapter, we described some recent advancement of nonlinear measurement techniques using phase object, aiming at characterizing the third-order optical nonlinearities of various materials. The experimental configurations and theoretical principles are presented in some detail, and the demonstrations of recent experimental results were given to stress the potential applications of these measurement techniques. These novel measurement techniques will play a key role in understanding the properties and mechanisms responsible for the nonlinear optical effect. We hope that the systematic investigation of a ‘tool-material-application’ triangle for nonlinear optics will inspire more research efforts to address the current challenges and promote a better understanding of the physical processes taking place in nonlinear optical phenomena.

REFERENCES

[1] Franken, PA; Hill, AE; Peter, CW; Weinreich, G. “Generation of optical harmonics”, *Phys. Rev. Lett.*, 1961, 7, 118–122.

[2] Eckhardt, GM; Hellwarth, RW; Mcclung, FJ; Schwarz, SW; Weiner, D; Eoodbury, EJ. “Stimulated Raman scattering from organic liquids”, *Phys. Rev. Lett.*, 1962, 9, 455–461.

[3] Chiao, RY; Townes, CH; Stoicheff, BP. “Stimulated brillouin scattering and coherent generation of intense hypersonic waves”, *Phys. Rev. Lett.*, 1964, 12, 592-596.

[4] Chiao, RY; Garmire, E; Townes, CH. “Self-trapping of optical beams”, *Phys. Rev. Lett.*, 1964, 13, 479–483.

[5] Maker, PD; Terhune, RW; Savage, CM. “Intensity dependent changes in the refractive index in liquids”, *Phys. Rev. Lett.*, 1964, 12, 507-509.

[6] Kelley, PL. “Self-focusing of optical beams”, *Phys. Rev. Lett.*, 1965, 15, 1005-1008.

[7] Giordmaine, JA; Miller, RC. “Tunable coherent parametric oscillation in $LiNbO_3$ at optical frequencies”, *Phys. Rev. Lett.*, 1965, 14, 973–975.

[8] Ashkin, A; Boyd, GD; Dziedzic, JM; Smith, RG; Ballman, AA; Levinstein, JJ; Nassau, K. “Optically induced refractive index inhomogenities in $LiNbO_3$ and $LiTaO_3$”, *Appl. Phys. Lett.*, 1966, 9, 72–75.

[9] Kosonocky, WF; Harrison, SE. “Saturation of absorption and fluorescence in solutions of phthalocyanines”, *J. Appl. Phys.*, 1966, 37, 4789-4797.

[10] Veduta, AP; Kirsanoy, BP. "Variation of the refractive index of liquids and glasses in a high intensity field of a ruby laser", *Sov. Phys. JETP.*, 1968, 27, 736-738.

[11] Stolen, RH; Lin, C. "Self-phase-modulation in silica optical fibers", *Phys. Rev. A*, 1978, 17, 1448-1453.

[12] Yariv, A; Pepper, DM. "Amplified reflection, phase conjugation, and oscillation in degenerate four-wave mixing", *Opt. Lett.*, 1977, 1, 16-18.

[13] Yariv, A; Fekete, D; Pepper, DM. "Compensation for channel dispersion by nonlinear optical phase conjugation", *Opt. Lett.*, 1979, 4, 52-54.

[14] Bloembergen, N. "Nonlinear optics: past, present, and future", *IEEE J. Sel. Top. Quantum. Electron.*, 2000, 6, 876-880.

[15] Van Driel, H. "Celebrating the fiftieth anniversary of nonlinear optics: focus issue introduction", J. *Opt. Soc. Am. B*, 2011, 28, NLO1.

[16] Yariv, A. "The beginning of quantum nonlinear optics", *J. Opt. Soc. Am. B*, 2011, 28, A83-87.

[17] Nie, WJ. "Optical nonlinearity: phenomena, applications, and materials", *Adv. Mater.*, 1993, 5, 520-545.

[18] Clark, J; Lanzani, G. "Organic photonics for communications", *Nat. photonics*, 2010, 4, 438-446.

[19] Hartland, GV. "Optical studies of dynamics in noble metal nanostructures", *Chem. Rev.*, 2011, 111, 3858-3887.

[20] Gonzalez, JC; Grancini, G; Lanzani, G. "Pump-probe spectroscopy in organic semiconductors: monitoring fundamental processes of relevance in optoelectronics", *Adv. Mater.*, 2011, 23, 5468-5485.

[21] Cundiff, ST. "Coherent spectroscopy of semiconductors", *Opt. Express*, 2008, 16, 4639-4664.

[22] Goodson, III TG. "Optical excitations in organic dendrimers investigated by time-resolved and nonlinear optical spectroscopy", *Acc. Chem. Res.*, 2005, 38, 99-107.

[23] Sibbett, W; Lagatsky, AA; Brown, CTA. "The development and application of femtosecond laser systems", *Opt. Express*, 2012, 20, 6989-7001.

[24] Cundiff, ST; Ye, J. "Colloquium: femtosecond optical frequency combs", *Rev. Mod. Phys.*, 2003, 75, 325-342.

[25] Mathias, S; Vorakiat, CLO; Grychtol, P; Granitzka, P; Turgut, E; Shaw, JM; Adam, R; Nembach, HT; Siemens, ME; Eich, S; Schneider, CM; Silva, T; Aeschlimann M; Murnane MM; Kapteyn, HC. "Probing the timescale of the exchange interaction in a ferromagnetic alloy", *Proc. Natl. Acad. Sci. USA*, 2012, 109, 4792-4797.

[26] Ulbricht, R; Hendry, E; Shan, J; Heinz, TF; Bonn, M. "Carrier dynamics in semiconductors studied with time-resolved terahertz spectroscopy", *Rev. Mod. Phys.*, 2011, 83, 543-586.

[27] Kirilyuk, A; Kimel, AV; Rasing, T. "Ultrafast optical manipulation of magnetic order", *Rev. Mod. Phys.*, 2010, 82, 2731-2784.

[28] Sheik-bahae, M; Said, AA; Van Stryland, WE. "High-sensitivity single-beam n_2 measurements", *Opt. Lett.*, 1989, 14, 955-957.

[29] Sheik-bahae, M; Said, AA; Wei, TH; Hagan, DJ; Van Stryland, WE. "Sensitive measurement of optical nonlinearities using a single beam", *IEEE J. Quantum. Electron*, 1990, 26, 760-769.

[30] Wang, J; Sheik-Bahae, M; Said, AA; Hagan, DJ; Van Stryland, WE. "Time-resolved z-scan measurements of optical nonlinearities", *J. Opt. Soc. Am. B*, 1994, 11, 1009-1017.

[31] Hughes, S; Burzler, JM. "Theory of z-scan measurements using Gaussian-Bessel beams", *Phys. Rev. A*, 1995, 56, R1103-R1106.

[32] Sheik-Bahae, M; Wang, J; DeSalvo, R; Hagan, DJ; Van Stryland, WE. "Measurement of nondegenarate nonlinearities using a two-color z-scan", *Opt. Lett.*, 1992, 17, 258–260.

[33] Kozich, VP; Marcano AO; Hernandez, FE; Castillo, JA. "Dual-beam time-resolved z-scan in liquids to study heating due to linear and nonlinear light absorption", *Appl. Spec.*, 1994, 48, 1506–1512.

[34] Kawazoe, T; Kawaguchi, H; Inoue, J; Haba, O; Ueda, M. "Measurement of nonlinear refractive index by time-resolved z-scan technique", *Opt. Commun.*, 1999, 160, 125–129.

[35] Castillo, J; Kozich, VP; Marcano, OA. "Thermal lensing resulting from one and two-photon absorption with a two-color time-resolved z-scan", *Opt. Lett.*, 1994, 19, 171–173.

[36] Zhao, W; Palffy-Muhoray, P. "Z-scan technique using top-hat beams", *Appl. Phy. Lett.*, 1993, 63, 1613-1615.

[37] Petrov, DV; Gomes, ASL; De Arabjo, BC. "Reflection z-scan technique for measurements of optical properties of surfaces", *Appl. Phy. Lett.*, 1994, 65, 1067-1069.

[38] Martinelli, M; Bian, S; Leite, JR; Horowicz, RJ. "Sensitivity-enhanced reflection z-scan by oblique incidence of a polarized beam", *Appl. Phy. Lett.*, 1998, 72, 1427-1429.

[39] Balu, M; Hales, J; Hagan, DJ; Van Stryland, WE. "White-light continuum z-scan technique for nonlinear materials characterization", *Opt. Express*, 2004, 12, 3820-3826.

[40] Chapple, PB; Staromlynska, J; McDuff, RG. "Z-scan studies in the thin and the thick-sample limits", *J. Opt. Soc. Am. B*, 1994, 11, 975-982.

[41] Zang, WP; Tian, JG; Liu, ZB; Zhou, WY; Song, F; Zhang, CP. "Analytic solutions to z-scan characteristics of thick media with nonlinear refraction and nonlinear absorption", *J. Opt. Soc. Am. B*, 2004, 21, 63-66.

[42] Kuzyk, MG; Dirk, CW. *Characterization techniques and tabulations for organic nonlinear materials* (1st edition), Marcel Dekker, 1998, 655-692.

[43] Righini, R. "Ultrafast optical kerr effect in liquids and solids", *Science*, 1993, 262, 1386-1390.

[44] Smith, N; Meech, SR. "Optically-heterodyne-detected optical Kerr effect (OHD-OKE): applications in condensed phase dynamics", *Int. Rev. in Phys. Chem.*, 2002, 21, 75-100.

[45] Hunt, NT; Jaye, AA; Meech, SR. "Ultrafast dynamics in complex fluids observed through the ultrafast optically-heterodyne-detected optical-Kerr-effect (OHD-OKE)", *Phys. Chem. Chem, Phys.*, 2007, 9, 2167-2180.

[46] Zhong, Q; Fourkas, JT. "Optical kerr effect spectroscopy of simple liquids", *J. Phys. Chem. B*, 2008, 112, 15529-15539.

[47] Katayama, K; Yamaguchi, M; Sawada, T. "Lens-free heterodyne detection for transient grating experiments", *Appl, Phy. Lett.*, 2003, 17, 2775-2777.

[48] Wang, Y; Lin, CY; Nikolaenko, A; Raghunathan, V; Potma, EO. "Four-wave mixing microscopy of nanostructures", *Adv. Opt. Photon.*, 2010, 3, 1-52.

[49] Dao, LV; Lincoln, CN; Lowe, RM; Do, MTT; Eckle, P; Hannaford, P. "Spectrally resolved two-color femtosecond photon echoes: a new multidimensional technique for studying molecular dynamics", *Laser Phys.*, 2005, 15, 1257-1268.

[50] Khalil, M; Demirdoven, N; Tokmakoff, A. "Coherent 2D IR spectroscopy: molecular structure and dynamics in solution", *J. Phys. Chem. A*, 2003, 107, 5258-5279.

[51] Hochstrasser, RM. "Two-dimensional spectroscopy at infrared and optical frequencies", *Proc. Natl. Acad. Sci. USA*, 2009, 104, 14190 –14196.

[52] Boyd, RW. *Nonlinear optics* (3rd edition), Academic Press, 2007, 1-34.

[53] Leuthold, J; Koos, C; Freude, W. "Nonlinear silicon photonics", *Nat. Photonics*, 2010, 4, 535-544.

[54] Tokmakoff, A. "Orientational correlation functions and polarization selectivity for nonlinear spectroscopy of isotropic media I: Third order", *J. Chem. Phys.*, 1996, 105, 1-12.

[55] Richards, OW. "Phase difference microscopy", *Nature*, 1944, 154, 672.

[56] Linfoot, EH. "Phase difference microscopy", *Nature*, 1945, 155, 76.

[57] Zernike, F. "How I discovered phase contrast", *Science*, 1955, 121, 345-349.

[58] Holzner, C; Feser, M; Vogt, S; Hornberger, B; Baines, S; Jacobsen, C. "Zernike phase contrast in scanning microscopy with X-rays", *Nat. Phys.*, 2010, 6, 883-887.

[59] Martin, LC. "Phase-contrast methods in microscopy", *Nature*, 1947, 159, 827-830.

[60] Boudebs, G; Chis, M; Bourdin, JP. "Third-order susceptibility measurements by nonlinear image processing", *J. Opt. Soc. Am. B*, 1996, 13, 1450-1456.

[61] Boudebs, G; Chis, M; Monteil, A. "Contrast increasing by third-order nonlinear image processing: a numerical study for microscopic rectangular objects", *Opt. Commun.*, 1998, 150, 287-296.

[62] Cherukulappurath, S; Boudebs, G; Monteil, A. "4*f* coherent image system and its application to nonlinear optical measurements", *J. Opt. Soc. Am. B*, 2004, 21, 273–279.

[63] Boudebs, G; Cherukulappurath, S. "Nonlinear optical measurements using a 4*f* coherent imaging system with phase objects", *Phys. Rev. A*, 2004, 69, 053813.

[64] Boudebs, G; Fedus, K; Cassagne, C; Leblond, H. "Degenerate multiwave mixing using z-scan technique", *Appl. Phys. Lett.*, 2008, 93, 021118-021120.

[65] Godet, JL; Derbal, H; Cherukulappurath, S; Boudebs, G. "Optimization and limits of optical nonlinear measurements using imaging technique", *Eur. Phys. J. D*, 2006, 39, 307–312.

[66] Boudebs, G; Cherukulappurath, S. "Nonlinear refraction measurements in presence of nonlinear absorption using phase object in a 4*f* system", *Opt. Commun.*, 2005, 250, 416-420.

[67] Boudebs, G; De Araujo, CB. "Characterization of light-induced modification of the nonlinear refractive index using a one-laser-shot nonlinear imaging technique", *Appl. Phys. Lett.*, 2004, 85, 3740–3742.

[68] Li, YB; Zhang, XR; Wang, YX; Yang, K; Song, YL. "Optimization of phase objects in 4*f* coherent imaging system for nonlinear refraction measurements", *Opt. Commun.*, 2006, 266, 686-690.

[69] Li, YB; Song, YL; Wei, TH; He, CY; Zhang, XR; Wang, YX; Yang, K; Shen, M; Yang, JY. "Measurements of third-order nonlinearity of inhomogeneous samples using nonlinear-imaging technique with a phase object", *Appl. Phys. B*, 2008, 91, 119-122.

[70] Li, YB; Song, YL; Yang, JY; Yang, K; Zhang, XR; Wang, YX. "Measurement of third-order nonlinear refraction using a nonlinear-imaging technique with positive–negative phase object", *Appl. Phys. B*, 2008, 92, 253-257.

[71] Shui, M; Luo, XY; Zhang, XR; Jin, X; Li, CW; Yang, JY; Li, ZG; Shi, G; Yang, K; Wang, YX; Song, YL. "Phase filtering in the nonlinear-imaging technique with a phase object", *J. Opt. Soc. Am. A*, 2010, 27, 2514-2523.

[72] Bredas, JL; Adant, C; Tackx, P; Persoons, A. "Third-order nonlinear optical response in organic materials: theoretical and experimental aspects", *Chem. Rev.*, 1994, 94, 243-278.

[73] Ghebbemichael, F; Kuzyk, MG; Lackritz, HS. "Nonlinear optics and polymer physics", *Prog. Polym. Sci.*, 1997, 22, 1147-1201.

[74] Khoo, IC. "Nonlinear optics of liquid crystalline materials", *Phys. Rep.*, 2009, 471, 221-267.

[75] Innocenzi, P; Lebeau, B. "Organic-inorganic hybrid materials for non-linear optics", *J. Mater. Chem.*, 2005, 15, 3821-3831.

[76] Banfi, GP; Degiorgio, V. "Nonlinear optical properties of semiconductor nanocrystals", *Adv. Phys.*, 1998, 47, 447-510.

[77] Marder, SR; Kippelen, B; Jen, AKY; Peyghambarian, N. "Design and synthesis of chromophores and polymers for electro-optic and photorefractive applications", *Nature*, 1997, 388, 845-851.

[78] Haque, S; Nelson, J. "Toward organic all-optical switching", *Science*, 2010, 327, 1466-1467.

[79] Hales, JM; Matichak, J; Barlow, S; Ohira, S; Yesudas, K; Bredas, JL; Perry, JW; Marder, SR. "Design of polymethine dyes with large third-order optical nonlinearities and loss figures of merit", *Science*, 2010, 327, 1485-1488.

[80] Li, CW; Wang, YX; Shui, M; Yang, JY; Jin, X; Zhang, XR; Yang, K; Song, YL. "Characterization of signs change of nonlinear refraction in ZnSe based on a modified double 4*f* imaging system with a phase object", *Opt. Commun.*, 2009, 283, 1124-1128.

[81] Yang, JY; Zhang, XR; Wang, YX; Shui, M; Li, CW; Jin, X; Song, YL. "Method with a phase object for measurement of optical nonlinearities", *Opt. lett.*, 2009, 34, 2513-2515.

[82] Yang, JY; Song, YL. "Direct observation of the transient thermal-lensing effect using the phase-object z-scan technique", *Opt. lett.*, 2009, 34, 157-159.

[83] Christodoulides, ND; Khoo, IC; Salamo, JG; Stegeman, IG; Van Stryland, WE. "Nonlinear refraction and absorption: mechanisms and magnitudes", *Adv. Opt. Photon.*, 2010, 2, 60-200.

[84] Krausz, F; Ivanov, M. "Attosecond physics", *Rev. Mod. Phys.*, 2009, 81, 163-234.

[85] Li, YB; Pan, GF; Yang, K; Zhang, XR; Wang, YX; Wei, TH; Song, YL. "Time-resolved pump-probe system based on the nonlinear-imaging technique with a phase object", *Opt. Express*, 2008, 16, 6251-6259.

[86] Yang, JY; Song, YL; Wang, YX; Li, CW; Jin, X; Shui, M. "Time-resolved pump-probe technology with phase object for measurements of optical nonlinearities", *Opt. Express*, 2009, 17, 7110-7116.

[87] Zhong, Q; Zhu, X; Fourkas, JT. "Antiresonant-ring Kerr spectroscopy", *Opt. Express*, 2007, 15, 6561-6568.

[88] Koos, C; Vorreau, P; Vallaitis, T; Dumon, P; Bogaerts, W; Baets, R; Esembeson, B; Biaggio, I; Michinobu, T; Diederich, F; Freude, W; Leuthold, J. "All-optical high-speed

signal processing with silicon-organic hybrid slot waveguides", *Nat. Photonics*, 2009, 3, 216-219.

[89] Fecko, CJ; Eaves, JD; Tokmakoff, A. "Isotropic and anisotropic Raman scattering from molecular liquids measured by spatially masked optical Kerr effect spectroscopy", *J. Chem. Phys.*, 2002, 117, 1139-1154.

[90] Gardecki, JA; Yu, G; Constantine, S; Peng, J; Zhou, Y; Ziegler, LD. "A unified treatment of ultrafast optical heterodyne detected and z-scan spectroscopies", *J. Chem. Phys.*, 2001, 114, 3586-3597.

[91] Ganeev, RA. "Single-shot reflection z-scan for measurements of the nonlinear refraction of nontransparent materials", *Appl. Phys. B*, 2008, 91, 273-277.

[92] Yang, JY; Song, YL; Wang, YX; Li, CW; Jin, X; Shui, M. "Reflection 4*f* coherent imaging technique for measurements of optical nonlinearity", *Opt. Commun.*, 2010, 283, 209-215.

[93] Fishman, DA; Cirloganu, CM; Webster, S; Padilha, LA; Monroe, M; Hagan, DJ; Van Stryland, WE. "Sensitive mid-infrared detection in wide-bandgap semiconductors using extreme non-degenerate two-photon absorption", *Nat. Photonics*, 2011, 5, 561-565.

[94] Konstantatos, G; Sargent, EH. "Nanostructured materials for photon detection", *Nat. Nanotech.*, 2010, 5, 391-400.

[95] Hu, XY; Jiang, P; Ding, CY; Yang, H; Gong, QH. "Picosecond and low-power all-optical switching based on an organic photonic-bandgap microcavity", *Nat. Photonics*, 2008, 2, 185-189.

[96] Shui, M; Jin, X; Li, ZG; Yang, JY; Shi, G; Nie, ZQ; Wu, XZ; Wang, YX; Yang, K; Zhang, XR; Song, YL. "Investigation of nonlinear dynamics in CdS by time-resolved nondegenerate pump–probe system with phase object", *J. Mod. Opt.*, 2011, 58, 973-977.

[97] Shui, M; Li, ZG; Jin, X; Yang, JY; Nie, ZQ; Shi, G; Wu, XZ; Yang, K; Zhang, XR; Wang, YX; Song, YL. "Measurements of dynamics of nondegenerate optical nonlinearity in ZnS with pulses from optical parameter generation", *Opt. Commun.*, 2011, 285, 1940-1944.

In: Nonlinear Performance and Characterization Methods … ISBN: 978-1-62808-093-3
Editor: Jingsong Wei

Chapter 2

Theory and Applications of the Z-Scan Technique through Beam Dimension Measurements

George Tsigaridas[1,2,3*]

[1]Department of Physics, National Technical University of Athens, Zografou, Athens, Greece
[2]Department of Informatics and Computer Technology, Technological and Educational Institute of Lamia, Lamia, Greece
[3]Department of Physics, University of Patras, Rion Patras, Greece

Abstract

In this chapter we describe a modification of the well-known z–scan technique, based on the direct measurement of the beam dimensions in the far field. More specifically, the quantity measured in the case of circular Gaussian beams is the beam radius, while in the case of elliptic Gaussian beams the measured quantities are the lengths of the principal semiaxes of the beam. The measurements can easily be performed through a CCD camera in connection with a laser beam profiler. The advantages of the proposed modification compared to the classic z-scan technique are the elimination of the sensitivity in beam pointing instability as well as the drastic reduction of the sensitivity in energy fluctuations. This technique can also be applied to the more general case of nearly-Gaussian beams, namely Gaussian beams perturbed by a small contribution of higher-order Hermite-Gaussian modes.

Samples exhibiting higher-order nonlinearities can also be measured by this technique. Further, the simultaneous presence of nonlinear refraction and absorption is discussed, and it is shown that the effects of nonlinear absorption can be suppressed by multiplying the closed and open aperture z-scan curves. Finally, a simplified approach for determining the nonlinear refractive index in the case of an astigmatic (elliptic) Gaussian beam is presented.

[*] Corresponding author e–mail:gtsig@physics.upatras.gr.

1. INTRODUCTION

The z-scan has been used to date as a simple and accurate technique for measuring optical nonlinearities [1-3]. However, the choice of the normalized transmittance [1-3], as the measured quantity, implies several disadvantages [3], such as high sensitivity to beam energy fluctuations and pointing instability, as well as difficulties in the extension of the technique to more general input beam profiles (for example, an elliptic Gaussian beam). Therefore, we have introduced the measuring of more fundamental quantities, namely the beam dimensions (and their variations) in the far field.

More specifically, in the case of a circular Gaussian beam the measured quantity is the beam radius in the far field, defined as the distance from the beam center to the points where the intensity reduces to a certain fraction q of its peak value, as shown in Figure 1.

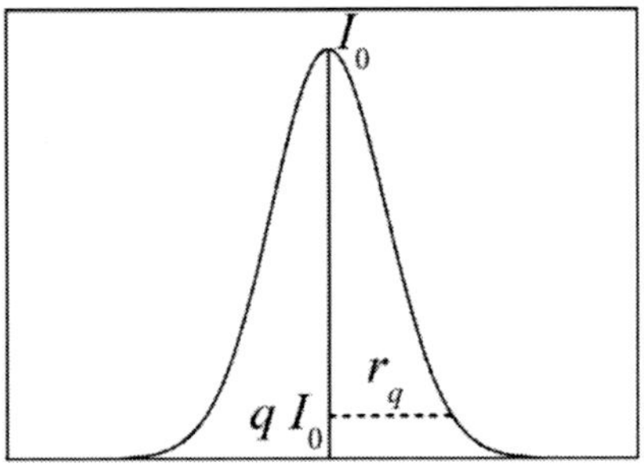

Figure 1. Definition of the beam radius.

It is calculated through the relation

$$I\left(r_q, D, t; \Delta\varphi_0(z_s, t)\right) = \frac{c\varepsilon_0}{2}\left|E\left(r_q, D, t; \Delta\varphi_0(z_s, t)\right)\right|^2 = qI\left(0, D, t; \Delta\varphi_0(z_s, t)\right) \tag{1}$$

where c is the speed of light in the vacuum, ε_0 the vacuum permittivity, $I[r_q, D, t;\ \Delta\varphi_0(z_s, t)]$ the intensity of the beam at a distance D from the exit plane of the sample ($z = z_s$), and $E[r_q, D, t;\ \Delta\varphi_0(z_s, t)]$ the corresponding electric field pattern. The latter can be analytically calculated using either diffraction theory [4-5] or Gaussian decomposition [2, 6-8]. $\Delta\varphi_0(z_s, t)$ is the nonlinear phase shift induced to the beam by the optical nonlinearities (either refractive or absorptive) of the material, at the sample position. For the case of a single first order refractive nonlinearity, it is related to the nonlinear refractive index n_2 by the formula [2]

$$\Delta\varphi_0(z_s, t) = -k\Delta n_0(z_s, t)L_{eff} = -kn_2 I_0(z_s, t)L_{eff} \tag{2}$$

where k is the wavenumber, $\Delta n_0(z_s, t)$ the on-axis refractive index change, L_{eff} the effective length of the sample [2] and $I_0(z_s, t)$ the on-axis value of the intensity of the incident beam at the sample plane. Fitting Equation (1) to the experimental z-scan measurements the phase parameter $\Delta\varphi_0(z_s, t)$ and consequently the nonlinear refractive index n_2 can be calculated.

The value of q should be chosen properly in order to maximize the signal to noise ratio. In more detail, as the value of q decreases the sensitivity increases, because the beam radius lies in the region of the beam wings and its variations become very large. However, for very low values of q the beam intensity is comparable to the noise level, introducing high uncertainty in the measurements. Therefore, an optimum level of q should be chosen, providing the highest signal-to-noise ratio.

In the more general case of an elliptic Gaussian incident beam, the measured quantities are the lengths of the principal semiaxes of the beam (x_q, y_q).

They are defined as the distances from the beam center to the points at which the intensity drops to a certain fraction q of its peak $(x = y = 0)$ value and are calculated through the formulas

$$I\left(x_q,0,D,t;\Delta\varphi_0(z_s,t)\right)=\frac{c\varepsilon_0}{2}\left|E\left(x_q,0,D,t;\Delta\varphi_0(z_s,t)\right)\right|^2=qI\left(0,0,D,t;\Delta\varphi_0(z_s,t)\right) \tag{3a}$$

$$I\left(0,y_q,D,t;\Delta\varphi_0(z_s,t)\right)=\frac{c\varepsilon_0}{2}\left|E\left(0,y_q,D,t;\Delta\varphi_0(z_s,t)\right)\right|^2=qI\left(0,0,D,t;\Delta\varphi_0(z_s,t)\right) \tag{3b}$$

where $I[x,y,D,t;\ \Delta\varphi_0(z_s,t)]$ the intensity of the beam at a distance D from the exit plane of the sample $(z = z_s)$, and $E[x,y,D,t;\ \Delta\varphi_0(z_s,t)]$ the corresponding electric field pattern. In both cases, the electric field pattern can be calculated quite straightforwardly through the Gaussian decomposition method, as described below.

2. Calculation of the Far Field Pattern for Circular and Elliptic Gaussian Beams

The electric field pattern of the beam can be analytically calculated through the Gaussian decomposition method, which is described in the current section. First, let us suppose that an elliptic Gaussian laser beam, including astigmatism, is focused by a lens (Figure 2(a)). The electric field of this beam at a distance z from the lens is given by theEquation [9, 10]

$$E(x,y,z,t)=E_0(t)\left(\frac{w_{0x}w_{0y}}{w_x(z)w_y(z)}\right)^{1/2}\exp\left(-\frac{x^2}{w_x^2(z)}-\frac{y^2}{w_y^2(z)}\right)$$
$$\times\exp\left(-\frac{ikx^2}{2R_x(z)}-\frac{iky^2}{2R_y(z)}\right)\exp\left[-ikz+i\theta(z)\right] \tag{4}$$

where $E_0(t)$ is a constant in space containing the temporal envelope of the electric field, k is the wave number, and w_{0x}, w_{0y} are the beam widths at the two waists of this astigmatic elliptic beam. The parameters $w_{x,y}(z)$, $R_{x,y}(z)$, $\theta(z)$ are the beam widths, radii of curvature and on axis phase shift of the beam, defined by the relations

$$w_{x,y}(z) = w_{0x,y}\left[1+\left(\frac{z-z_{0x,y}}{z_{Rx,y}}\right)^2\right]^{1/2} \tag{5}$$

$$R_{x,y}(z) = \left(z-z_{0x,y}\right)\left[1+\left(\frac{z_{Rx,y}}{z-z_{0x,y}}\right)^2\right] \tag{6}$$

$$\theta(z) = \frac{1}{2}\tan^{-1}\left(\frac{z-z_{0x}}{z_{Rx}}\right)+\frac{1}{2}\tan^{-1}\left(\frac{z-z_{0y}}{z_{Ry}}\right) \tag{7}$$

where $z_{0x,y}$ are the locations of the waists and $z_{Rx,y}$ the Rayleigh lengths of the incident beam for each principal direction (x, y). All these parameters are shown in Figures 2(b) and 2(c).

If this beam passes through a thin sample of a nonlinear material, a phase shift of the form

$$\Delta\varphi(x, y, z_s, t) = -k\Delta n(x, y, z_s, t)L_{eff} \tag{8}$$

is induced [2]. In the Equation (8), z_s is the sample distance from the focusing lens, $L_{eff} = [1-exp(-a_0L)]/a_0$ the effective propagation length inside the nonlinear material and $\Delta n(x, y, z_s, t) = n_2I(x, y, z_s, t)$ the refractive index change due to the nonlinearity of the material. Here, L is the sample length, a_0 its linear absorption coefficient, n_2 the nonlinear refractive index of the material, and $I(x, y, z_s, t)$ the irradiance profile of the incident beam. Thus, the electric field pattern at the exit plane of the sample takes the form

$$E_e(x, y, z_s, t) = E_{in}(x, y, z_s, t)\exp\left[i\Delta\varphi(x, y, z_s, t)\right]\exp\left(-a_0L/2\right) \tag{9}$$

where $E_{in}(x, y, z_s, t)$ is the electric field pattern of the incident beam, given by Equation (4), for $z = z_s$. According to the Gaussian decomposition method [2, 6-8], if we expand the induced phase shift $\Delta\varphi(x, y, z_s, t)$ into a Taylor series, the electric field pattern at the exit plane of the sample can be expressed as a linear combination of elliptic Gaussian beams. These can be propagated to any desired distance D from the exit plane of the sample and the electric field at this distance is given by the equation

$$E_e\left(x, y, D, t; \Delta\varphi_0(z_s, t)\right) = E_0(t)\left(\frac{w_{0x}w_{0y}}{w_x(z_s)w_y(z_s)}\right)^{1/2}\exp\left(-\frac{a_0L}{2}\right) \times \sum_{m=0}^{\infty}\frac{[i\Delta\varphi_0(z_s, t)]^m}{m!}E^{(m)}(x, y, D) \tag{10}$$

where $\Delta\varphi_0(z_s, t)$ is the on-axis ($x = y = 0$) phase shift, and $E^{(m)}(x, y, D)$ the electric field pattern of each derivative beam, given by the formula

$$E^{(m)}(x,y,D) = F^{(m)}\left(\frac{w_{0x}^{(m)}w_{0y}^{(m)}}{w_x^{(m)}(D)w_y^{(m)}(D)}\right)^{1/2} \exp\left[-\left(\frac{x}{w_x^{(m)}(D)}\right)^2 - \left(\frac{y}{w_y^{(m)}(D)}\right)^2\right]$$

$$\times\exp\left(-\frac{ikx^2}{2R_x^{(m)}(D)} - \frac{iky^2}{2R_y^{(m)}(D)}\right)\exp\left[-ik\left(z_s + D\right) + i\left(\theta^{(m)}(D) + \eta^{(m)}\right)\right] \tag{11}$$

where

$$w_{0x,y}^{(m)} = \left(\frac{2z_{Rx,y}^{(m)}}{k}\right)^{1/2} \tag{12}$$

$$w_{x,y}^{(m)}(D) = w_{0x,y}^{(m)}\left[1 + \left(\frac{d_{x,y}^{(m)} + D}{z_{Rx,y}^{(m)}}\right)^2\right]^{1/2} \tag{13}$$

$$R_{x,y}^{(m)}(D) = \left(d_{x,y}^{(m)} + D\right)\left[1 + \left(\frac{z_{Rx,y}^{(m)}}{d_{x,y}^{(m)} + D}\right)^2\right] \tag{14}$$

$$\theta^{(m)}(D) = \frac{1}{2}\tan^{-1}\left(\frac{d_x^{(m)} + D}{z_{Rx}^{(m)}}\right) + \frac{1}{2}\tan^{-1}\left(\frac{d_y^{(m)} + D}{z_{Ry}^{(m)}}\right) \tag{15}$$

$$F^{(m)} = \left(\frac{w_x^{(m)}(D=0)w_y^{(m)}(D=0)}{w_{0x}^{(m)}w_{0y}^{(m)}}\right)^{1/2} = \left[1 + \left(\frac{d_x^{(m)}}{z_{Rx}^{(m)}}\right)^2\right]^{1/4}\left[1 + \left(\frac{d_y^{(m)}}{z_{Ry}^{(m)}}\right)^2\right]^{1/4} \tag{16}$$

and

$$\eta^{(m)} = \theta(z_s) - \theta^{(m)}(D=0)$$

$$= \frac{1}{2}\tan^{-1}\left(\frac{z_s - z_{0x}}{z_{Rx}}\right) + \frac{1}{2}\tan^{-1}\left(\frac{z_s - z_{0y}}{z_{Ry}}\right) - \frac{1}{2}\tan^{-1}\left(\frac{d_x^{(m)}}{z_{Rx}^{(m)}}\right) - \frac{1}{2}\tan^{-1}\left(\frac{d_y^{(m)}}{z_{Ry}^{(m)}}\right) \tag{17}$$

The parameters $F^{(m)}$ and $\eta^{(m)}$ have been introduced for consistency with the electric field pattern at the exit plane of the sample. Here $d_{x,y}^{(m)}$ are the distances between the sample and the waists of each derivative beam for the two principal directions (x, y), defined positive if the sample plane is behind the beam waist.

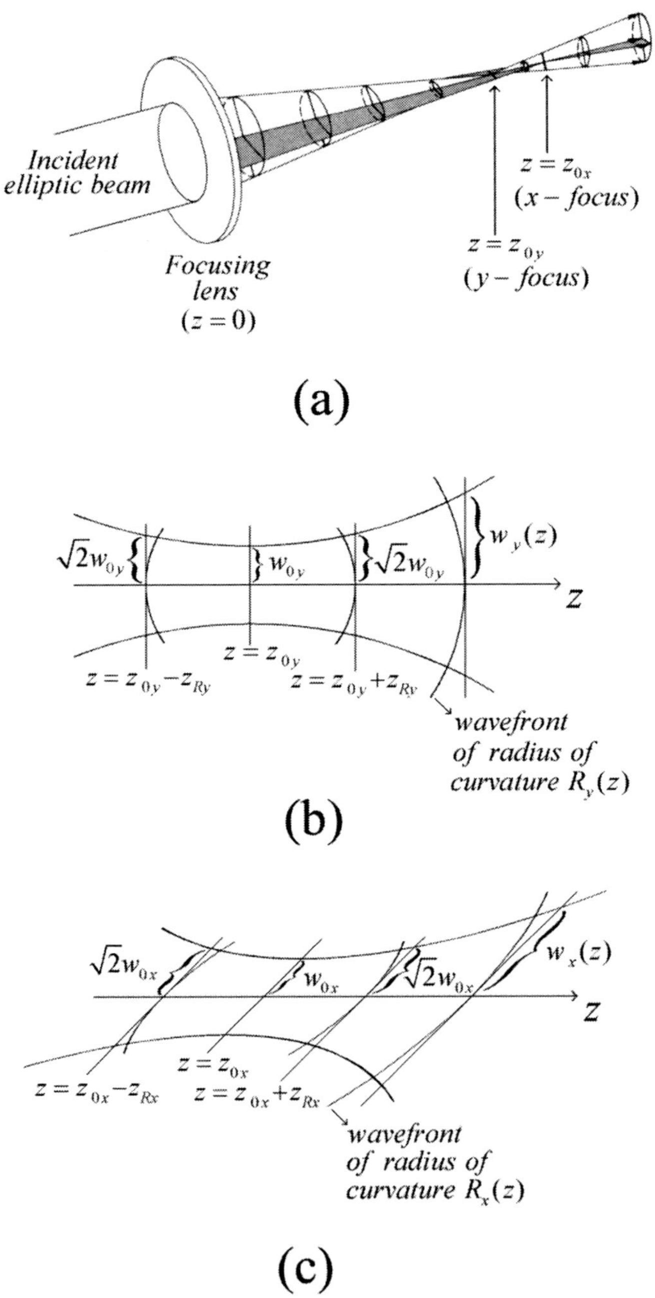

Figure 2. The basic parameters of a focused elliptic Gaussian beams: (a) Beam focusing, (b) details of the beam focus on the yz plane and (c) details of the beam focus on the *xz* plane.

Also $z_{Rx,y}^{(m)}$ are the Rayleigh lengths of each derivative beam and each principal direction. These parameters are given by the equations

$$d_{x,y}^{(m)}=\frac{\left(A_{x,y}^{(m)}\right)^2 R_{x,y}(z_s)}{\left(A_{x,y}^{(m)}\right)^2+R_{x,y}^2(z_s)} \tag{18}$$

and

$$z_{Rx,y}^{(m)}=\frac{A_{x,y}^{(m)} R_{x,y}^2(z_s)}{\left(A_{x,y}^{(m)}\right)^2+R_{x,y}^2(z_s)} \tag{19}$$

where

$$A_{x,y}^{(m)}=\frac{k}{2}\frac{w_{x,y}^2(z_s)}{2m+1} \tag{20}$$

Thus, the electric field pattern and consequently the irradiance distribution of an elliptic Gaussian beam passed through a thin nonlinear optical material have been determined in the far field.

In the special case of a circular Gaussian beam, where $w_{0x}=w_{0y}=w_0$ and $z_{0x}=z_{0y}=z_0$, Equations (10)–(20) become

$$E_e\left(r,D,t;\Delta\varphi_0(z_s,t)\right)=E_0(t)\frac{w_0}{w(z_s)}\exp\left(-\frac{a_0L}{2}\right)\sum_{m=0}^{\infty}\frac{[i\Delta\varphi_0(z_s,t)]^m}{m!}E^{(m)}(r,D) \tag{21}$$

Each derivative beam is given by the formula

$$E^{(m)}(r,D)=F^{(m)}\frac{w_0^{(m)}}{w^{(m)}(D)}\exp\left[-\left(\frac{r}{w^{(m)}(D)}\right)^2\right]$$

$$\times\exp\left(-\frac{ikr^2}{2R^{(m)}(D)}\right)\exp\left[-ik\left(z_s+D\right)+i\left(\theta^{(m)}(D)+\eta^{(m)}\right)\right] \tag{22}$$

where

$$w_0^{(m)}=\left(\frac{2z_R^{(m)}}{k}\right)^{1/2} \tag{23}$$

$$w^{(m)}(D) = w_0^{(m)} \left[1 + \left(\frac{d^{(m)} + D}{z_R^{(m)}} \right)^2 \right]^{1/2} \tag{24}$$

$$R^{(m)}(D) = \left(d^{(m)} + D \right) \left[1 + \left(\frac{z_R^{(m)}}{d^{(m)} + D} \right)^2 \right] \tag{25}$$

$$\theta^{(m)}(D) = \tan^{-1} \left(\frac{d^{(m)} + D}{z_R^{(m)}} \right) \tag{26}$$

$$F^{(m)} = \frac{w^{(m)}(D=0)}{w_0^{(m)}} = \left[1 + \left(\frac{d^{(m)}}{z_R^{(m)}} \right)^2 \right]^{1/2} \tag{27}$$

and

$$\eta^{(m)} = \theta(z_s) - \theta^{(m)}(D=0) = \tan^{-1} \left(\frac{z_s - z_0}{z_R} \right) - \tan^{-1} \left(\frac{d^{(m)}}{z_R^{(m)}} \right) \tag{28}$$

In this case

$$d^{(m)} = \frac{\left(A^{(m)} \right)^2 R\,(z_s)}{\left(A^{(m)} \right)^2 + R^2(z_s)} \tag{29}$$

and

$$z_R^{(m)} = \frac{A^{(m)} R^2(z_s)}{\left(A^{(m)} \right)^2 + R^2(z_s)} \tag{30}$$

where

$$A^{(m)} = \frac{k}{2} \frac{w^2(z_s)}{2m+1} \tag{31}$$

More general cases can also be treated by the Gaussian decomposition method, namely nearly Gaussian beams [7] or higher order nonlinearities [8]. The Gaussian decomposition method for high-order nonlinearities, which more usually occurs in practice, is described in some detail in the Appendix.

3. EXPERIMENTAL APPLICATIONS

Experimentally, the lengths of the principal semiaxes (x_q, y_q) can be measured through a laser beam analysis system, e.g. a CCD camera connected with a laser beam profiler. The use of this system entails several advantages, such as elimination of the problems due to beam pointing instability [3] as well as better supervision of the entire beam profile during the experiment. Furthermore, the choice of the beam radius defined by Equation (1) as the measured quantity reduces the sensitivity in beam energy fluctuations [11]. This occurs because the beam radius is determined only by the spatial distribution of the power (Equation (1) and not by its total amount. The experimental setup is shown in Figure 3.

In our case we employed a Spiricon, Inc. LBA – 300 PC Laser Beam Analyzer, in connection to a Cohu, Inc. 4915 RS–170 two–dimensional CCD camera having 768×494 pixels (pixel size $8.4\, um \times 9.8\, um$). The sample used was the standard nonlinear material CS_2 placed in a cuvette of 1 mm thickness. The laser employed was a high repetition rate (82 MHz) femtosecond system (TSUNAMI-Spectra Physics) pumped by a continuous wave (CW) doubled frequency Nd: YVO_4 laser (MILLENNIA – Spectra Physics). The duration of the pulses was $\sim 85\, fs$ while their central wavelength was at $800\, nm$. The beam was focused through a lens of focal length $12.5\, cm$.

For the case of a circular Gaussian input beam, the average power of the laser beam was $\sim 770\, mW$ corresponding to pulse energy of $\sim 9.4\, nJ$. The beam radius at the waist was estimated to be $w_0 \approx 55\, um$ leading to a peak value of the on-axis irradiance at focus $I_0 \approx 2.4\, GW/cm^2$. For the beam radius measurements the value of q was chosen to be 0.08, an optimum one under our experimental conditions achieving the highest sensitivity for a given noise level. The results are shown in Figure 4.

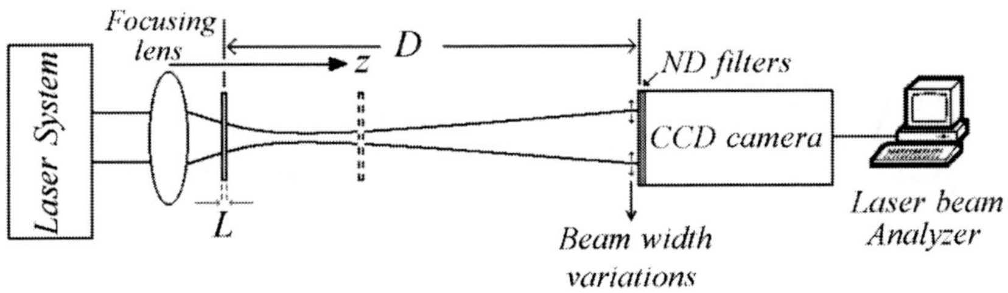

Figure 3. Experimental setup for the z-scan technique through beam dimension measurements.

The solid curve represents a least square fitting of the beam radius as defined by Equation (1) to the experimental results. In this procedure, the Gaussian decomposition described in section 2 and more specifically Equations (21-31) are utilized for determining the far field pattern of the beam. It should be noted that because of the high repetition rate of the laser system (82 MHz) the thermal nonlinearities are dominant. Due to the much longer relaxation time of the induced thermal lens in CS_2 ($\sim 100\, msec$) [12] compared with the temporal separation of the pulses ($\sim 12.2\ nsec$) the sample is considered to be in a quasi–steady state. Under these conditions, only the average values of the nonlinear optical parameters can be calculated. The fitting procedure provided an average value of the nonlinear phase shift $\langle \Delta\Phi_0 \rangle = -\ 0.78$, which through Equation (2) corresponds to a refractive index change of $\langle \Delta n_0 \rangle = -9.9\times10^{-5}$. In order to compare our results with those reported in the literature [13] an equivalent refractive index $n_{2,eq} = \langle \Delta n_0 \rangle / I_0$ for the thermal nonlinearities has been used. In our case the value of $n_{2,eq}$ are $-4.1 \times 10^{-14}\ cm^2/W$.

For verifying the validity of the proposed technique a standard (transmittance z-scan) has been simultaneously performed. The transmittance measurements were made by setting a soft aperture with linear transmission *S*=0.4 in the software of the beam profiler and measuring the fraction of the beam energy inside the aperture as function of the sample position. The experimental results are shown in Figure 5.

It should be noted that the transmittance z-scan plot has an opposite configuration (peak-valley) compared to the radius z-scan plot (valley-peak) because the transmitted power through the aperture increases as the beam radius decreases and vice-versa.

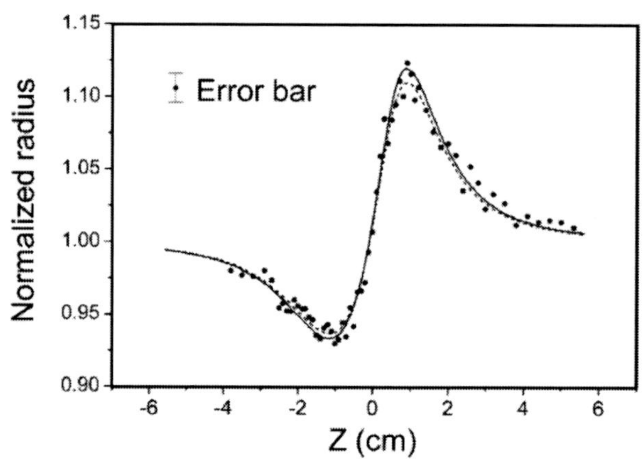

Figure 4. Radius z-scan plot for a circular Gaussian beam.

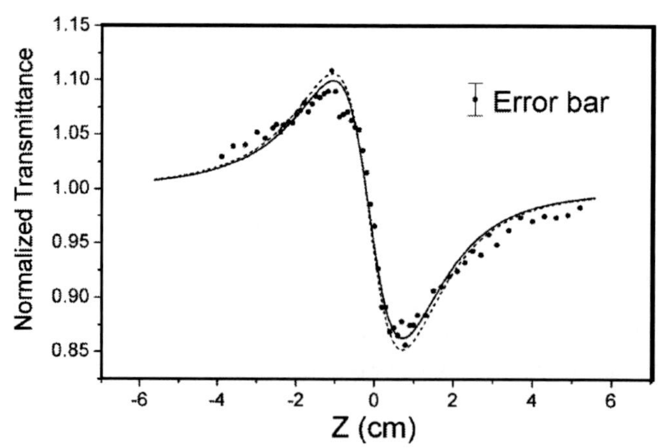

Figure 5. Transmittance z-scan plot for a circular Gaussian beam.

In this case the power transmittance is calculated through the relation

$$T(z)=\frac{\int_{-\infty}^{+\infty} P_T\left(z,\Delta\Phi_0(t)\right) dt}{S\int_{-\infty}^{+\infty} P_i(t)\, dt} \tag{32}$$

where $P_i(t)$ is the instantaneous input power (within the sample), S the aperture linear transmittance and $P_T[z,\Delta\Phi_0(t)]$ the transmitted power through the aperture obtained by spatially integrating the irradiance up to the aperture radius r_a, i.e.

$$P_T\left(z,\Delta\Phi_0(t)\right)=c\varepsilon_0\pi\int_0^{r_a}\left|E_e\left(r,D,t;\Delta\varphi_0(z_s,t)\right)\right|^2 rdr \tag{33}$$

where the electric field pattern is still calculated through Equations (21-31). The least square fitting of Equation (32) to the experimental transmittance results gave the value $\langle\Delta\Phi_0\rangle = -0.68$ for the on-axis phase shift, corresponding to $\langle\Delta n_0\rangle = -8.7\times10^{-5}$ and $n_{2,eq} = -3.6\times10^{-14} cm^2/W$. The reasonable agreement of the γ_{eq} values for the two cases verifies the validity of the suggested modification of the z–scan technique. Also, these values are comparable with the ones reported in the literature [13], within the experimental error arising mainly from the uncertainty in the determination of the value of the on-axis intensity $I_0(z_s,t)$.

Extensive numerical simulations of z-scan experiments based on the variations of the beam radius have been performed. They have shown that for $|\Delta\Phi_0|\leq 1$ the difference between the normalized peak and valley radii Δr_{P-V} in a radius z–scan plot depends almost linearly on $|\Delta\Phi_0|$. Specifically, we have found that Δr_{P-V} depends linearly on $|\Delta\Phi_0|$ and the explicit formula connecting the two quantities is [11]

$$\Delta r_{P-V} = 0.154\, q^{-0.214}\left|\Delta\Phi_0\right| \tag{34}$$

which is valid within an accuracy of ±2%. The above equation can be used to readily estimate the on-axis phase shift $\Delta\Phi_0$ nonlinear refractive index n_2 by simply measuring the normalized peak–valley difference. Also, Equation (34) provides a measure of the sensitivity of this technique, defined as [11]

$$p = \Delta r_{P-V} / \left|\Delta\Phi_0\right| = 0.154\, q^{-0.214} \tag{35}$$

Its dependence on the parameter q is shown in Figure 6. As shown in theFigure (6), decreasing the value of q, the sensitivity of the method increases dramatically and it finally becomes considerably higher than that of the original z–scan technique. This occurs because for low values of q the beam radius lies in the region of the beam wings and its variations become very large. However, for very low values of q the beam intensity is comparable to the noise level, introducing high uncertainty into the measurements, which is also evident by the large error bars in the experimental values. Therefore, the value of q must be kept to an optimum level in order to achieve a reliable signal to noise ratio. In our case this value is 0.08 as mentioned above.

For the measurements based on an elliptic Gaussian incident beam we incorporated the same laser system. The ellipticity has been induced through a slight misalignment of the laser resonator [14].

In this case the average power was $\sim 650\ mW$ corresponding to pulse energy $\sim 8\ nJ$. The duration of the pulses was again $\sim 85\ fs$, while their central wave length $\sim 800\ nm$. The value of q chosen for the measurement of the lengths of the principal semiaxes was again 0.08, balancing between high sensitivity and low noise. The experimental results, together with their fitting curves, are shown in Figure 7.

It should be noted that in this case two z-scan curves are produced, one for each principal semiaxis. The horizontal separation between the two curves is approximately equal to the distance between the two focal points of the beam. In this case, the curves are fitted by Equation (3), where the far-field pattern is given by the Equations (10)-(20).

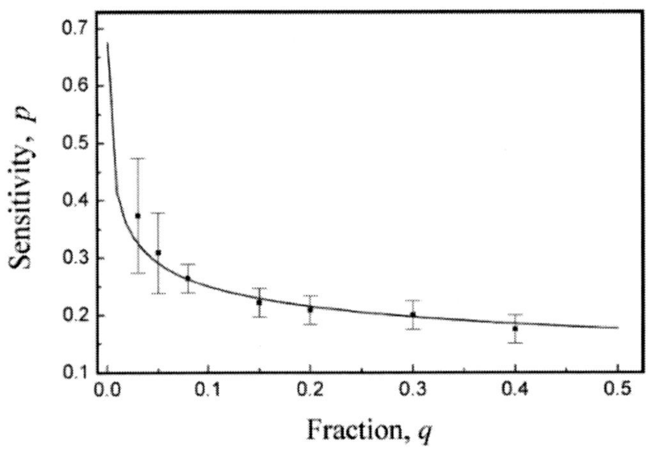

Figure 6. The sensitivity of the radius z-scan technique as function of the fraction q.

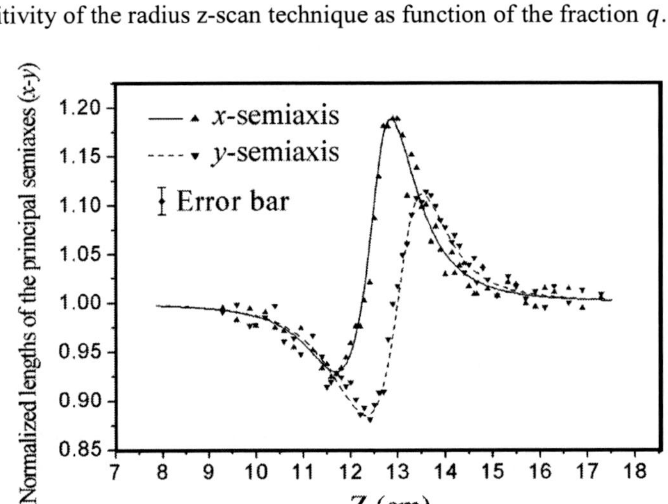

Figure 7. Z-scan curves for an elliptic Gaussian beam.

The thermal nonlinearities dominate again. Therefore, only the average values of the nonlinear parameters can be calculated. From the fitting we find that $n_{2,eq} = -4.2\times10^{-18}m^2/W$, in excellent agreement with the value obtained in the case of the circular Gaussian beam, verifying the validity of the proposed approach.

Finally, the fitting procedure provides the values of the fundamental parameters of the incident beam, namely the locations of the beam waists z_{0x}, z_{0y}, and the beam widths at the waists w_{0x}, w_{0y}. Consequently, the elliptic z-scan technique enables the characterization of the beam emitted by the laser system, as a by-product of the fitting process.

4. Further Advances

The above analysis assumes that the sample is only characterized by nonlinear refraction.

However, in many cases nonlinear absorption is also present. In more detail, the absorption coefficient of the material depends linearly on the intensity, as given by the relation

$$a = a_0 + \beta I \tag{36}$$

where a_0 is the linear and β the nonlinear absorption coefficient of the material. In this case the Gaussian decomposition method developed in section 2 is still valid provided that the factor $[i\Delta\varphi_0(z_s,t)]^m/m!$ in Equations (10) and (21) should be replaced by

$$\frac{F^{(m)}\left[Q_0(z_s,t)\right]^m}{m!} \tag{37}$$

where

$$Q_0(z_s,t) = \beta I_0\left(z_s,t\right)L_{eff} \tag{38}$$

and

$$F^{(m)} = \begin{cases} 1 & \text{for m=1} \\ \prod_{r=1}^{m}\left(\dfrac{ikn_2}{\beta} - r + \dfrac{1}{2}\right) & \text{for m>1} \end{cases} \tag{39}$$

Performing extensive numerical simulations of z-scan experiments where nonlinear refraction and absorption are present [15], we have found that the main influence of nonlinear absorption is that induces asymmetry in the z-scan curves, both in the cases of circular and elliptic (astigmatic) Gaussian beams, as shown in Figure 8.

Specifically, in the case of circular Gaussian beams, we define the asymmetry in the z-scan curves as

$$A = (V_{max} - 1) - (1 - V_{min}) = V_{max} + V_{min} - 2 \tag{40}$$

where V_{max}, V_{min} are their maximum and minimum values respectively. We have also found that for small values of the nonlinearities the asymmetry A grows exponentially with the nonlinear phase shift $\Delta\varphi_0 = -kn_2I_0L_{eff}$. Some characteristic examples for $q = 0.05$, $q = 0.1$ and $q = 0.15$ are shown in Figure 9. It is evident that the asymmetry increases as the values of q decreases.

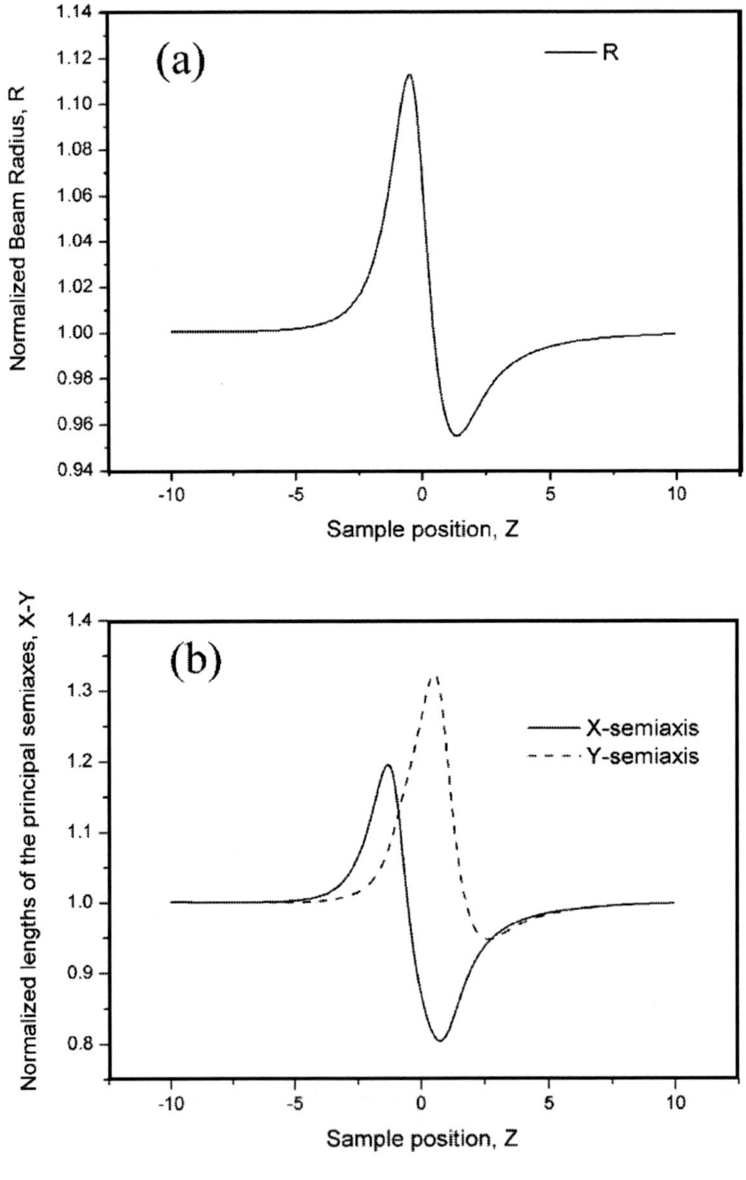

Figure 8. Simulated z-scan curves for circular (a) and elliptic (b) Gaussian beams. In both cases nonlinear refraction and absorption are present simultaneously.

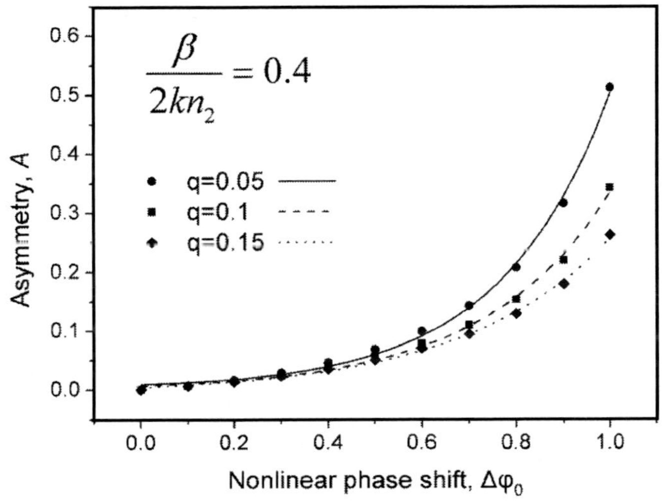

Figure 9. Exponential growth of the asymmetry with the nonlinear phase shift.

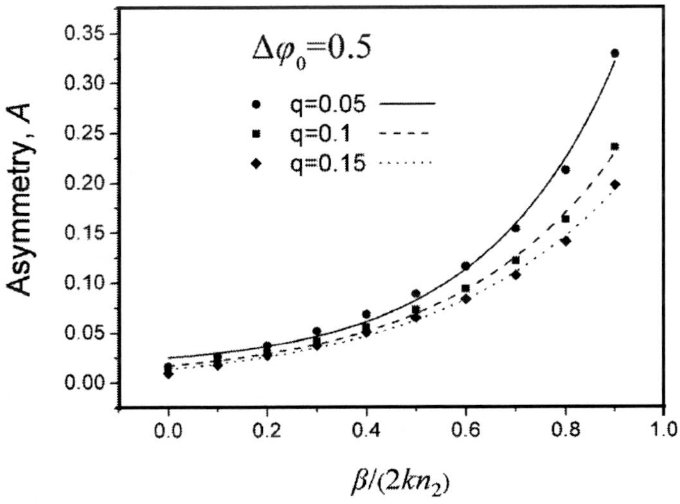

Figure 10. Exponential growth of the asymmetry with the nonlinear absorption coefficient.

The asymmetry also grows exponentially with nonlinear absorption, as shown in Figure 10. Again, the asymmetry increases as the values of q decreases. It should be noted that in all simulations the value of $q_0 = \beta I_0 L_{eff}$, where β is the nonlinear absorption coefficient of the material, was kept below unity in order to be within the limits of the Gaussian decomposition theory. However, the most important finding from a practical point of view is that the nonlinear refractive index can be calculated through the multiplication of the radius z-scan curve with the open z-scan curve, as shown in Figure 11.

Specifically, we have found that the normalized peak-valley difference Δr_{p-v} of the resulting curve is very close to that obtained by a purely refractive material ($\beta = 0$) within a good level of accuracy ±10%.

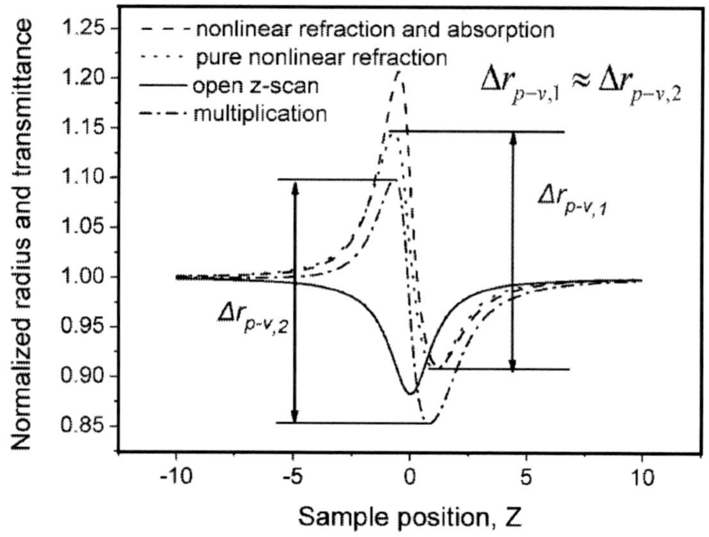

Figure 11. The effects of nonlinear absorption can be suppressed if the radius z–scan curve is multiplied with the corresponding open z–scan curve.

This result is valid for $0.1 \leq |\Delta\varphi_0| \leq 1.2$ and $\beta/2kn_2 \leq 0.25$. In this case, the simplified Equation (34) which has been obtained in the case of purely refractive z-scan [11] is still valid and can be used for calculating the nonlinear phase shift $\Delta\varphi_0$. Then, the nonlinear refractive index n_2 can be obtained through Equation (2). Thus, it is possible to extract the nonlinear refractive index without performing an exact fitting of the theoretical relations to the experimental results, which is a complex and time-consuming procedure. It should be noted that this procedure is analogous to that employed in the case of transmittance z-scan [2], where the effects of nonlinear absorption are suppressed through division of the close z-scan curve by the open z-scan plot.

Finally, we have found that even in the case of an astigmatic (elliptic) Gaussian beam it is still possible to determine the nonlinear refractive index through a simplified procedure. However, this approach is valid only for pure refractive nonlinearities.

In more detail, through extensive numerical simulations we have found that the normalized peak-valley difference of the z-scan curve corresponding to the first focus of the astigmatic beam is proportional to the nonlinear phase shift. Further, the proportionality factor is related to the value of the fraction q where the beam dimensions are defined through a double exponential decay function of the form

$$\Delta x_{p-v} = \left(y_0 + A_1 e^{-q/t_1} + A_2 e^{-q/t_2}\right)|\Delta\varphi_0| \tag{41}$$

where $\Delta\varphi_0 = -kn_2 I_0 L_{eff}$ is the value of the nonlinear phase shift at focus, or more specifically at the center between the two focal points of the beam. Δx_{p-v} is graphically defined in Figure 12.

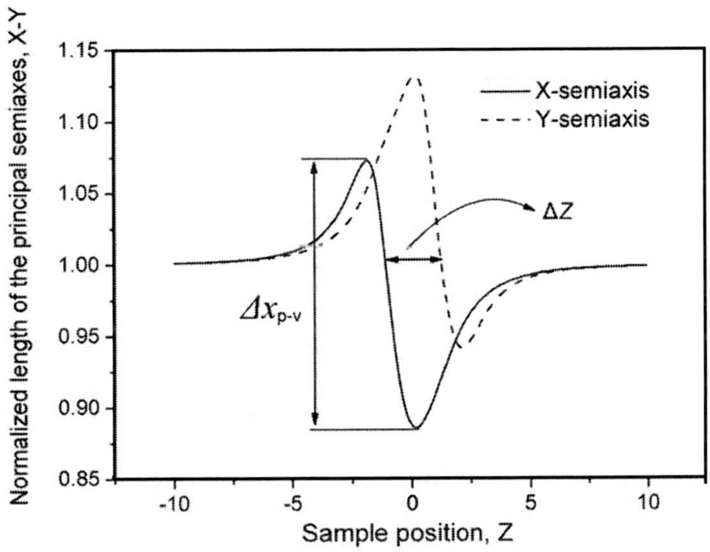

Figure 12. Normalized peak-valley difference Δx_{p-v}.

The values of the parameters y_0, A_1, t_1, A_2, t_2 for several values of the astigmatic difference ΔZ-defined as the distance between the two focuses of the beam normalized to the Rayleigh length are given in the following table 1.

In this way, the nonlinear phase shift – and finally the nonlinear refractive index – can be calculated quite straight forward through the normalized peak-valley difference of the z-scan curve, eliminating the need for extensive numerical calculations.

Table 1. The values of the parameters y_0, A_1, t_1, A_2, t_2 as function of the astigmatic difference

ΔZ	$y0$	$A1$	$t1$	$A2$	$t2$
0.0	0.158	0.123	0.205	0.177	0.026
0.1	0.158	0.121	0.195	0.180	0.025
0.2	0.159	0.121	0.208	0.180	0.025
0.3	0.170	0.127	0.189	0.177	0.024
0.4	0.182	0.125	0.203	0.181	0.025
0.5	0.197	0.130	0.207	0.184	0.024
0.6	0.218	0.132	0.193	0.185	0.024
0.7	0.239	0.130	0.201	0.193	0.024
0.8	0.262	0.130	0.192	0.203	0.024
0.9	0.288	0.128	0.178	0.216	0.023
1.0	0.311	0.123	0.180	0.227	0.024
1.1	0.334	0.117	0.105	0.243	0.025
1.2	0.358	0.119	0.169	0.258	0.025
1.3	0.384	0.125	0.147	0.266	0.025
1.4	0.409	0.136	0.128	0.272	0.025
1.5	0.433	0.166	0.108	0.276	0.022

Conclusion

In conclusion, we have presented a new z-scan technique based on the direct measurement of the beam dimensions in the far field. This entails several advantages such as elimination of the sensitivity in beam pointing instability and energy fluctuations as well as better supervision of the evolution of the beam profile throughout the z-scan experiment. Further, this technique can easily be extended to elliptic (astigmatic) Gaussian beams. The experimental application of this technique has been demonstrated both for circular and elliptic Gaussian beam, providing excellent results. A simplified relation for calculating the nonlinear refractive index directly from the radius z-scan curves has been presented. Further, we have shown that the effects of nonlinear absorption can be almost eliminated by multiplying the radius z-scan curve with the corresponding open z-scan curve. Finally, a simplified procedure for the straightforward calculation of the nonlinear refractive index even in the case of an elliptic (astigmatic) Gaussian beam has also been presented.

Appendix: Gaussian Decomposition for High-Order Nonlinearities

We suppose that this astigmatic elliptic beam is incident on a thin sample of a material characterized by both the first two orders of refractive nonlinearity. Namely, the refractive index n of the material is given by the relationship

$$n = n_0 + n_2^{(3)} I + n_2^{(5)} I^2 \tag{A1}$$

where n_0 is the linear refractive index of the material, $n_2^{(3)}$ and $n_2^{(5)}$ the third-and fifth-order nonlinear refractive indices respectively, and I the intensity of the incident beam. The effect of the refractive nonlinearities is to induce a phase shift of the form

$$\Delta\varphi(x,y,z_s,t) = -kn_2^{(3)} I(x,y,z_s,t) L_{eff}^{(3)} - kn_2^{(5)} I^2(x,y,z_s,t) L_{eff}^{(5)} \tag{A2}$$

where z_s is the location of the sample and

$$L_{eff}^{(3)} = \left[1 - \exp(-a_0 L)\right] / a_0 \,, \quad L_{eff}^{(5)} = \left[1 - \exp(-2a_0 L)\right] / (2a_0) \tag{A3}$$

are the effective lengths corresponding to the third-and fifth-order nolnearities respectively. Here L is the physical length of the sample and a_0 the linear absorption coefficients. Consequently the electric field pattern $E_e(x,y,z_s,t)$ of the beam at the exit plane of the sample becomes

$$E_e(x,y,z_s,t) = E_{in}(x,y,z_s,t) \exp\left[i\Delta\varphi(x,y,z_s,t)\right] \exp(-a_0 L/2) \tag{A4}$$

where $E_{in}(x,y,z_s,t)$ is the electric field pattern of the incident beam at the sample plane ($z = z_s$). Taking into account that $I(x,y,z_s,t) = c\varepsilon_0 n_0 |E(x,y,z_s,t)|^2/2$, the induced phase shift $\Delta\varphi(x,y,z_s,t)$ can also be written as

$$\Delta\varphi(x,y,z_s,t) = \Delta\varphi_0^{(3)}(z_s,t)\exp\left(-\frac{2x^2}{w_x^2(z_s)} - \frac{2y^2}{w_y^2(z_s)}\right) + \Delta\varphi_0^{(5)}(z_s,t)\exp\left(-\frac{4x^2}{w_x^2(z_s)} - \frac{4y^2}{w_y^2(z_s)}\right) \tag{A5}$$

where $\Delta\varphi_0^{(3)}(z_s,\mathrm{t}) = -kn_2^{(3)} I_0(z_s,t) L_{eff}^{(3)}$ and $\Delta\varphi_0^{(5)}(z_s,\mathrm{t}) = -kn_2^{(5)} I_0^2(z_s,t) L_{eff}^{(5)}$ are the on-axis phase shifts due to the third-and fifth-order nonlinearities respectively. Here $I_0(z_s,t)$ is the on-axis intensity of the incident beam.

Expanding the phase term $exp[i\Delta\varphi(x,y,z_s,t)]$ into a Taylor series and assuming that $\Delta\varphi_0^{(5)}(z_s,\mathrm{t}) \ll \Delta\varphi_0^{(3)}(z_s,t)$, the electric field at the exit plane of the sample becomes

$$\begin{aligned} E_e(x,y,z_s,t) = {} & E_0(t)\left(\frac{w_{0x}w_{0y}}{w_x(z_s)w_y(z_s)}\right)^{1/2} \exp\left(-\frac{a_0 L}{2}\right) \exp\left[-ikz_s + i\theta(z_s)\right] \sum_{m=0}^{\infty} \left\{\frac{\left[i\Delta\varphi_0^{(3)}(z_s,t)\right]^m}{m!}\right. \\ & \times\exp\left[-\frac{(2m+1)x^2}{w_x^2(z_s)} - \frac{(2m+1)y^2}{w_y^2(z_s)} - \frac{ikx^2}{2R_x(z_s)} - \frac{iky^2}{2R_y(z_s)}\right] + \frac{m\, i^m}{m!}\left[\Delta\varphi_0^{(3)}(z_s,t)\right]^{m-1} \\ & \left.\times\Delta\varphi_0^{(5)}(z_s,t)\exp\left[-\frac{(2m+3)x^2}{w_x^2(z_s)} - \frac{(2m+3)y^2}{w_y^2(z_s)} - \frac{ikx^2}{2R_x(z_s)} - \frac{iky^2}{2R_y(z_s)}\right]\right\} \end{aligned} \tag{A6}$$

Each term in the sum over m can be associated with two elliptic Gaussian beams with principal semiaxes $w_{x,y}^{(m1)} = w_{x,y}(z_s)/\sqrt{2m+1}$ for the beam associated to the first subterm and $w_{x,y}^{(m2)} = w_{x,y}(z_s)/\sqrt{2m+3}$ for the beam associated to the second subterm. The radii of curvature for these beams are the same with those of the incident beam, namely $R_{x,y}^{(m1,2)} = R_{x,y}(z_s)$. These relations, together with Equations (A2) and (A3) can be used for determining the location of the waists and the Rayleigh lengths for each individual Gaussian beam, which are

$$z_{0x,y}^{(m1,2)} = z_s - \frac{\left(A_{x,y}^{(m1,2)}\right)^2 R_{x,y}(z_s)}{\left(A_{x,y}^{(m1,2)}\right)^2 + R_{x,y}^2(z_s)}$$

and

$$z_{Rx,y}^{(m1,2)} = \frac{k\left[w_{0x,y}^{(m1,2)}\right]^2}{2} = \frac{A_{x,y}^{(m1,2)} R_{x,y}^2(z_s)}{\left(A_{x,y}^{(m1,2)}\right)^2 + R_{x,y}^2(z_s)} \tag{A7}$$

where $A_{x,y}^{(m1)} = kw_{x,y}^2(z_s)/(4m+2)$ and $A_{x,y}^{(m2)} = kw_{x,y}^2(z_s)/(4m+6)$. Equation (A7) fully define each Gaussian beam, which can be easily propagated to any desired distance D from the exit plane of the sample according to Equations (A1)-(A4). Finally, their sum gives the total electric field pattern of the laser beam at this distance. A more detail presentation of the above analysis can be found in [8].

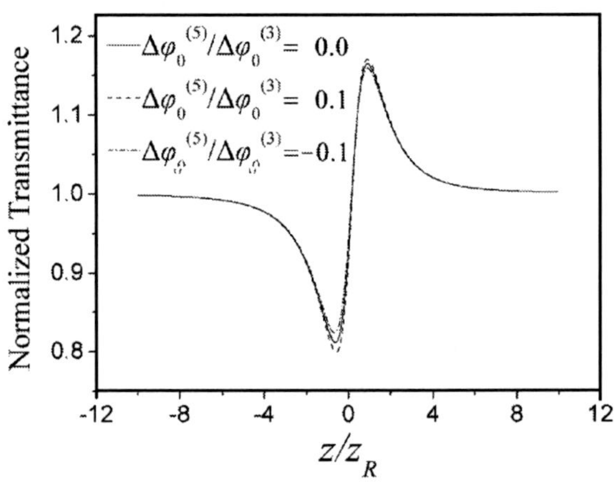

Figure A1. Simulations of the transmittance variations in the special case of a circular Gaussian beam.

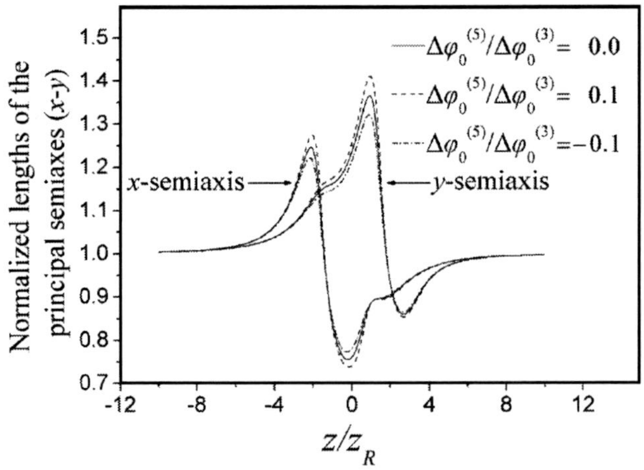

Figure A2. Simulations of the variations of the beam's principal semiaxes in the case of an astigmatic elliptic beam.

The above results can be used for simulating z-scan experiments based on either the measurement of the transmittance through an aperture or the direct measurement of the beam dimensions in the far field. In the following we depict transmittance z-scan plots for the case of a circular Gaussian beam and dimension z-scan plots in the case of an elliptic (astigmatic) beam. From the Figures A1 and A2 it is evident that the main effect of the higher-order nonlinearities is to modify the distance between the peaks and valleys of the curves. Consequently, it is not easy to decide from a single set of experimental data whether higher order nonlinearities are involved or not. Thus, several z–scan experiments should be performed for different values of the beam intensity. Based on these data the values of $n_2^{(3)}$ as function of the beam intensity should be calculated, using the standard theory for third-order nonlinearity [1, 11, 14]. If all these values are constant, then only third-order nonlinearities are present and no further elaboration is needed. However if above a certain value of the intensity, the $n_2^{(3)}$ values begin to increase or decrease, then fifth-order nonlinearities are also involved and Equations (A6) and (A7) should be used in order to determine the far field pattern of the beam.

REFERENCES

[1] Sheik–Bahae, M; Said, AA; Van Stryland, EW.“High-sensitivity, single-beam n_2 measurement”, *Opt. Lett.*, 1989, 14, 955-957.

[2] Sheik–Bahae, M; Said, AA; Wei, TH; Hagan DJ;Van Stryland,EW. “Sensitive measurement of optical nonlinearities using a single beam”, *IEEE J. Quantum Electron.*, 1990, 26, 760-769.

[3] Chapple, PB; Staromlynska, J; Hermann, JA;McKay, TJ;McDuff, RG. “Single beam z – scan: measurement techniques and analysis”, *J. Nonlinear Opt. Phys. Mater.*,1997, 6, 251-293.

[4] Siegman AE. *Lasers,* University Science Books, Mill Valley, California, 1986.

[5] Samad, RE; VieiraJr, ND.“Analytical description of z-scan on-axis intensity based on the Huygens – Fresnel principle”, *J. Opt. Soc. Am. B*,1998, 15, 2742-2747.

[6] Weaire, D; Wherrett, BS; Miller, DAB; Smith, SD.“Effect of low –power nonlinear refraction on laser – beam propagation in InSb”, *Opt. Lett.*,1979, 4,331-333.

[7] Tsigaridas, G; Fakis, M; Polyzos, I; Tsibouri, M; Persephonis, P; Giannetas, V. “Z-scan analysis for near-Gaussian beams through Hermite-Gaussian decomposition”, *J. Opt. Soc. Am. B*, 2003, 20, 670–676.

[8] Tsigaridas, G; Fakis, M; Polyzos, I; Persephonis, P; Giannetas, V. “Z-scan analysis for high-order nonlinerities through Gaussian decomposition”, *Opt. Commun.*,2003 225, 253-268.

[9] Siegman, AE. *Lasers*, University Science Books, Mill Valley, California, 1986.

[10] Yariv, A. *Quantum Electronics*, 3rd edition, John Wiley and Sons, 1989.

[11] Tsigaridas, G; Fakis, M; Polyzos, I; Persephonis, P; Giannetas, V. “Z-scan technique through beam radius measurements”, *Appl. Phys. B*, 2003,76, 83-86.

[12] Hayes, JN. “Thermal blooming of laser beams in fluids”, *Appl. Opt.*, 1972, 2, 455-461.

[13] Kawazoe, T; Kawaguchi, H; Inoue, J; Haba, O; Ueda, M. "Measurement of nonlinear refractive index by time-resolved z-scan technique", *Opt. Commun.*, 1999, 160, 125-129.

[14] Tsigaridas, G; Fakis, M; Polyzos, I; Persephonis, P; Giannetas, V. "Z-scan technique for elliptic Gaussian beams", *Appl. Phys. B*, 2003, 77, 71-75.

[15] Tsigaridas, G; Persephonis, P; Giannetas, V. "Effects of nonlinear absorption on the z-scan technique through beam dimension measurements", *Mater. Sci. Eng. B*, 2009, 165, 182-185.

In: Nonlinear Performance and Characterization Methods ... ISBN: 978-1-62808-093-3
Editor: Jingsong Wei

Chapter 3

Z-SCAN TECHNIQUE USING FRESNEL–KIRCHHOFF DIFFRACTION METHOD

Ehsan Koushki[1,2*]
[1]Physics Department, Kharazmi University, Tehran, Iran
[2]Physics Department, Hakim Sabzevari University, Sabzevar, Iran

ABSTRACT

In this chapter, Fresnel–Kirchhoff diffraction method is used to relate the electric field of the passing laser beam through the nonlinear sample and the electric field of the far-field where the effects or modulations of the nonlinear sample on the laser beam should be detected or studied. Fresnel–Kirchhoff diffraction integral is a general method that could adopt with different kinds and magnitudes of nonlinear phase changes. As observed later, it's a useful and precise tool for the simulation of the nonlinear deviations in the beam shape, beam radius and intensity distribution (that is directly used in Z-scan method). However, this method has some shortcomings such as disability in short distances from the sample. Here, we compare this method with other usual methods and study the virtues. Furthermore, the faults of each method and the reliable region of each one are discussed.

Keywords: Fresnel–Kirchhoff diffraction method, Kerr medium, Z-scan technique

1. INTRODUCTION

M. Sheik Bahaei et al. introduced Z-scan method to obtain nonlinear optical coefficients of materials using a single laser beam in 1989 [1] and enveloped it in 1990 [2]. In this method, a thin sample with only nonlinear refraction was considered and the phase shift of the electric field of the transmitted beam was studied in the far-field by using the Gaussian decomposition method [3,4]. Nonlinear refraction index could be either a result of electron transition

* Corresponding author e-mail:ehsan.koushki@yahoo.com.

nonlinearity, reorientation effect or thermal lensing nonlinearity, or a mixture of them [5,6]. In the next step, they represented a more complete phase shift of the electric field containing both the nonlinear refraction and two photon absorption indices [2], but the electric field was proceded till reaching the outer surface of the sample and wasn't evaluated in far field. In the subsequent papers [7, 8] Fresnel–Kirchoff integral method was used to relate the far-field electric field to the transmitted beam through the sample. In [8], based on this method, a perfect formalism for the close aperture Z-scan of high refractive nonlinearities was obtained and Fresnel–Kirchoff integral method was also used for the simulation of the diffraction ring patterns of the passing beam in far field. This method could be used for any kind of beam phase shifts and doesn't confine to the phase shift chosen. In [9] a more complete model, it was offered that the electric field in far-field was related to the passing beam from the surface of the sample and the phase shift contains both the refraction and absorption nonlinearities. This model applied the basic relations of the two photon absorption presented in [2]. In the present study, we have offered a complete discription of the Fresnel–Kirchoff integral method and its applications in description of the passing laser beam from a Kerr medium in far field. Using this method, some applicable methods such a Z-scan method, irradiance destribution of the passing beam in the far-field, Z-scan technique through beam radius measurement, beam profile passing through the sample and the radius of curvature of the passing wavefront were modified or improved, which could help us to conduct precise measurement of nonlinear coefficients of materials.

2. THEORITICAL DESCRIPTION OF THE METHODS

The electric field of a TEM_{00} Gaussian beam with beam waist radius w_0 traveling in the z direction could be written as [2]:

$$E(z,r) = E_0 \frac{w_0}{w(z)} \exp\{-i(kz - \tan^{-1}(\frac{z}{z_0}))\} \exp\{-r^2(\frac{1}{w^2(z)} + \frac{ik}{2R(z)})\} \tag{1}$$

and

$$w(z) = w_0\sqrt{1+\frac{z^2}{z_0^2}},$$

$$R(z) = z(1+ {z_0^2}/{z^2}),$$

$$z_0 = {kw_0^2}/{2}$$

where I_0, $w(z)$, $R(z)$, z_0 are the irradiance intensity at beam waist of the Gaussian beam, the beam radius at position z, the diffraction length of the beam and the curvature radius of the wavefront at position z, respectively. The beam waist (focal point) is located at $z = 0$. The parameters w and R of a Gaussian beam change during the propagation of the beam along the z-axis.

The equations describing the propagation of the optical field inside the nonlinear material take the form:

$$\frac{d\Delta\varphi}{dz'} = -\Delta n(I)k \tag{2}$$

$$\frac{dI}{dz'} = -\alpha(I)I \tag{3}$$

We consider that a Gaussian beam passes through a thin third-order nonlinear optical sample with nonlinear refraction n_2 (MKS) located at z [1,2]. In the case of a cubic nonlinearity and negligible nonlinear absorption, Equation (2) is solved to give the phase shift $\Delta\varphi$ at the exit surface of the sample which simply follows the radial variation of the incident irradiance at a given sample position z. Thus, the phase shift of the beam is given by [1,2]:

$$\Delta\varphi(z,r) = \frac{\Delta\varphi_0}{1+(\frac{z}{z_0})^2}\exp(-2r^2 / w^2(z)) \tag{4}$$

where $\Delta\varphi_0$ is defined as the on-axis nonlinear phase shift at the beam focus, given by:

$$\Delta\varphi_0 = kn_2I_0L_{eff} = k\Delta nI_0 \tag{5}$$

Here, z, $L_{eff} = (1 - e^{-\alpha L})/\alpha$, α, L and $\Delta n = n_2I_0$ are the distance between the sample and the beam waist, the effective length of the sample, the linear absorption coefficient, the sample thickness, and the refraction change, respectively. The sample could be assumed as a thin sample if its thickness L is significantly smaller than z_0.

Using the Fresnel–Kirchhoff diffraction theory, the electric field at the exit surface of the sample is obtained as [2,10]

$$E(r,z) = E_0\frac{w_0}{w(z)}\exp\{-i[kz - Arc\tan(z/z_0)]\}\times \exp[\frac{-ikr^2}{2R(z)} - \frac{r^2}{w^2(z)} - i\Delta\phi(z,r) - \alpha L/2] \tag{6}$$

The electric field in far-field is obtained by means of free-space Fresnel-Kirchhoff diffraction integral at the Fraunhofer approximation [10,12]. For a radial symmetric field, it equals to a Hankell function of Fourier-Bessel transform of the field, namely:

$$E(\rho, z+D)=\frac{1}{i\lambda D}\exp(\frac{i\pi\rho^2}{\lambda D})\int_0^\infty J_0(k\theta r)E(r,z)\exp(\frac{i\pi r^2}{\lambda D})2\pi r dr \tag{7}$$

where $J_0(x)$, θ, ρ and D are the zero-order Bessel function of the first kind, the far-field diffraction angle and the radial coordinate in the far-field observation plane and the distance from the exit plane of the sample to the far-field observation plane, respectively. In the paraxial approximation, the distance from the exit plane of the medium to the far-field observation plane (D) is related to the radial coordinate and the diffraction angle by $\rho = D\theta$. Therefore, it could be written as follow:

$$E(\rho, z+D)=(\frac{2\pi}{i\lambda D})\times E_0\frac{w_0}{w(z)}\exp(-i(kz-\arctan(z/z_0)))e^{-\alpha L/2}$$
$$\times\int_0^\infty J_0(\frac{k\rho r}{D})\exp(-\frac{ikr^2}{2R}-\frac{r^2}{w^2}-i\Delta\varphi(z,r))rdr \tag{8}$$

Using Equation (8), the distribution pattern of irradiance could be written as:

$$I(\rho, z+D)=I'\left|\int_0^\infty J_0(\frac{k\rho r}{D})\exp(-\frac{ikr^2}{2R(z)}-\frac{r^2}{w^2}-i\Delta\varphi(z,r))rdr\right|^2 \tag{9}$$

where

$$I'=\frac{cn_0\varepsilon_0}{2}\times(\frac{1}{\lambda D})^2\times(2\pi E_0)^2(\frac{w_0}{w(z)})^2e^{-\alpha L} \tag{10}$$

The magnitude of E_0 could be calculated simply through $(2/w_0)\sqrt{P_0/c\varepsilon_0 n_0\pi}$, where P_0 is the laser output power. Equations (9) and (10) provide a Gaussian beam with a circular beam spot shape. In case of an elliptic Gaussian beam passing through a cubic nonlinear refractive medium (with only n_2) the electric field is presented as [13]:

$$E(x,y,z,t)=E_0(t)\sqrt{\frac{w_{0x}w_{0y}}{w_x(z)w_y(z)}}\times\exp(-\frac{x^2}{w_x^2(z)}-\frac{y^2}{w_y^2(z)})\times$$
$$\exp(-\frac{ikx^2}{2R_x(z)}-\frac{iky^2}{2R_y(z)})\times\exp(-ikz+i\theta(z)) \tag{11}$$

where w_{0x} and w_{0y} are the beam width at the two waists of this astigmatic elliptic beam, respectively. The parameters $w_{x,y}(z)$, $R_{x,y}(z)$, and $\theta(z)$ are the beam width, radii of curvature and on-axis phase shift of the beam, respectively, which are defined as follows:

$$w_{x,y}(z) = w_{0x,y}[1+(\frac{z-z_{0x,y}}{z_{Rx,y}})^2]^{1/2} \tag{12}$$

$$R_{x,y}(z) = (z-z_{0x,y})[1+(\frac{z_{Rx,y}}{z-z_{0x,y}})^2] \tag{13}$$

$$\theta(z) = \frac{1}{2}\tan^{-1}(\frac{z-z_{0x}}{z_{Rx}}) + \frac{1}{2}\tan^{-1}(\frac{z-z_{0y}}{z_{Ry}}) \tag{14}$$

where $z_{0x,y}$ are the locations of the waists and $z_{Rx,y}$ the Rayleigh lengths of the incident beam for each principal direction (x, y).

As the beam passes through a nonlinear thin sample placed at z, it would experience a phase shift with the following form

$$\Delta\varphi(x,y,z,t) = \Delta\varphi_0(t)\times\exp(-2(\frac{x^2}{w_x(z)^2}+\frac{y^2}{w_y(z)^2}))\times$$
$$\{(1+(\frac{z-z_{0x}}{z_{Rx}})^2)(1+(\frac{z-z_{0y}}{z_{Ry}})^2)\}^{-\frac{1}{2}} \tag{15}$$

The electric field at the far-field is usually obtained using Gaussian decomposition method in [13]. Here we offer the electric field of far-field using Fresnel-Kirchhoff diffraction method [14].

Multiplying the left side of Equation (11) by $exp[-i\Delta\varphi(x,y,z,t)]$, the electric field at the exit plane of the sample $E_e(x,y,z)$ is obtained. Substituting into Equation (7) there would be (only in Cartesian coordinate system definition):

$$E(x',y',z+D) = \frac{1}{i\lambda D}\exp(\frac{i\pi(x'^2+y'^2)}{\lambda D})\times$$
$$\int_{-\infty}^{\infty}\int_{-\infty}^{\infty} J_0(k\frac{\sqrt{x^2+y^2}\sqrt{x'^2+y'^2}}{D})E_e(x,y,z)\exp(\frac{i\pi(x^2+y^2)}{\lambda D})dxdy \tag{16}$$

where the related parameters refer to coordinates at far field.

The above-mentioned formulation would be useful when the nonlinear absorption is neglected or at least the absorption effect is removed from the close aperture curve; and there is almost a pure close aperture curve. However, when the nonlinear absorption is considerable, the close aperture curve includes both the nonlinear absorption and refraction effects. Thus, we need to provide a more complete model that includes both the nonlinear refraction and absorption coefficients. For the above cases, Equations (2) and (3) should be reexamined after the following substitution [2]

$$\alpha(I) = \alpha + \beta I \tag{17}$$

This yields to the irradiance distribution and phase shift at the exit surface of the sample as

$$I_e(z,r,t) = \frac{I(z,r,t)e^{-\alpha L}}{1 + q(z,r,t)} \tag{18}$$

and

$$\Delta\varphi(z,r,t) = \frac{kn_2}{\beta}\ln(1 + q(z,r,t)) \tag{19}$$

where $q(z,r,t) = \beta I(z,r,t)L_{eff}$. Combining Equations (18) and (19), we obtain the complex field at the exit surface of the sample as follows:

$$E_e = E(z,r,t)e^{-\alpha L/2}(1+q)^{(-\frac{ikn_2}{\beta}-\frac{1}{2})} \tag{20}$$

By using Equation (7) there would be for a circular beam:

$$E(\rho,z) = (\frac{2\pi}{i\lambda D}) \times E_0 \frac{w_0}{w(z)} \exp(-i(kz - \arctan(z/z_0)))e^{-\alpha L/2}$$
$$\times \int_0^\infty J_0(\frac{k\rho r}{D})\exp(-\frac{ikr^2}{2R} - \frac{r^2}{w^2})(1+q)^{-(\frac{ikn_2}{\beta}+\frac{1}{2})} rdr \tag{21}$$

and

$$I(\rho,z) = I'\left|\int_0^\infty J_0(\frac{k\rho r}{D})\exp(-\frac{ikr^2}{2R(z)} - \frac{r^2}{w^2})(1+q)^{-(i\frac{kn_2}{\beta}+\frac{1}{2})} rdr\right|^2 \tag{22}$$

where I' is defined in Equation (10). Equation (22) is a good relation for electric field of a Gaussian beam passing through a cubic nonlinear sample, and has some virtues in comparison with other previous formulations. This technique is based on the Fresnel-Kirchhoff method and could evaluate the passing electric field in the region far from the axis. Furthermore, Equation (22) is suitable for high optical nonlinearity case which may include diffraction ring patterns [8]. The main virtue is that the effects of all linear and nonlinear optical coefficients are presented within one formulation simultaneously. Thus, Equation (22) could be utilized for the simulation of both the open and close aperture Z-scan curves.

3. Applications in the Measurement of Third Order Nonlinear Optical Indexes

The calculation of the electric field of the passing beam needs to be carried out through many measurement methods. Some applicable ones concerned with evaluating nonlinear optical indices are discussed using the Fresnel-Kirchhoff method. The results and analysis could be expanded to both the self-focusing and self-defocusing cases. The studies were performed for both the circular and elliptic Gaussian beams. For simplicity, sample is considered with only nonlinear refraction; and the nonlinear absorption would be analyzed if necessary for all cases.

3.1. Z-Scan Experiments

Z-scan is a well known technique to calculate the nonlinear refraction and absorption coefficients of samples. This technique has been well developed in the past 30 years [2,4,8]. Close aperture theory was firstly presented by M. Sheik Bahae et al. by using Gaussian-decomposition method. In this close-aperture Z-scan, a sample is scanned along the optical axis (z-direction) of a focused Gaussian laser beam around the focus. The high intensity of the electromagnetic field in this region induces a non-linear lens of variable focal length on the sample. This causes a refractive divergence or convergence of the laser beam, depending on the sample position relative to the beam focus and the sign of the non-linearity. In the original version of the z-scan technique, the measured quantity is the normalized transmittance of the irradiance, through a finite aperture in the far field [2].

Now, assuming that a finite aperture of linear transmittance $S = 1 - exp(2r_a^2/w_a^2)$, ($r_a$ and w_a are the aperture radius and the beam waist at the aperture plane, respectively) is placed at the far field and the aperture centre is located at the beam center. The transmitted power through the aperture could be simply calculated by:

$$P(z) = \int_0^{r_a} I(\rho, z) 2\pi\rho d\rho$$
$$I(\rho, z) = \frac{C\varepsilon_0 n_0}{2} |E(\rho, z)|^2 \tag{23}$$

Moreover, the normalized transmittance could be calculated as well:

$$T(z) = \frac{\int_0^{\infty} P_T(z, \Delta\varphi_0(t))dt}{\int_0^{\infty} P_T(z, \Delta\varphi_0 = 0)dt} \tag{24}$$

Two of the virtues of Gaussian decomposition method include the facility and rapidity in numerical computing. Nonetheless, when the nonlinearity is high, Fresnel-Kirchhoff method is more reliable, and the close aperture Z-scan technique was developed using Fresnel-Kirchhoff method [7,8]. Figure 1 shows the fitting experimental curve of close aperture Z-scan experiment for a solution of Erioglaucine component [8]. All the curves obey a well-known peak-valley configuration due to the close aperture Z-scan setup and are in agreement with GD curves.

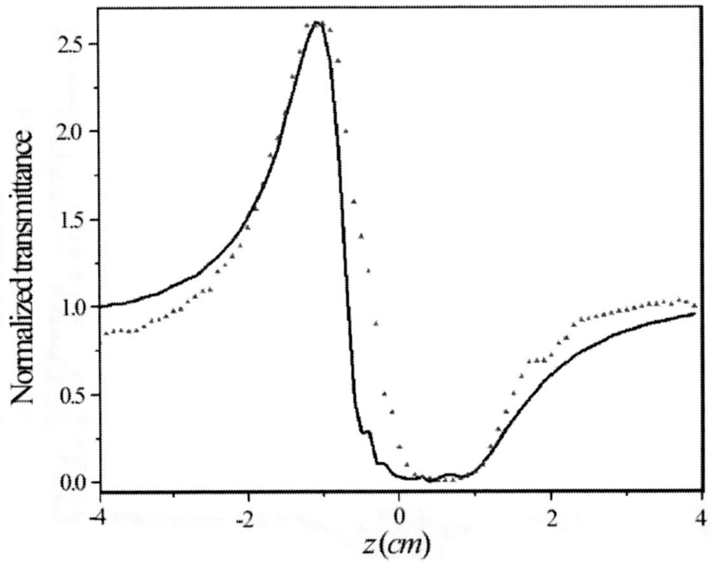

Figure 1. Normalized close aperture Z-scan plot of Erioglaucine ($\Delta\varphi_0 = -18rad$) [8]. The solid line is the theoretical fitting.

The more the nonlinear phase shift increases, the more Fresnel-Kirchhoff method reliable results would give. In this method, the numerical calculation would take a long time, which is the main shortcoming of this method. Numerical evaluation with the Fresnel-Kirchhoff method gives an applicable relation for experimental applications as follows [8]:

$$\Delta T_{p-v} = 3.06(1-\exp(\frac{-|\Delta\varphi_0|}{5.81}))(1-s)^{0.25} \quad for \quad |\Delta\varphi_0| < 25 \tag{25}$$

In real cases, the sample has both the nonlinear refraction and nonlinear absorption. Using Equations (22)-(24) a proper model for close aperture Z-scan experiment could be obtained. Dividing the close aperture curve by the open aperture one gives a pure close aperture curve that could be defined as well within Equations (5) and (9) (shown in Figure 2)[9].

Open aperture Z-scan measurement is a good method to evaluate the nonlinear absorption coefficient of the materials. Open aperture Z-scan measurement could neutralize the nonlinear refraction effect. In [2], using Equation (2) and substituting $\alpha(I) = \alpha + \beta(I)$the irradiance distribution at the exit surface of the sample has been obtained. The integrating of the intensity over the exit surface of the sample could give the transmitted power $P_e(z,t)$ as follows:

$$P_e(z,t) = P_i(t)e^{-\alpha L}\frac{\ln[1+q(z,r=0,t)]}{q(z,r=0,t)} \tag{26}$$

The $P_i(t)$ and $P_e(t)$ are the initial incident power onto the sample and the exit power from the sample, respectively. In this derivation it is assumed that the powermeter is placed immediately behind the sample.

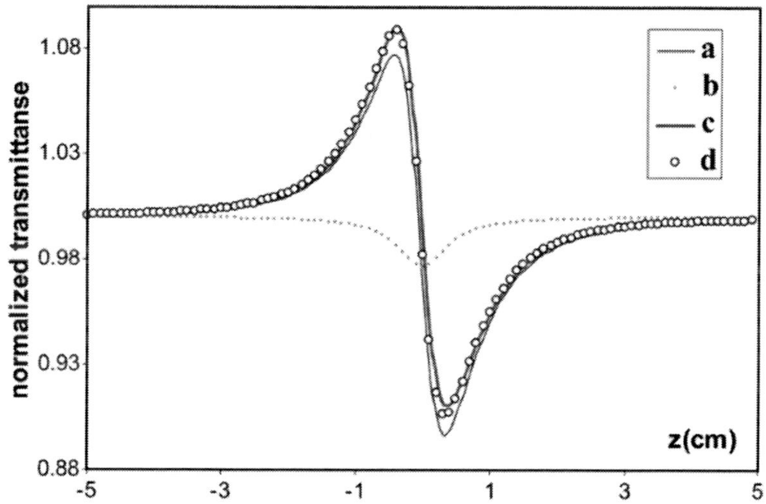

Figure 2. Curve a: theoretical close aperture Z-scan curve using Equation (22);curve b:theoretical open aperture Z-scan curve; curve c: pure close aperture z-scan curve obtained by dividing curve a by curve b; curved: pure close aperture Z-scan curve using Equation (9) [9].

In [9] the open apeture curve is obtained using Equations (22) and (23) with $r_a \to \infty$. In this method, the powermeter is located at distance D from the sample and the whole beam spot energy is received. The value of n_2 doesn't have any effects on the results; because the convergence (ordivergence) of the beam induced by the nonlinear refraction doesn't have any effects on the transmitted power as $r_a \to \infty$. In Figure 3, the two models are compared through curve fitting for an open aperture Z-scan curve of TiO_2 nanoparticles [9].

In case of elliptic Gaussian beam, integrating the intensity presented with Equation (16) over the aperture surface at the far field provides the close aperture z-scan curve. By considering a focused laser beam, although with an astigmatism ($z_{0x} = 1\ cm$ and $z_{0y} = 0$) and with semi-axes of waist spot $w_{0x} = 50\ um$ and $w_{0y} = 30um$ is passed through a nonlinear sample (with $\Delta\varphi_0 = -1$), Figure 4 presents the curve where an astigmatism is between 0 and 1 cm. In addition, Gaussian decomposition method (GD) gave similar results for elliptic beams, which could be found in [13]. For general cases, the nonlinear absorption effect should be considered in close aperture z-scan measurement, which could be obtained by substituting Equation (19) into Equation (15) instead of using Equation (5).

3.2. Irradiance Distribution of the Passing Beam in the Far-Field

Irradiance distribution of the passing beam at far field could be obtained easily using Equations (9) and (22). Fitting the observed distribution pattern with the theoretical curves could give the nonlinear optical phase shift and the nonlinear refractive index. It needs to be noted that the accuracy of this method is less than the Z-scan method. This method could be applicable for high linearity where the diffraction ring pattern could be observed [8], as shown in Figure 5.

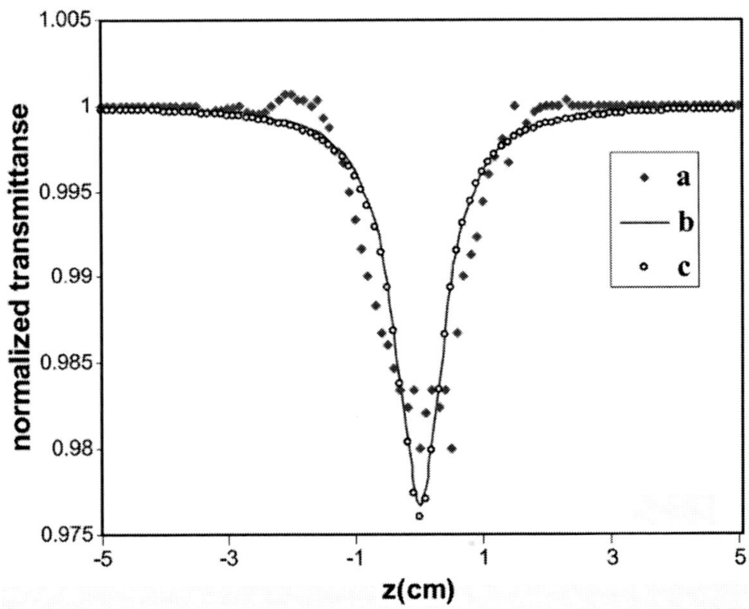

Figure 3. Curve a: experimental data of open aperture Z-scan measurement of TiO_2 nanoparticles [9]; curve b: theoretical fitting curve using Equation (26); curve c: theoretical fitting curve using Equations (22) and (23) with $r_a \to \infty$.

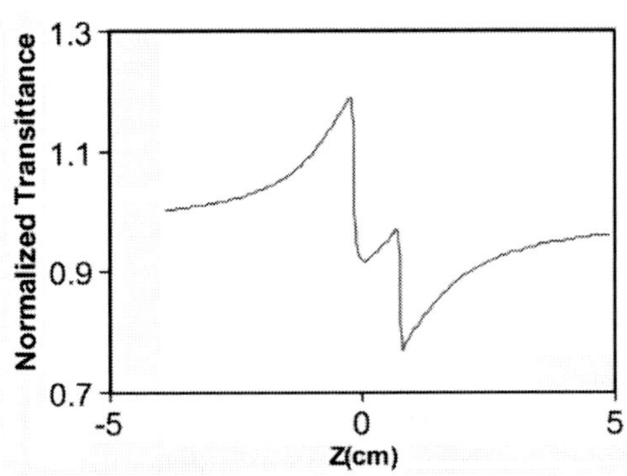

Figure 4. Close aperture Z-scan curve for an elliptic Gaussian beam passing through the sample.

For a divergent beam in self-defocusing medium ($R > 0$, $\Delta\varphi < 0$) and a convergent beam in self-defocusing medium ($R < 0$, $\Delta\varphi > 0$), a well-defined ring pattern could be observed and the phase shift is proportional to the number of bright ring (N) through the following equation:

$$|\Delta\varphi(z, r = 0)| = 2\pi N \tag{27}$$

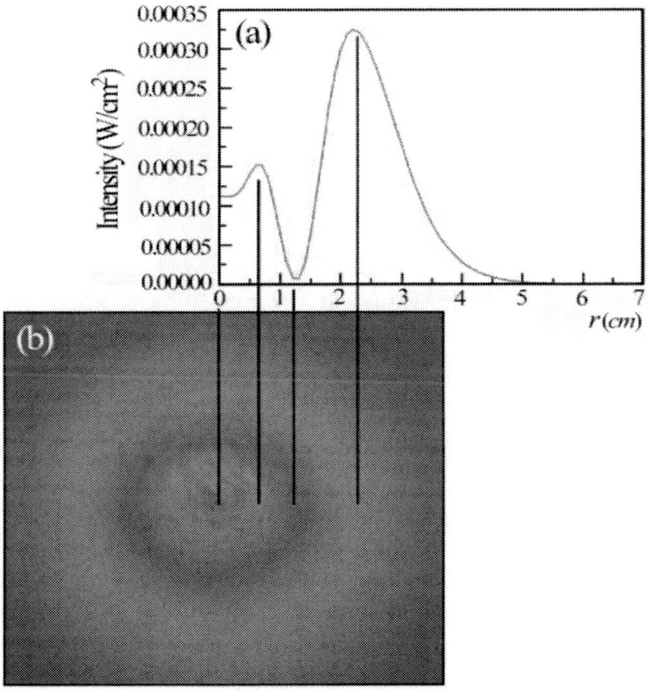

Figure5. Far-feld diffraction pattern of 1 millimole concentration of Erioglaucine with $\Delta\varphi(z = +z_0, r = 0) = -10$, (a) theoretical result using Fresnel–Kirchoff integral method, (b) observed diffraction ring pattern [8].

The measurement accuracy with this method is in the order of π. In order to obtain the optical nonlinearity accurately, the sample should be put near $z = +z_0$ (in the case of $n_2 < 0$) or $z = -z_0$ (for $n_2 > 0$). The theoretical curve could also be obtained using Gaussian decomposition method, but for high order nonlinearity ($\Delta\varphi > 8$) the Fresnel–Kirchoff integral method is more suitable and precise.

3. 3. Z-Scan Technique through Beam Radius Measurement

The measured quantity is the beam radius r_q, which is defined as the distance from the beam center to the point where the irradiance reduces to a specific fraction q of on-axis ($r = 0$) irradiance value. If the value of q is equal to $1/e^2 = 0.135$, r_q will correspond to the commonly defined radius $w(z)$ of the Gaussian beam. The radius is numerically calculated through the Equation (28) [14,15]

$$I(z,r_q,t,\Delta\varphi_0(t)) = \frac{C\varepsilon_0}{2}\left|E(z,r_q,t,\Delta\varphi_0(t))\right|^2 = qI(z,0,t,\Delta\varphi_0(t)) \quad (28)$$

The value of the nonlinear phase shift $\Delta\varphi_0(t)$, the refractive index change $\Delta n(t)$ as well as the nonlinear refractive index n_2 are calculated by fitting the radius fromthe experimental z-scan measurement with Equation (28). Obviously, the numerical calculations involved in this method are much simple compared with the common z-scan technique, which is mainly because the spatial integration of Equation (23) is avoided. When r is equal to the beam radius ($r_q = w(z)$), the ratio of the intensity at $r_q = w(z)$ to the intensity at $r = 0$ is $q = 1/e^2 = 0.135$. In this section, the nonlinearity larger than π doesn't be analyzed, which is because when the diffraction ring pattern appears the Z-scan technique through beam radius measurement is unmeaning.

In Figure 6 the dependences of normalized passing beam radius curves on the sample position $z$are plotted using both the Gaussian decomposition method and Fresnel–Kirchhoff diffraction method at $q = 0.135$ for $\Delta\varphi_0 = -1$ and $\Delta\varphi_0 = -2$. The curvespresent a valley-peak configuration and show that when a self-defocusing sample is placed behind the focus (in a convergent beam), the beam radius in the far field is smaller than the case where the sample is placed after focus, which is in good agreement with results of Z-scan.

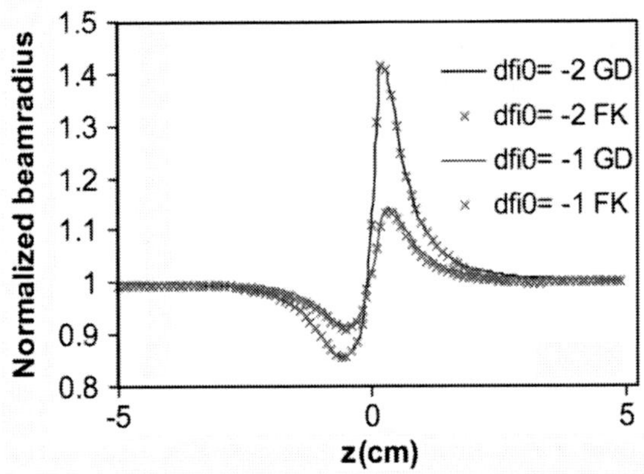

Figure 6. Normalized beam radius ($q = 0.135$) via sample position$z$ using Gaussian decomposition method and Fresnel–Kirchhoff diffraction method, $\Delta\varphi_0 = -1$ and $\Delta\varphi_0 = -2$.

3.4. Beam Profile Passing through the Sample

The parameter $w(z)$ guides us to stimulate the beam profile before and after the sample. This could be realized by moving a vertical transverse detection plane to detect the beam radius in a far-field position when the sample is fixed at z. This method offers a numerical simulation of the beam profile before and after the sample [14,15]. The beam shape before the sample could be easily estimated by the relation of Gaussian beam radius ($w(z) = w_0[1 + (z/z_0)^2]^{0.5}$). After the sample, the beam profile could be estimated by using Equation (28) by numerical calculation.

We again consider a thin self-defocusing sample placed at $z = -z_0$. Using GD method, the observable beam shape (at $q = 0.135$) is evaluated in Figure 7 (solid curve). As shown in Figure 7, the curve is similar to a Gaussian beam and the sample makes the beam converge at far field, which results in increase of the transmitted power from a finite aperture. This result is in agreement with common Z-scan measurement.

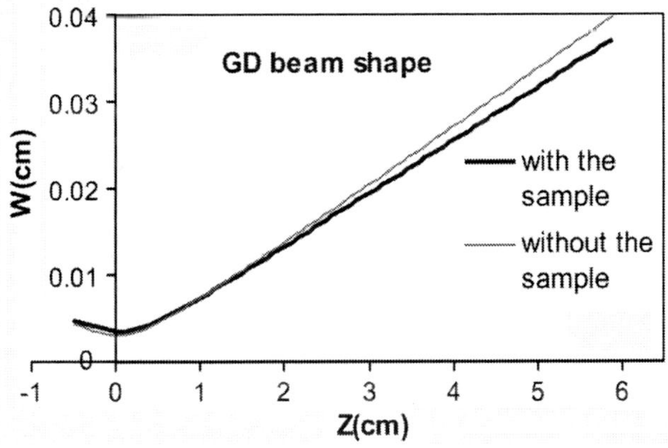

Figure 7. Transmitted beam profile using GD method for $\Delta\varphi_0 = -1$ (sample at $z = -z_0$).

The same curves are plotted using FK method (Figure 8). Figure 8 shows that the curve is not similar to the Gaussian beam profile. Also in the sample position at $z < 5.5\ cm$, the beam radius is larger than without sample case, which is not physically acceptable. However, the beam radius is smaller than the without sample case at $z > 5.5\ cm$. In the other words, only at far field, FK method has met our physical expectations, which is due to the range validity of the FK method. When the distance from the exit plane of the medium to the far-field observation plane (D) is large enough, the FK method is reliable and the beam profile is physically acceptable.

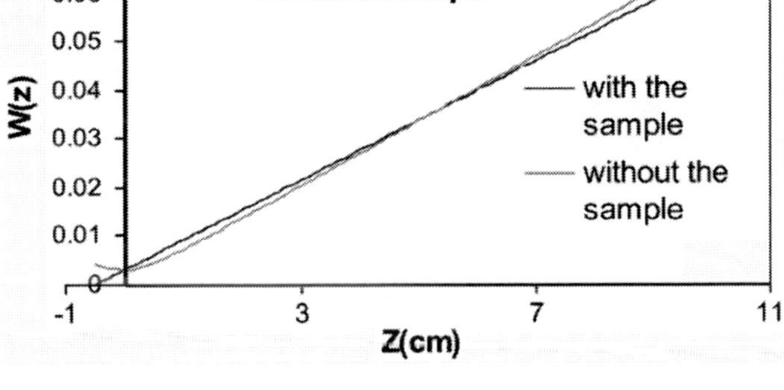

Figure 8. Transmitted beam profile using FK method for $\Delta\varphi_0 = -1$ (sample at $z = -z_0$).

3.5. The Curvature Radius of the Transmitted Wavefront

The phase of Gaussian beam could be calculated from Equation (1), which is given by:

$$\varphi(r,z) = kz - \arctan(z/z_0) + \frac{kr^2}{2R(z)} \tag{29}$$

This phase is constant for a specified wavefront. If a circular Gaussian beam passes through the sample, the phase of an arbitrary point of the space could be calculated using the argument of the electric field that is a complex value. In order to get $R(z)$, the phase magnitude of the beam is firstly calculated on the propagation axis at z. The wavefront is a paraboloidal surface and the forehead is at point (z_a, $z = 0$) while the phase of this point is marked as φ_a. We need to find another point of the surface at $z_a = -\Delta z$ (Δz must be smaller than $\lambda/2$). A vertical line will cut the same wavefront surface at $z_a = -\Delta z$ if $\Delta z \ll \lambda/2$ (Figure 9). To find this point one needs to move on the vertical line from the point ($z_a = -\Delta z$, $r = 0$) with a small step (Δr) and check the phases of every point until find the point with a phase equaling to the phase of the wavefront (φ_a). Thus the coordinate of another point of the wavefront is ($z_b = z_a - \Delta z$, r_b) and then from Equation (29) one can get:

$$R(z_b) = \frac{r_b^2}{2} \frac{1}{\left(\frac{\arctan(z_b/z_0) - \arctan(z_a/z_0)}{k}\right) + z_a - z_b} \tag{30}$$

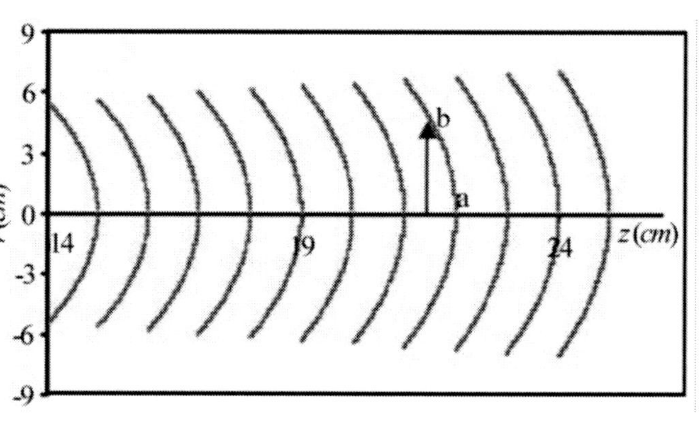

Figure 9. Finding another point on the wavefront.

Figure 10 shows the curvature radius of the wavefront versus z assuming that the sample is placed near $\pm z_0$ where the deformation maximum in the beam profile comes out. Gaussian decomposition method has been used to evaluate the phase of the electric field in Figure 10. As a result, the curves obey $R(z) \approx z$ for $z > z_0$ region. Having repeated this method for some different values of nonlinear refraction, the slops of the lines remain constant and don't depend on the position or quality of the sample.

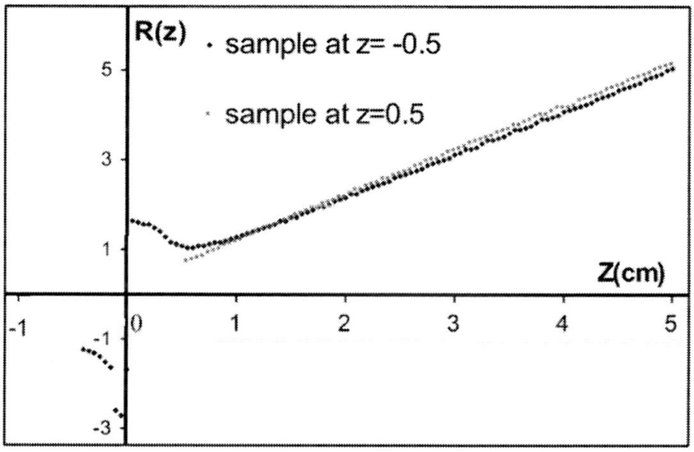

Figure 10. The curvature radius of the transmittedwavefront versus z for $\Delta\varphi_0 = -1$.

When the sample is a self-defocusing medium ($\gamma < 0$), R in the far-field decreases, in case the sample is placed at a convergent beam ($z < 0$) and increases if the sample is placed at a divergent beam ($z > 0$). Our results are still consistent with Z-scan results.

FK method has two main shortcomings in this case. The first one is the dependence of the phase value of an arbitrary point on the selected step dr in encountering the integral over r (Equations (8) and (9)). For example, in the above sections, numerical evaluations by Fresnel–Kirchhoff diffraction method were performed with the step of $dr = 10^{-4}cm$ that is larger than $\lambda/2 = 316 \times 10^{-7}cm$. To reach a credible the phase magnitude, the integration step should be smaller than $\lambda/2$, that may cause noticeable increase in the computer time. Different dr values may lead to different curves. Gaussian decomposition method is no integral over the radius of the sample area and is credible in this condition. The second shortcoming is the disability of Fresnel–Kirchhoff diffraction method in calculating the electric field in the region near the sample as mentioned in the above section.

Conclusion

Fresnel–Kirchhoff diffraction method is a good method to describe the electric field of a Gaussian beam passing through a Kerr medium at far field. Numerical calculations show that, as the nonlinear phase shift increases, the accuracy of this method exceeds to Gaussian decomposition method. Nevertheless there are two shortcomings in measurements and curve fittings with this method: one is longer time duration required to perform computer calculations, and the other one is the disability in describing the electric field in very short distances. However, for a long distance from the sample, Fresnel–Kirchhoff diffraction method is more accurate and adaptable compared to the other method and formula of nonlinear phase shift.

REFERENCES

[1] Sheik-bahae, M; Said, AA; Van Stryland, EW. "High-sensitivity, single-beam n_2 measurements", *Opt. Lett.*,1989, 14, 955-957.

[2] Sheik-bahae, M; Said, AA; Wei, TH; Hagan, DJ; Van Stryland, EW. "Sensitive measurement of optical nonlinearity using a single beam", *IEEE J. Quantum Electron.*,1990, 26, 760-769.

[3] Weaire, D; Wherrett, BS; Miller, DAB; Smith, SD. "Effect of low-powernonlinear refraction on laser beam propagation in InSb," *Opt. Lett.*,1974, 4, 331-333.

[4] Tsigaridas, G; Fakis, M; Polyzos, I; Persephonis, P; Giannetas, V. "Z-scan analysis for high order nonlinearities through Gaussian decomposition",*Opt. Commun.*,2003, 225, 253-268.

[5] Palfalvi, L; Hebling, J. "Z-scan study of the thermo-optical effect",*Appl. Phys. B*,2004, 78, 775-780.

[6] Boyd, RW. "*Nonlinear Optics*", 3thedition, Academic Press, New York, 2007.

[7] Wang, J; Sheik-Bahae, M; Said, AA; Hagan, DJ; Van Stryland, EW. "Time-resolved z-scan measurements of optical nonlinearities", *J. Opt. Soc. Am. B*, 1994,11, 1009-1017.

[8] Koushki, E; Farzaneh, A; Mousavi, SH. "Closed aperture z-scan technique using the Fresnel–Kirchhoff diffraction theory for materials with high nonlinear refractions", *Appl. Phys. B*, 2010,99, 565–570.

[9] MajlesAra, MH; Koushki, E. "Data analysis of z-scan experiment using Fresnel–Kirchoff integral method in colloidal TiO_2 nanoparticles", *Appl. Phys. B*, 2012, 107, 429–434.

[10] Nascimento, CM; Alencar, MARK; Chavez-Cerda, S; da Silva, MGA; Meneghetti, MR; Hichmann, JM. "Experimental demonstration of novel effects on the far-field diffraction patterns of a Gaussian beam in a Kerr medium",*J. Opt. A: Pure Appl. Opt.*, 2006, 8, 947-951.

[11] Koushki, E; Mousavi, SH; MajlesAra, MH; Koushki, A. "Numerical simulation of a Gaussian beam profile passing through a Kerr medium", *Current Appl. Phys.*,2009, 9, 1347-1350.

[12] Goodman, Joseph W."*Introduction to fourieroptics*", 2ndedition, McGraw-Hill, New York, 1988

[13] Tsigaridas, G; Fakis, M; Polyzos, I; Persephonis, P; Giannetas, V. "Z-scan technique for elliptic Gaussian beams", *Appl. Phys. B*, 2003, 77, 71-75.

[14] Koushki, E; MajlesAra, MH. "Comparison of the Gaussian-decomposition and the Fresnel–Kirchhoff diffraction methods in circular and elliptic Gaussian beams", *Opt. Commun.*,2011, 284, 5488–5494.

[15] Tsigaridas, G; Fakis, M;Polyzos, I; Giannetas.V; Persephonis, P. "Z-scan technique through beam radius measurements", *Appl. Phys. B*,2003, 76, 83-86.

In: Nonlinear Performance and Characterization Methods … ISBN: 978-1-62808-093-3
Editor: Jingsong Wei

Chapter 4

MEASUREMENTS OF NONLINEAR OPTICAL PARAMETERS OF TRANSPARENT AND NONTRANSPARENT MATERIALS USING SINGLE-SHOT TECHNIQUES

R. A. Ganeev*
Imperial College London, Prince Consort Road, London, UK

ABSTRACT

Single-shot techniques for the measurements of the nonlinear optical refraction of the materials possessing limited and high transparency are presented. These methods allow for analyzing the nonlinear refraction using the tilted samples placed from both sides of the focal plane of cylindrical lens. The proposed techniques were examined in the cases of opaque materials (GaAs and Si) and transparent liquid (carbon disulfide) at a wavelength of 790 nm. We compare these methods with the standard z-scan technique.

Keywords: Nonlinear refraction index, single-shot z-scan, single shot rz-scan

1. INTRODUCTION

All of the modifications of the z-scan are based on propagation of the sample through the focal plane of focused laser radiation and analysis of the phase and amplitude characteristics of transmitted (or reflected) radiation in the far field. A drawback of this technique is the need for multiple measurements during propagation of the sample through the focal plane. Instability of laser intensity requires averaged measurements of normalized transmittance when, at the same sample position, one has to carry out multiple measurements. One has to repeat these measurements at least a few hundred times to reproduce the normalized

* Corresponding author e-mail: r.ganeev@imperial.ac.uk.

transmittance of the sample in the far field in the cases of both open aperture and closed aperture z-scan schemes.

These repeatable measurements can (in some cases when the nonlinear optical response of the medium is weak and the required laser intensity is too strong) lead to damage of the sample. Furthermore, the wedge shape of the sample can cause the laser beam to move out of the axis of the far-field aperture. All these drawbacks of the z-scan technique have encouraged the development of some new approaches. Below we demonstrate new single-shot techniques for the measurements of the nonlinear optical refraction of the materials possessing limited and high transparency. These methods allow for analyzing the nonlinear refraction using the tilted samples placed from both sides of the focal plane of cylindrical lens. The proposed techniques were examined in the cases of opaque materials (GaAs and Si) and transparent liquid (carbon disulfide) at a wavelength of $790\,nm$. We compare these methods with the standard z-scan technique.

2. Single-shot Reflection Z-Scan for Measurements of the Nonlinear Refraction of Non-transparent Materials

The z-scan technique [1] for the measurements of the nonlinear optical parameters of media, which currently can be considered as most versatile one, was realized as a conventional transmission z-scan, off-axis z-scan, time-resolved z-scan, eclipsing z-scan, and reflection z-scan). Most of these modifications, excluding the latter one, are based on the propagation of the transparent sample through the focal plane of focused laser radiation and the analysis of the phase and amplitude characteristics of transmitted radiation in the far field.

The reflection z-scan (rz-scan) is used in the case of the opaque materials or the samples with limited transparency. The principles of this method are based on the analysis of the phase and amplitude characteristics of the beam reflecting from the non-transparent sample during its propagation through the focal plane of the focused radiation [2-6]. A drawback of this technique is a necessity of multiple measurements during the reflection of strong laser pulse from the surfaces of the materials propagating through the focal plane. The instability of the output intensity of lasers requires the averaged measurements of the reflectance of the samples under investigation, when, at the same position of sample, one has to carry out the multiple measurements. These averaged measurements have to be repeated at least few tens of times to define the normalized reflectance of the sample. This can lead, in some cases, to the damage of the sample, since one has to apply the strong laser fields to achieve the measurable nonlinear optical response from the surface. In that case one has to repeat all these measurements at the fresh position of the sample.

These drawbacks of rz-scan technique could be diminished by using some new approaches. The application of single-shot scans would allow overcoming the above difficulties. In this section, a new rz-scan method for defining the nonlinear optical refraction of opaque materials during single shot of laser radiation is discussed [7]. The method is based on the analysis of the spatial shape of laser beam after the reflection from the tilted sample, which is placed at the focal plane of the cylindrical lens. GaAs and Si plates were used to validate this approach. The proposed method is advantageous due to lower stress imposed on the fragile materials, since a single shot of laser pulse is sufficient for the definition of their nonlinear refractive indices. Another advantage is that it is no longer necessary to average a large number of pulses. It also needs no displacement of the medium under investigation, unlike the case of conventional rz-scan method.

The experimental scheme of this method is extremely simple. A broad ($18\ mm$ diameter FWHM $1/e^2$) laser beam (Ti: sapphire laser, $\lambda = 790\ nm,\ t = 120\ fs$) was focused by a cylindrical lens ($f = 150\ mm$) on a sample tilted at the angle of 45^0 (Figure 1). The focusing occurs in such a manner that the laser beam size does not change along the plane of drawing of this scheme. A sample was placed at such a position of z-axis when the central part of sample approximately coincided with the focal plane of the cylindrical lens, as shown in the picture. A CCD camera registered the shape of reflected radiation.

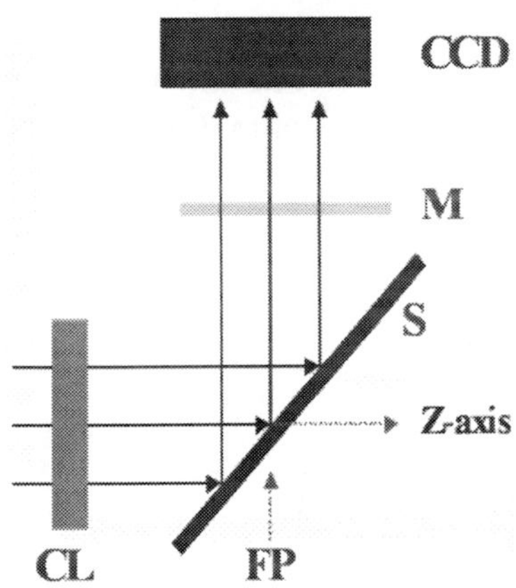

Figure 1. Experimental scheme for the SSSR. CL, cylindrical lens; S, sample; M, 99.7% reflection mirror; CCD, charge-coupled device; FP, focal plane of the cylindri- cal lens.

Two sets of acquisitions are needed for these measurements. Firstly, a dense filter (or a 99.7% reflecting mirror, which can leak a small amount of radiation) is placed in front of the cylindrical lens. The obtained image of the laser beam leaked through this strong filter (or mirror) is used just for confirmation of the approximately homogeneous radiation distribution along the z-axis, especially close to the focal plane. Secondly, a strong filter, or mirror was moved from the previous position toward the position in front of the CCD camera (as shown in Figure 1). In that case, the intensity of focused radiation becomes strong enough to induce the nonlinear optical processes influencing the shape of reflected beam along the z-axis.

This method can be dubbed as “Single-shot scan by reflection from the tilted samples” (SSSR). The proposed scheme requires, after confirming the homogeneity of the beam shape reflecting from the sample at low-intensity conditions, only one shot to define the variations of the spatial distribution along the z axis, which further could be processed to define the nonlinear refractive index (γ) of the sample.

GaAs and Si plates were used for the definition of their γ at the wavelength of $\lambda = 790\ nm$ using the SSSR. This wavelength is below the band gaps of the GaAs ($890\ nm$) and Si ($1170\ nm$), so they were fully non-transparent at this spectral range and could not be analyzed using the conventional transmission z-scan technique. The $2\ mm$ thick plate of intrinsic, polished crystalline Si cut with normal axis along the $\langle 001 \rangle$ crystal direction and the 0.5 mm thick plate of polished GaAs were used in these studies.

Figure 2 shows the beam shapes reflected from the GaAs plate in the cases of weak and strong irradiation of the focused pulse. At weak irradiation (i.e., when the 99.7% reflecting mirror was placed in front of the cylindrical lens), the spatial distribution along the z-axis remains unchanged with regard to the input radiation (Figure 2(a)). Another pattern appeared in the case of strong irradiation ($I_0 = 3.6 \times 10^{11}\ W\ cm^{-2}$) on the surface of GaAs when the mirror was moved towards the CCD camera. At the central part of the beam (i.e., at maximum intensity), the nonlinear growth of reflectance led to the increase of reflected intensity. This growth was caused by the positive adding to the refractive index of GaAs and corresponding increase of the Fresnel reflection (Figure 2(b)). The normalized reflectance $R(z)$ depicted from the shape of reflected beam is presented in Figure 2(c). One can see a considerable increase of normalized reflectance at the central part of this SSSR curve. When one returns the mirror to the previous position, a featureless pattern analogous to the Figure 2(a) appeared again, without the indication of any damage of the surface of GaAs.

To calculate the nonlinear refraction of non-transparent samples, the following expression for the normalized reflectance along the z-axis was used for the fitting with the experimental data [8]:

$$R(z,\Theta) = 1 - \frac{4\frac{R_1(\Theta)}{R_0} I_0 k_2 x}{(x^2+9)(x^2+1)} + \frac{2\frac{R_1(\Theta)}{R_0} I_0 \gamma (x^2+3)}{(x^2+9)(x^2+1)} + \frac{\left(\frac{R_1(\Theta)}{R_0}\right)^2 I_0^2 (\gamma^2 + k_2^2)}{(x^2+9)(x^2+1)}. \quad (1)$$

Here I_0 is the incident radiation intensity at the focal plane; k_2 is the nonlinear extinction coefficient, R_0 is the coefficient of linear reflection: $x = \frac{z}{z_0}$, $z_0 = 0.5\ kw^2$ is the diffraction length of the beam, w is the radius of the beam waist, $k = 2\pi/\lambda$ is the wave number,

$$R_1(\Theta) = \frac{2n_0^3 \cos\Theta - 4n_0 \cos\Theta \sin^2\Theta}{n_0^4 \cos^2\Theta - n_0^2 + \sin^2\Theta} \left[n_0^2 - \sin^2\Theta\right]^{-0.5} \ [9],$$

n_0 is the linear refractive index, and Θ is the angle of radiation incidence. In Equation (1), the first term is responsible for nonlinear absorption, the second term characterizes nonlinear refraction, and the third term characterizes their joint influence.

A characteristic feature of the rz-scan is the use of an open aperture for investigating the nonlinear refraction, while for the conventional transmitting procedure; an open aperture makes it possible to analyze the nonlinear absorption. It should be noted that the scheme without the aperture in the rz-scan allows one to exclude the effect of nonlinear absorption on the results of experiments. Taking into account an insignificant role of the nonlinear absorption with regard to the nonlinear refraction in the case of GaAs, which is seen from the experimental $R(z)$ dependence showing the absence of the valley, one can use the second term of Equation (1) for the fitting with the experimental data. A thin curve in the Figure 2 (c) shows the fit with the experimentally observed normalized reflectance. The corresponding fitting value of the nonlinear refractive index was $\gamma = (1.7 \pm 0.3) \times 10^{-11} cm^2 W^{-1}$.

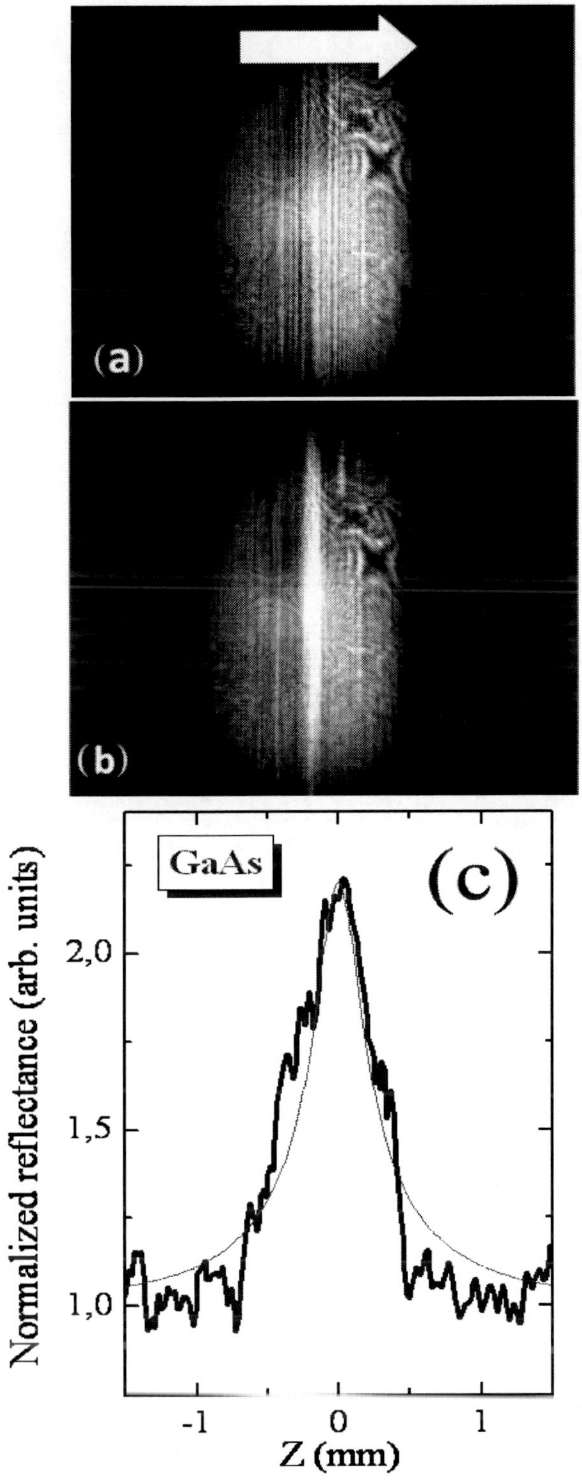

Figure 2. Reflected beam shapes in the cases of (a) weak and (b) strong intensities of the femtosecond pulse focused on the surface of GaAs plate, (c) normalized SSSR curve of the high-intensity reflected beam registered by a CCD camera. Thin curve in (c) is a fit of the second term of Equation (1) with the experimental result (thick curve).

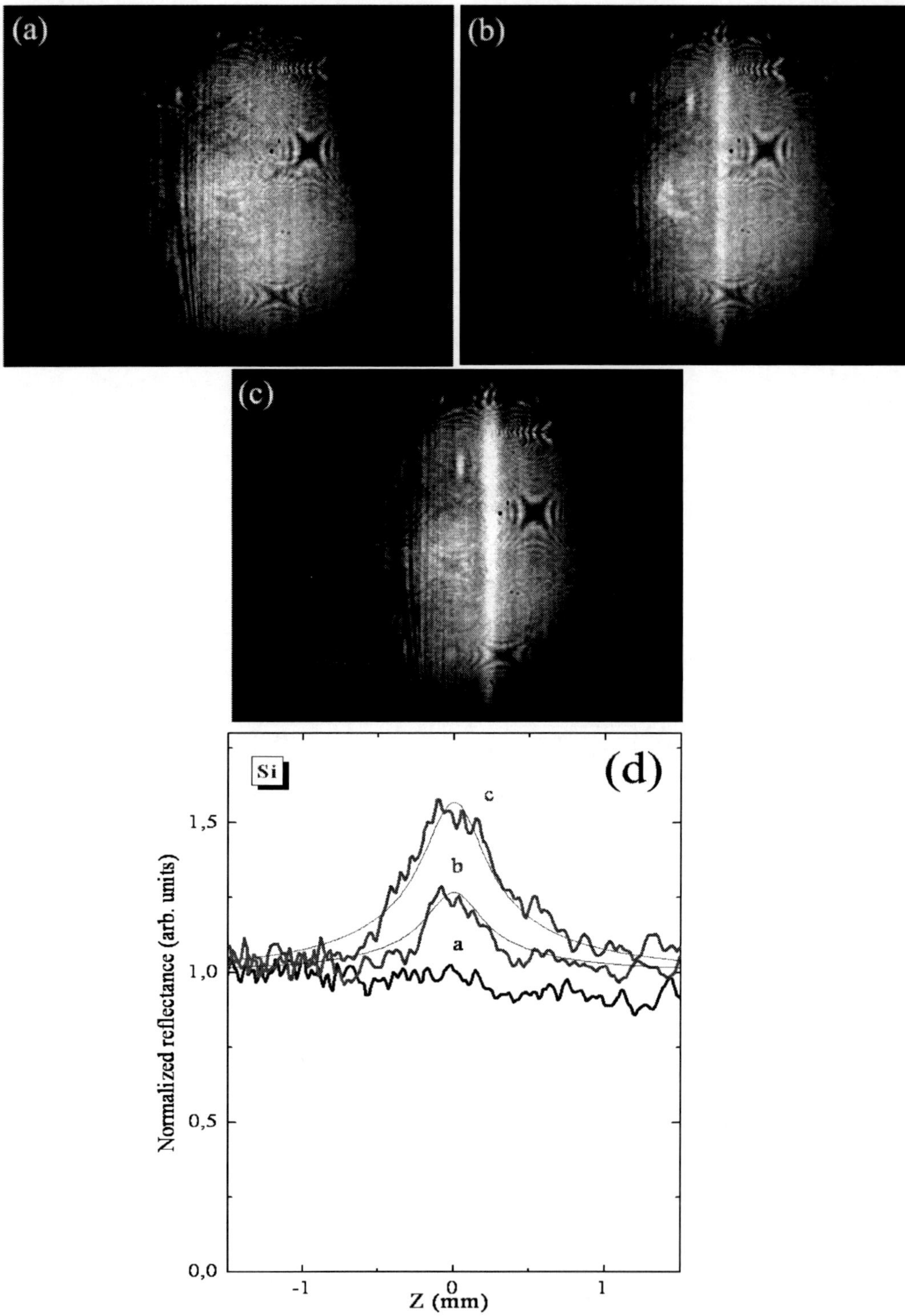

Figure 3. Reflected beam shapes in the cases of (a) weak ($9\times10^{8}\ W\ cm^{-2}$), (b) moderate ($1.2\times10^{11}\ W\ cm^{-2}$), (c) higher ($1.8\times10^{11}\ W\ cm^{-2}$) intensities of the laser beam irradiated the Si plate, and (d) normalized SSSR curves of the reflected laser beam in above-presented three cases. Thin curves in (d) are the fits of the second term ofEquation (1) with the experimental results (thick curves).

The data of the nonlinear refractive indices of GaAs in the mid-infrared range ($3\times10^{-13}\ cm^2W^{-1}$, $\lambda = 1.7\ um$) measured using the conventional transmission z-scan are presented in [10]. However, close and below the edge of transmission band, one can expect the increase of the optical Kerr effect. Indeed, the value of γ increases drastically when one approaches toward the band gap energy. Theoretical predictions show that the nonlinear refraction in this region also changes from a positive value (for $E < 0.65\ E_g$, E and E_g are the photon energy and band gap energy, respectively) to a negative one. Close to the band gap energy of GaAs ($E_g = 1.42\ eV$), the nonlinear refractive index was reported to be $-2.1\times10^{-6}\ cm^2W^{-1}$ [11]. In the meantime, previous studies of the γ of GaAs in the spectral range below its band gap ($538\ nm$, $532\ nm$ [12]) measured using the rz-scans have shown the positive sign of nonlinear refraction. In particular, a value of $\gamma = 7.8\times10^{-8}\ cm^2W^{-1}$ was reported when the dye laser radiation at $538\ nm$ was used.

An accurate knowledge of the nonlinear optical parameters of silicon is essential for various applications of this material (cross-phase modulation in silicon waveguides, wavelength conversion, Raman lasing and amplification, etc). However, as in the case of GaAs, current knowledge of the nonlinear optical parameters of silicon is limited to the near-infrared region. A steady decrease of the nonlinear refractive index of Si from $3\times10^{-14}\ cm^2W^{-1}$ ($\lambda = 1.9\ um$) to $1\times10^{-14}\ cm^2W^{-1}$ ($\lambda = 1.9\ um$) in the mid-infrared region was reported by Lin *et al.* [13].

The SSSR measurements carried out using the silicon plate have shown a drastic increase of the γ for shorter wavelength ($790\ nm$). Figures 3(a), 3(b), and 3(c)) present the shapes of the beam reflected from the surface of Si at different intensities of the femtosecond pulse. The normalized reflectance of sample in these three cases is presented in Figure 3. Here also shown the fitting procedure using the Equation (1). Thin curves represent the fits for (b) $1.2\times10^{11}\ W\ cm^{-2}$ and (c) $1.8\times10^{11}\ W\ cm^{-2}$ intensities of laser radiation at the surface of silicon plate. The fitting nonlinear refractive index of silicon was found to be $\gamma = (2.6\pm0.5)\times10^{-11} cm^2W^{-1}$.

In rz-scan, the nonlinear refraction is analyzedby the measurement of the wavefront distortion of a laser beam due to photoinduced modification in the reflection coefficient. One can assume considerably smaller modifications of the wavefront compared with the conventional transmission z-scan, since in the latter case one has an advantage of analyzing the accumulated nonlinear optical process using whole sample's length. In most cases, themodification of refractive index is below the measurement sensitivity, while the increase of laser intensity is undesirable due to optical damage of the sample through the accumulative effect of multiple shots.

One of potential applications of SSSR is the analysis of the nonlinear refraction of semiconductor saturable mirrors (SESAM) currently used for mode-locking as a standard technique for self-starting femtosecond lasers. The manufacturing of SESAM is a complicated and expensive process requiring accurate irradiation of this structure to prevent the surface damage. Previously, the application of the oblique incidence of a polarized beam allowed increasing the sensitivity of rz-scan technique at the Brewster angle. The tilting at Brewster angles could enhance the sensitivity of the SSSR method, as well. At the same time, one has to examine the applicability of rz-scan relations and used approximations for the case of the tilted samples at these conditions.

3. Single-shot Y-Scan for Characterization of the Nonlinear Optical Parameters of Transparent Materials

So far, a few single-shot methods based on the analysis of the laser beam shape in the far field after propagation of the material were offered. A single-shot technique using a phase object at the entry of a $4f$ (quarter-phase) coherent imaging system to characterize the value of nonlinear refractive index was reported in [14, 15]. Instead z-scanning the material they analyzed the transverse modification in the image acquired with a CCD camera. Another method is based on the use of two crossed cylindrical lenses to obtain an astigmatic beam [16]. The sample is placed between the focal planes where the beam has a round intensity profile. By evaluating the far-field beam profile it is possible to estimate the nonlinear phase shift. Third method uses the conventional open aperture z-scan and extracts the two-dimensional transverse modifications of the whole far-field pattern by processing all the pixels of the CCD camera [17]. The key element of the fourth single-shot method is a spatial intensity variation of an elliptical Gaussian beam in conjunction with an array detector [18]. In their method, one has to record the changes in the relative intensity of the pump or the probe beams. However, some of these methods did not define the sign of nonlinearity, which is a drawback of the above single-shot techniques.

In this section, a new approach for analysis of the nonlinear optical parameters of transparent materials using single shot is discussed [19]. The method is based on the analysis of the spatial shape of laser beam after propagation through the tilted sample, which is placed at the focal plane of cylindrical lens. The use of the tilted sample allows different parts of the sample to be simultaneously irradiated at different distances from the focusing lens, thus allowing the standard z-scan-like data to be acquired in a single shot based on the intensity variation in z-axis followed the phase variations along the sample's y-axis. The CS_2 was used to validate the approach in the case of the media demonstrating the nonlinear refraction and absorption.

The $14\,mm$ diameter (FWHM $1/e^2$) laser beam was focused by a cylindrical lens ($f = 150\,mm$) on a sample tilted at the angle of 45^0, as it shown in Figure 4. The light source was a Ti: sapphire laser ($\lambda = 790\,nm$). For these experiments, the two ranges of pulse duration were used ($120\,fs$ and $210\,ps$ of compressed and chirped radiation respectively). The beam waist dimensions at the focal plane of cylindrical lens were $14\,mm \times 25\,um$. The laser was operated at a 10 Hz pulse repetition rate. The shutter was used for separation of single shot. The focusing of laser radiation was carried out in such a manner that the size of laser beam in the plane of Figure 4 (i.e. along the y-axis) remained unchanged during propagation through this scheme. A sample was placed at a position of z-axis when the central part of sample coincided with the focal plane of cylindrical lens. A CCD camera registered the shape of propagated radiation in the far field (i.e. at the distance of $600\,mm$ from the samples).

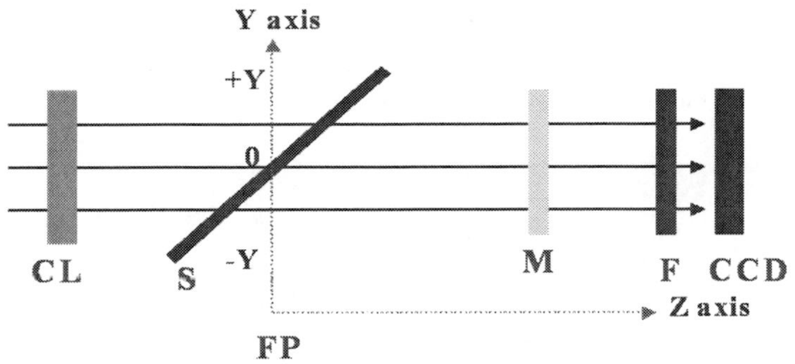

Figure 4. Experimental setup for the single-shot measurements of the nonlinear optical characteristics of transparent materials. CL: cylindrical lens; S: sample; FP: focal plane; M: 99.7% reflecting mirror (or dense filter); F: filter; CCD: charge-coupled device camera.

Initially, a dense filter (or 99.7% reflecting mirror, which can leak a small (~0.3%) amount of radiation) was placed in front of the cylindrical lens. The obtained image of the laser beam leaked through the dense filter (or mirror) was used to calibrate the spatial shape of radiation along the y-axis. Then a strong filter (or mirror) was moved from the previous position toward the CCD camera (as shown in Figure 4). In that case, the intensity of focused radiation became strong enough to cause the appearance of the nonlinear optical processes influencing the beam phase modulation along the y-axis, which was further transformed in the spatial shape variation in the far field.

The method is related with the single-shot z-scan of the sample. However, since the changes of laser intensity along the z-axis cause the variation of beam shape along the y-axis, the method can be dubbed as a "Single-shot y-scan of tilted sample" (SSYSTS). This method requires only one laser shot to define the changes in the spatial distribution of laser radiation along the y-axis in the far field, which further could be processed to define the nonlinear refractive index and nonlinear absorption coefficient of the sample.

The example of SSYSTS is presented in Figure 5. A $1\,mm$ thick silica glass cell was filled with the carbon disulfide, which is a commonly used reference medium for the z-scans and pump-probe measurements of the nonlinear optical parameters of media. Its nonlinear refractive index was precisely defined during multiple measurements in a broad spectral range and showed no considerable variation in the femtosecond timescale ($\gamma \sim 3.5\times10^{-15}\,cm^{2}W^{-1}$), while the ten-fold increase of this parameter was reported for the pulses longer than 1.5 ps [20], which was caused by the influence of laser-induced molecular reorientation (see also previous section). The high transparency down to $\lambda = 380\,nm$ makes CS_2 an excellent choice for the demonstration of the advantages of the proposed method in different spectral ranges.

Figure 5(a) shows the raw image of the beam shape after the propagation of a weakprobe radiation($t = 120\,fs$, $\lambda = 790\,nm$) through the cell, when the 99.7% reflecting mirror was placed in front of the cylindrical lens. This image demonstrates a featureless Gaussian-like intensity distribution along the y-axis. The change in the position of mirror toward the CCD camera led to considerably stronger irradiation of the sample, which in turn caused the phase variations of different parts of the beam propagating through the sample. Most significant changes occurred close to the focal plane, which were further transformed in the intensity variations in the far field (Figure 5(b)). One can see a considerable change in the spatial distribution along the y-axis, when the bottom part of the beam (close to its central part) was weakened (thick arrow), while the upper part of the beam showed some enhancement of its intensity (thin arrow) compared to the weak irradiation.

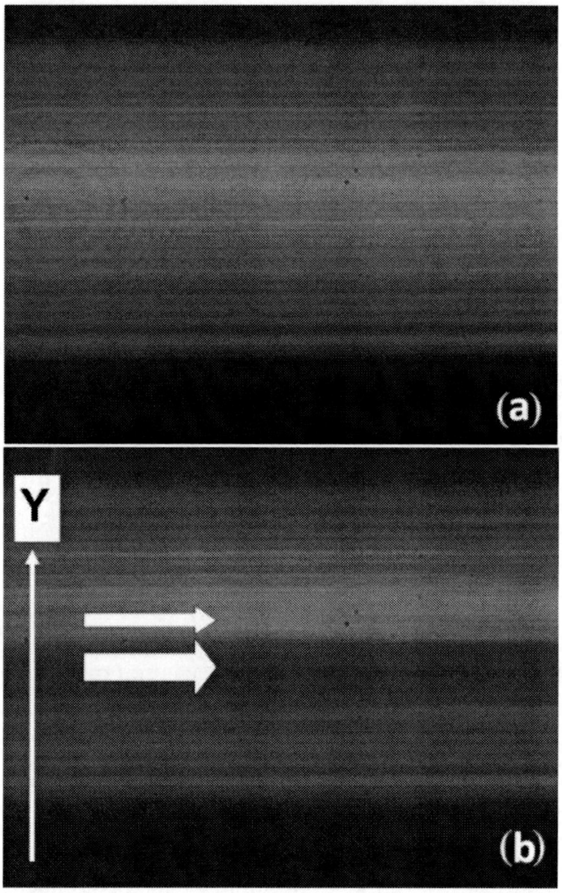

Figure 5. Beam shape images obtained in the cases of (a) weak and (b) strong irradiation using the 120 fs pulse propagating through the tilted CS_2-containing cell. Vertical arrows show the directions at which the analysis of spatial distributions was carried out. These data were further used for preparation of the normalized SSYSTS plots.

The initial processing of these data consisted of the definition of the spatial distribution of the propagated beam along the y-axis for these two cases. Then the spatial distribution curve defined from Figure 5(b) was normalized by dividing to the curve obtained from the Figure 5(a) to exclude the influence of the heterogeneities of laser beam shape caused by occasional diffraction patterns originated from the damaged parts of optical elements, as well as the heterogeneities of beam shape itself. This processresulted in the dependence of the normalized intensity on the position of laser beam along the y-axis ($I(y)$, Figure 6(a), thick line). One can see that this curve resembles the normalized transmittance dependence $T(z)$ of the standard closed aperture z-scan of the medium possessing the positive nonlinear refraction and strong nonlinear absorption (when the valley considerably exceeds the peak). The standard closed aperture z-scan measurements were performed using the spherical lens and observed analogous pattern in the case of 120 fs pulses at approximately identical irradiation conditions (i.e., using the laser intensity at the focal plane in the range of $I_0 = 1\times10^{10}\ W\ cm^{-2}$, Figure 6(b)), when a deep valley followed by a relatively small peak. Different ratios between the valley and peak in these two cases (Figures 6(a) and 6(b)) can be attributed to some difference in the peak intensities at the focal planes of cylindrical and spherical lenses, as well as to different focusing conditions.

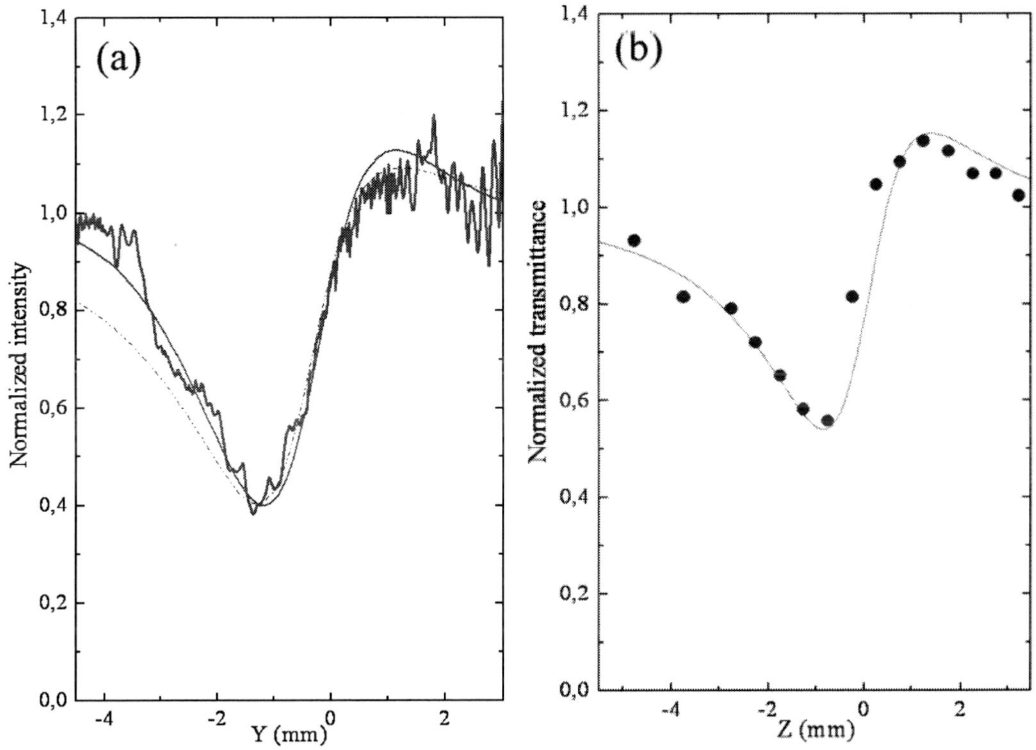

Figure 6. (a) SSYSTS and (b) standard closed aperture z-scan of the 1 mm thick cell filled with the CS_2 in the case of irradiation by the 120 fs pulses. Thin dotted line in (a) and thin solid line in (b) show the theoretical fits for the SSYSTS and z-scan. Thin solid line in (a) is the fitting curve defined from the [19] in the case of cylindrical focusing.

The standard z-scan was obtained by measuring the normalized transmittance at 15 positions of the CS_2-containing cell moving along the z-axis. Each point in the Figure 6(b) is the averaged measurement of normalized transmittance obtained during twenty laser shots. Here one has to take into account the shot-to-shot instability of laser intensity. The conventional technique of the fitting of these results allowed for defining the nonlinear optical parameters of CS_2 ($\gamma = 4\times10^{-15}\ cm^2W^{-1}$, $\beta = 3\times10^{-13}\ cm\ W^{-1}$), which were close to previously reported data measured at analogous pulse duration and intensity ($t = 110\ fs, I_0 = 3\times10^{10}\ W\ cm^{-2}$, $\gamma = 3\times10^{-15}\ cm^2W^{-1}$, $\beta = 4.2\times10^{-13}\ cm\ W^{-1}$ [20, 21]). One can see a nece- ssity in multiple experimental measurements and cumbersome actions during the definition of the sample's nonlinearities in the case of z-scan.

Analogous fitting procedure was applied in the case of SSYSTS and showed lack of coincidence between the experimental data obtained at cylindrical focusing conditions and the theory based on the spherical focusing conditions (thin dotted curve in Figure 6(a)). The application of the relations [1, 22] commonly used for fitting of the open and closed aperture z-scans seems inappropriate for analysis of the SSYSTS. At the same time the analysis quantifying the beam propagation characteristics of the cylindrical beam interacting with the nonlinear medium compared with the symmetrical beam used in a z-scan will be useful for the definition of the applicability of the proposed technique at different focusing configurations. The fitting using the theory of SSYSTS developed in [19] is presented in Figure 6 as a solid line.

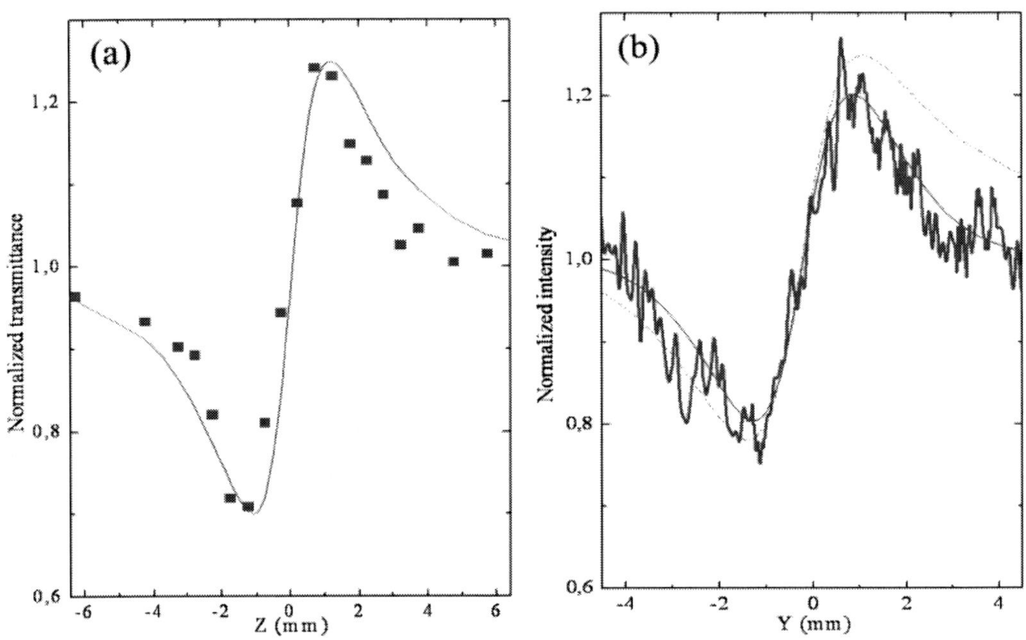

Figure 7. (a) SSYSTS and (b) standard closed aperture z-scan of the 1 mm thick cell filled with the CS_2 in the case of irradiation by the 210 ps pulses. Thin dotted line in (a) and thin solid line in (b) show the theoretical fits for these two methods using the relations of conventional z-scan. Thin solid line in (a) shows the fitted curve defined from [19].

To diminish the role of nonlinear absorption in these measurements, one can decrease the pulse energy at the focal plane. Another way is the application of longer pulses, when one can obtain relatively symmetric pattern of both the SSYSTS and z-scan in the case of CS_2. As it was mentioned, for the carbon disulfide, an increase of pulse duration above a few picoseconds leads to appearance of additional nonlinear refractive mechanism (i.e. molecular reorientation), which causes a considerable increase innonlinear refraction [20]. To examine this approach, the 210 ps pulses of the uncompressed radiation of Ti: sapphire laser ($\lambda = 790\ nm$) were used for both the SSYSTS and z-scan of CS_2. Figure 7 shows these scans, which were obtained at considerably smaller intensity compared to the previous case. One can see that the curves in Figure 7 are relatively symmetric with regard to initial normalized transmittance. The fitted value of z-scan at $t = 210\ ps$ was $\gamma = 4{\times}10^{-14}\ cm^2W^{-1}$, which is close to the data reported for this range of probe pulse durations. In the case of SSYSTS, the calculations [19] gave the value of nonlinear refractive index of carbon disulfide approximately equal to the γ of CS_2 defined from the z-scan measurements

($\gamma \approx 3.5 \times 10^{-14}\ cm^2 W^{-1}$). At these conditions, the SSYSTS curve of carbon disulfide is more suitable for comparison with other samples.

The results of fitting using the relations [19] are presented in Figure 7(a) by the thin solid line. Here the fitting using relations are also shown taking into account the spherical focusing of the probe radiation (thin dotted line). One can see the discrepancy between the fitting curve and experimental data in the latter case (analogous to Figure 6(a)) and the reasonable coincidence in the former case.

In described studies, the SSYSTS appear noisy due to some defects on the sample, mirrors, and filters. One can remind that this is a single-shot method. The averaging of the curves of single-shot scan gives relatively smooth pattern. One can also compare this result with the averaged results obtained during standard z-scan procedure. About few hundred shots were used for the results presented in Figure 7(b).

CONCLUSION

A single-shot technique for the measurements of the nonlinear optical refraction of the highly absorptive materials was analyzed using $790\ nm$, $120\ fs$ laser pulses and GaAs and Si plates. This SSSR method is based on the analysis of the spatial shape of a laser beam after the reflection from the tilted sample, which is placed at the focal plane of the cylindrical lens. The proposed method is advantageous and imposes a lower stress in the cases of fragile materials, since a single shot of laser pulse is enough for the definition of their nonlinear refractive indices. With this technique, it is no longer necessary to average a large number of pulses as well as move the medium under investigation, unlike the case of the rz-scan method.

We also discuss another new single-shot technique for definition of the nonlinear optical characteristics of transparent materials based on the measurements of spatial distribution of the beam propagating through a tilted sample. The method allows one to analyze the nonlinear refraction and nonlinear absorption using tilted samples placed at the focal plane of cylindrical lens. The use of a tilted sample allows different parts of the sample to be simultaneously irradiated at different distances from the focusing lens, thus allowing standard z-scan-like data to be acquired in a single shot based on the intensity variation in the z-axis following the phase variations along the sample's y-axis. Examples of such measurements are compared with the conventional z-scan technique in the case of carbon disulfide.

ACKNOWLEDGMENTS

The author acknowledges support from the Marie Curie International Incoming Fellowship Grant within the 7th European Community Framework Programme (grant No. PIIF-GA-2009-253104). The author also thanks H. Kuroda, I. A. Kulagin, M. Suzuki and M. Baba for their help during these studies.

References

[1] Sheik-Bahae, M; Said, AA; Wei, TH; Hagan, DJ; van Stryland, EW. "Sensitive measurement of optical nonlinearities using a single beam", *IEEE J. Quantum Electron.*,1990, 26, 760-769.

[2] Petrov, DV; Gomes, ASL; de Araujo, CB. "Reflection z-scan technique for measurements of optical properties of surfaces", *Appl. Phys. Lett.*,1994, 65, 1067-1069.

[3] Sheik-Bahae M; Ebrahimzadeh, M. "Measurements of nonlinear refraction in the second-order $\chi^{(2)}$ materials $KTiOPO_4$, $KNbO_3$, β-BaB_2O_4, and LiB_3O_5", *Opt. Commun.*,1997, 142, 294-298.

[4] Martinelli, M; Bian, S; Leite, JR; Horowicz, RJ. "Sensitivity-enhanced reflection z-scan by oblique incidence of a polarized beam", *Appl. Phys. Lett.*, 1998, 72, 1427-1429.

[5] Martinelli, M; Gomes, L; Horowicz, RJ. "Measurement of refractive nonlinearities in GaAs above bandgapenergy", *Appl. Opt.*,2000, 39, 6193-6196.

[6] Wang, K; Xing, QR; Li, HY; Li, JP; Zhang, ZG; Zhang, N; Chai, L; Wang, QY. "Optical characteristics of semiconductor saturable absorber mirrors investigated by the reflection *z*-scan technique", *Opt. Commun.*, 2006, 265, 369-372.

[7] Ganeev, RA. "Single-shot reflection *z*-scan for measurements of the nonlinear refraction of nontransparent materials", *Appl. Phys. B*, 2008, 91, 273-277.

[8] Ryasnyanskii, AI. "Nonresonant nonlinearity of silver nanoparticles", *J. Appl. Spectrosc.*, 2005, 72, 712-715.

[9] Bristow, AD; Rotenberg, N; van Driel, HM. "Two-photon absorption and Kerr coefficients of silicon for 850–2200 nm", *Appl. Phys. Lett.*, 2007, 90, 191104.

[10] Hurlbut, WC; Le, YS; Vodopyanov, KL; Kuo, PS; Fejer, MM. "Multiphoton absorption and nonlinear refraction of GaAs in the mid-infrared", *Opt. Lett.*, 2007, 32, 668-670.

[11] Fang JW; Zhang, SY. "Modeling for laser-induced surface thermal lens in semiconductors", *Appl. Phys. B*, 1998, 67, 633-639.

[12] Ganeev RA; Ryasnyansky, AI. "Reflection z-scan measurements of opaque semiconductor thin films", *Phys. Stat. Sol. A*, 2005, 202, 120-125.

[13] Lin, Q; Zhang, J; Piredda, G; Boyd, RW; Fauchet, PM; Agrawal, GP. "Dispersion of silicon nonlinearities in the near infrared region", *Appl. Phys. Lett.*, 2007, 91, 021111.

[14] Boudebs, G; Cherukulappurath, S. "Nonlinear optical measurements using a 4*f* coherent imaging system with phase objects", *Phys. Rev. A*, 2004, 69, 053813.

[15] Cherukulappurath, S; Boudebs, G; Monteil, A. "4*f* coherent imager system and its application to nonlinear optical measurements", *J. Opt. Soc. Am. B*, 2004, 21, 273-279.

[16] Hein, J; Helbig, M; Rentsch, S. "Measurements of a nonlinear refractive index with a single laser pulse", *Appl. Opt.*, 1997, 36, 1173-1176.

[17] Marcano, A; Maillotte, H; Gindre, D; Métin, D. "Picosecond nonlinear refraction measurement in single-beam open z-scan by charge-coupled device image processing", *Opt. Lett.*, 1996, 21, 101-103.

[18] Jayabalan, J; Singh, A; Oak, SM. "Single-shot measurement of nonlinear absorption and nonlinear refraction", *Appl. Opt.*, 2006, 45, 3852-3858.

[19] Ganeev RA; Kulagin, IA. "Single-shot *Y*-scan for characterization of the nonlinear optical parameters of transparent materials", *J. Opt. A*, 2009, 11, 085001.

[20] Ganeev, RA; Ryasnyansky, AI; Kuroda, H. "Nonlinear optical characteristics of carbon disulfide", *Opt. Spectrosc.*, 2006, 100, 108-118.

[21] Ganeev, RA; Baba, M; Ryasnyansky, AI; Suzuki, M; Turu, M; Kuroda, H. "Two- and three-photon absorption in CS_2", *Opt. Commun.*, 2004, 231, 431-436.

[22] Liu, X; Guo, S; Wang, H; Hou, L. "Theoretical study on the closed-aperture z-scan curves in the materials with nonlinear refraction and strong nonlinear absorption", *Opt. Commun.*, 2001, 197, 431-437.

In: Nonlinear Performance and Characterization Methods … ISBN: 978-1-62808-093-3
Editor: Jingsong Wei

Chapter 5

PSEUDO-NONLINEARITY AND ELIMINATION IN Z-SCAN MEASUREMENT

Rui Wang*[1,2], *Hui Yan*[1,2] *and Jingsong Wei*

[1]Shanghai Institute of Optics and Fine Mechanics,
Chinese Academy of Sciences, Shanghai, China
[2]University of Chinese Academy of Sciences, Beijing, China

ABSTRACT

The optical nonlinearity was an intrinsic characteristic of the materials, and usually occurs under a strong laser pulse irradiation. The z-scan is a commun measuring method for the optical nonlinearity. In this chapter, we introduce three kind of pseudo-nonlinearity in the z-scan measurement and the elimination methods, accordingly. The first is from the incident angle change in the z-scan measurement; the second is from the laser device instability; the third is from the sample surface morphology damage or structural transformation. The incident angle change induced pseudo-nonlinearity can be eliminated by adopting a low numerical aperture converging lens or adjusting the incident light polarization state; the laser device instability induced pseudo-nonlinearity can be avoided through the combination of placing an opto-isolator and systemic noise detector in the z-scan setup and advanced data processing methods. The sample damage needs to be ruled out during the z-scan measurement, which can be realized by a real-time imaging unit in the z-scan setup.

Keywords: Optical nonlinearity, z-scan, pseudo-nonlinearity

1. INTRODUCTION

The optical nonlinearity is one of the fundamental properties of optical materials. All kind of methods were proposed to measure the optical nonlinearity in past years, and the

* Corresponding author e-mail: weijingsong@siom.ac.cn.

accuracy varies with the methods employed for the measurements. Among others, the z-scan technique proposed by Sheik-Bahae [1, 2] has been widely used due to its simplicity and high sensitivity. A variety of new methods based on the original version of the z-scan method have also been proposed in recent years, such as the time-resolved z-scan method [3, 4] and the reflection z-scan technique [5]. As for the incident beams, a number of different beam profiles have been so far tested, such as the round Gaussian [6], the elliptic Gaussian [7], the Gaussian–Bessel [8], the flat-topped [9], the top-hat [10], and the polarization beams [11, 12], etc. In addition, the sensitivity and accuracy of the measurements are also improved with various data processing and fitting methods [13-16].

In a basic z-scan measurement, a collimated laser beam is focused through a lens, and the sample is scanned along the z-axis (optical axis) through the focal point. If there is significant nonlinearity in the sample, the light intensity at different z-scan position either increases or decreases depending on the positive or negative sign of the nonlinearity. The recorded z-scan curves are compared with the theoretically determined fitting curves, and the nonlinear index is subsequently extracted. In this chapter, we introduce three kind of pseudo-nonlinearity in the z-scan measurement and the elimination methods, accordingly. In section 2, the incident angle change in the z-scan measurement induced pseudo-nonlinearity and the corresponding elimination method are presented. In section 3, we give the laser device instability induced pseudo-nonlinearity. The sample surface morphology damage or structural transformation also will cause the pseudo-nonlinearity, which are presented in section 4. Finally, the work is concluded in Section 5.

2. Incident Angle Change Induced Pseudo-nonlinear Absorption in the Z-Scan Measurement

Generally speaking, in the case of pure linear absorption, for the idea of Gaussian beams, i.e. within the near-axis approximation, the light intensity should keep constant during the z-scan, which is a well-known basic principle for nonlinear index measurements. However, based on theoretical analyses and numerical simulations, it is demonstrated the existence of pseudo-nonlinear absorption in z-scan measurements even for samples of pure linear absorption materials due to the incident angle change. Since the shape of the discovered pseudo-nonlinear absorption curves appears similar to that of the nonlinear absorption curves of an open-aperture z-scan measurement, false nonlinearities may be recorded especially for focusing lenses with a relatively large numerical aperture. Numerical results are also presented to reveal that the plausible nonlinear effects stem from the changes in the incident and polarization azimuth angles of the focused laser beam. The incident angle change induced pseudo-nonlinearity can be eliminated by adopting a low numerical aperture converging lens or adjusting the incident light polarization state.

2.1. Theoretical Analysis [17]

2.1.1. Incident Angle as a Function of Z-Scan Position

Assuming that a TEM_{00} Gaussian beam with a beam waist radius w_0 travels in z direction, the electric field as a function of z and off-axis radial position r can be written as

$$E(z,r) = E_0 \frac{w_0}{w(z)} \exp\left(-\frac{r^2}{w^2(z)}\right) \exp\left[-i\left(\frac{kr^2}{2R(z)} - \Phi(z)\right)\right] \tag{1}$$

where $w(z) = w_0 \times \sqrt{1 + (z/z_0)^2}$ and $R(z) = z_0[z/z_0 + z_0/z]$ are the beam and curvature radii of the wave front, respectively. w_0 is the beam waist radius at $z = 0$. $z_0 = kw_0^2/2$ is the diffraction length of the beam, $k = 2\pi/\lambda$ is the wavevector, and λ is the laser wavelength. E_0 is the electric field at the focal point and $\Phi(z) = tan^{-1}(z/z_0)$ is the phase factor at position z. In apractical z-scan measurement, a collimated laser beam is focused by a converging lens with an angular semi-aperture of θ_0. In Figure 1, schematic pictures are presented to show how the light beam travels along the z-axis according to Equation (1). The relation between the axis and radial amplifications obeys the following hyperbolic equation:

$$\frac{w^2(z)}{w_0^2} - \frac{z^2}{z_0^2} = 1 \tag{2}$$

The angular semi-aperture θ_0 is only determined by the converging lens. In a z-scan measurement, the incident angle is $\theta_i = \theta_0$ when the sample is located far from the beam waist position (position 1 in Figure 1(a)). The incident angle θ_i decreases with the sample movement from positions 1 to 2 along the optical z-scan axis (Figure 1(b)). The incident angle reaches $\theta_i = 0$ when the sample moves to the beam waist position (position 3 in Figure 1(c)).

The dependence of the incident angle θ_i on the sample position z during z-scan measurement can be directly obtained from the derivative of Equation (2),

$$\theta_i = \tan^{-1}\left[\frac{\partial w(z)}{\partial z}\right] = \tan^{-1}\left[\frac{z}{\sqrt{z_0^2 + z^2}} \frac{\lambda}{\pi w_0}\right]. \tag{3}$$

At $z = 0$, $\theta_i = 0$ based on Equation (3), which corresponds to the beam waist position (position 3 in Figure 1(c)) and at $z \to +\infty$, $\theta_i = \theta_0 = \lambda/\pi w_0$, which corresponds to the angular semi-aperture (position 1 in Figure 1(a)). Figure 2 shows an example when the incident angle θ_i changes with the z-scan position z, where the laser wavelength and beam waist radius are chosen as $\lambda = 633\ nm$ and $w_0 = 1\ um$, respectively. In the z-scan process from $z \to -\infty$ to $z = 0$, the incident angle decreases from $\theta_i = \theta_0 = 11.5^o$ to $\theta = 0$, and then increases from $\theta_i = 0$ to $\theta_i = \theta_0 = 11.5^o$ when the z-scan changes from $z = 0$ to $z \to +\infty$.

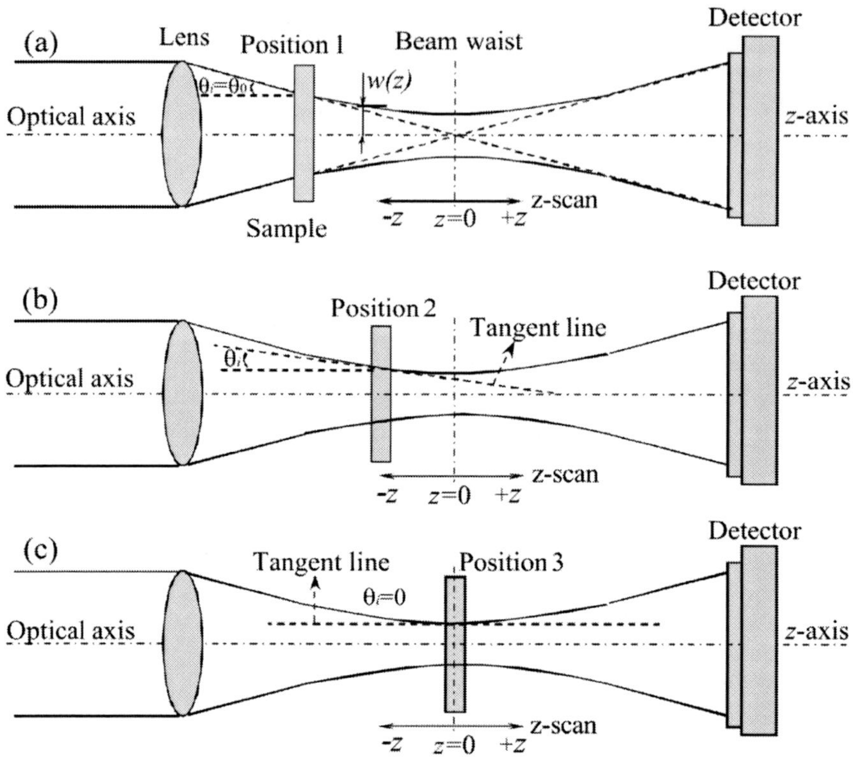

Figure 1. Schematic of z-scan measurement, sample position (a) far away from focal point, (b) close to focal point, and (c) at focal point [17].

In the z-scan measurement, the beam waist radius and angular semi-aperture change with the numerical aperture of the converging lens. Letting NA be the numerical aperture of the converging lens, the angular semi-aperture θ_0 can be calculated as:

$$\theta_0 = \sin^{-1}\frac{NA}{n_0}. \tag{4}$$

Generally speaking, n_0 is set to be 1 because the z-scan is usually carried out in an atmospheric environment. Figure 3 shows the relationship between θ_0 and NA where θ_0 increases with NA. This means that in a z-scan measurement, the angular semi-aperture becomes large if a converging lens with a large NA is used. As shown in Figure 3, the angular semi-aperture θ_0 determines the maximum incident angle of the z-scan measurement.

The above reasoning establishes that the field variation along the z-axis depends closely on the incident angle, and the significance of the derivation is determined by the angular semi-aperture and eventually by the numerical aperture of the focusing lens. In turn, the field variation along the z-axis is practically measured by detecting the transmittance through the sample in z-scan measurements. Since the numerical aperture of the focusing lens is a device parameter and has nothing to do with the intrinsic property of the sample, the measured variation in the transmittance during z-scan cannot be due to the nonlinear absorption in the sample and shall therefore be interpreted as some additional plausible effects, namely the pseudo-nonlinear absorption, which will be shown with numerical results in the next Section.

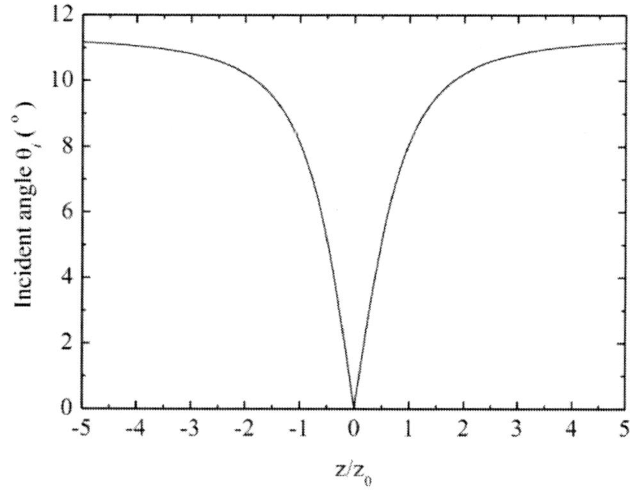

Figure 2. Changes of calculated z-scan incident angle with position z ($\lambda = 633\ nm, w_0 = 1\ um$) [17].

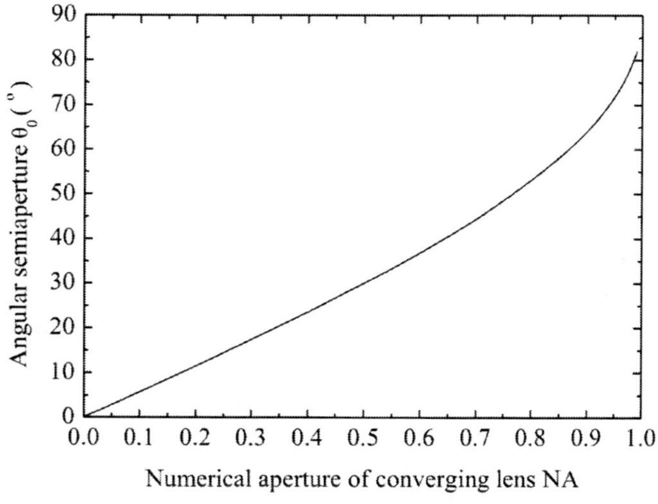

Figure 3. Relationship between angular semi-aperture and numerical aperture of lens [17].

2.1.2. Transmittance against Incident and Polarization Azimuth Angles

In this sub-section, the dependence of the transmittance and reflectance on the incident and polarization azimuth angles is explicitly shown. For the polarization, two special cases of s- and p-polarizations are taken into account. Figure 4 shows the schematic of the transmission and reflection when a laser beam irradiates the sample surface. Note in passing that although the whole laser beam is generalized as a Gaussian beam with a propagation direction along the optical axis, i.e. in perpendicular to the sample surface, the field components of the beam remain vectors and the polarization of each component can always be decomposed into s- and p-polarizations according to the local surface and the inputting direction of the component, i.e. the polarization azimuth angles. Based on the Fresnel formula, the sample's interface reflection depends on beam polarization azimuth angles.

The light is incident from medium 1 with a refractive index of n_1 to medium 2 with a refractive index of n_2. The Poynting vectors of incident and refraction lights are marked as $\mathbf{s}^{(i)}$ and $\mathbf{s}^{(t)}$, respectively. The incident and refraction angles are θ_i and θ_t, respectively. The refraction angle can be calculated by

$$\theta_t = \sin^{-1}\frac{n_1 \cdot \sin\theta_i}{n_2}. \tag{5}$$

Figure 4(a) shows the p-polarization incident, where the electric vector of the incident light $E_{\parallel}^{(i)}$ is parallel to the incident plane, and Figure 4(b) shows the *s*-polarization incident, where the electric vector of the incident light $E_{\perp}^{(i)}$ is perpendicular to the incident plane. Based on the Fresnel formula, the electric field of the transmitted light can be expressed as:

$$E_{//}^{(t)} = \frac{2n_1\cos\theta_i}{n_2\cos\theta_i + n_1\cos\theta_t}E_{//}^{(i)}$$

$$E_{\perp}^{(t)} = \frac{2n_1\cos\theta_i}{n_1\cos\theta_i + n_2\cos\theta_t}E_{\perp}^{(i)} \tag{6}$$

In the general case of an incident electric field with a polarization azimuth angle α, it is always possible to decompose the field $E^{(i)}$ into s-polarization and p-polarization components, as schematically shown in Figure 4(c), such that $E_{\parallel}^{(i)} = E^{(i)}\cos\alpha, E_{\perp}^{(i)} = E^{(i)}\sin\alpha$.

The horizontal component of the transmittance, defined as the ratio of the horizontal component of the transmitted light intensity to the horizontal component of the incident light intensity is then,

$$T_{//}(\theta_i) = \frac{n_2\cos\theta_t\left|E_{//}^{(t)}\right|^2}{n_1\cos\theta_i\left|E_{//}^{(i)}\right|^2} = \frac{4n_1n_2\cos\theta_i\cos\theta_t}{\left(n_1\cos\theta_t + n_2\cos\theta_i\right)} \tag{7}$$

and the vertical component of the transmittance, defined as the ratio of the vertical component of the transmitted light intensity to the vertical component of the incident light intensity can be written as,

$$T_{\perp}(\theta_i) = \frac{n_2\cos\theta_t\left|E_{\perp}^{(t)}\right|^2}{n_1\cos\theta_i\left|E_{\perp}^{(i)}\right|^2} = \frac{4n_1n_2\cos\theta_i\cos\theta_t}{\left(n_2\cos\theta_t + n_1\cos\theta_i\right)} \tag{8}$$

Finally, the total transmittance can be calculated by,

$$T(\theta_i, \alpha) = T_{//}(\theta_i)\cos^2\alpha + T_{\perp}(\theta_i)\sin^2\alpha \tag{9}$$

Figure 5 shows the calculated transmittance dependence on the incident angle for light with various polarization azimuth angles. For $\alpha = 0$, the transmittance increases with the incident angle at $\theta_i < 56°40'$. The transmittance is 1 at $\theta_i = 56°40'$, and decreases with the incident angle at $\theta_i > 56°40'$. For $\alpha < 45^\circ$, the transmittance initially increases to a certain value, and then decreases with the incident angle. For $\alpha > 45^\circ$, the transmittance always decreases with the incident angle.

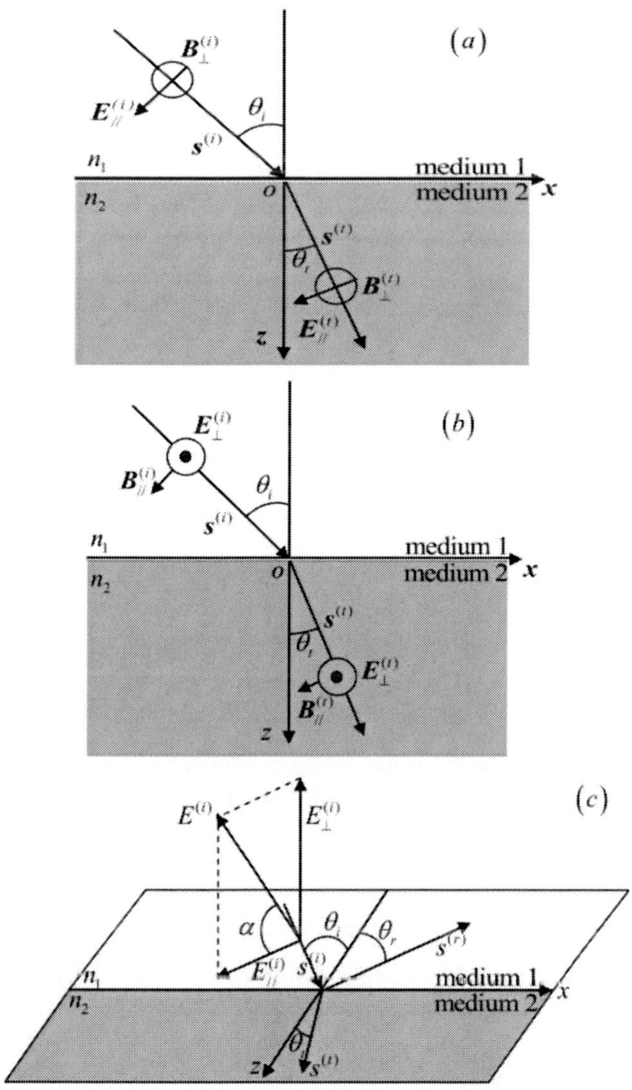

Figure 4. Schematic of transmission: (a) p-polarization incident, (b) s-polarization incident, and (c) arbitrary polarization azimuth angle [17].

Therefore, it is shown that the transmittance in z-scan depends further on the polarization azimuth angle, which implies that additional pseudo-nonlinear effects could also be caused, apart from the incident angle, by the change of the incident polarization state. In the next Section, numerical results will be provided to confirm the assertion.

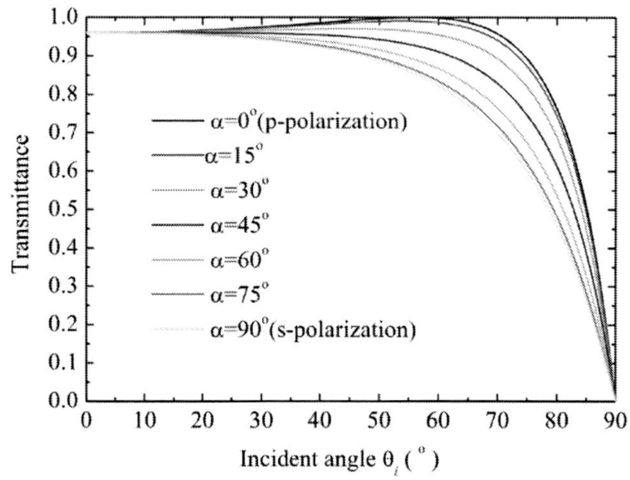

Figure 5. Transmittance dependence on the incident angle with different polarization azimuth angle [17].

2.1.3. Pseudo-nonlinear Absorption Analysis

The actually detected light transmittance is integration along the radial coordinate r. For a given r, according to Equation (1), the incident laser beam intensity $I = E \times E^*$ can be written as:

$$I^{(i)}(z,r) = I_0 \frac{w_0^2}{w^2(z)}.\exp(-\frac{2r^2}{w^2(z)}) \tag{10}$$

$$w_0 = \frac{\lambda}{\pi \tan(\theta_0)} \tag{11}$$

In the z-scan measurement, the sample is moved between the –z and +z positions. For open-aperture transmittance z-scan, the Fourier transform of the amplitude of launched beam $E_a^{(i)}(z,f)$ (from Equation (1)) is called the spatial spectrum $E^{(i)}(z,f)$, where f is the spatial frequency [18]:

$$E_a^{(i)}(z,r) = \int_{-\infty}^{+\infty} \tilde{E}^{(i)}(z,f)\exp(i2\pi fr)df \tag{12}$$

$$E_a^{(i)}(z,r) = E_0 \frac{w_0}{w(z)}\exp[-\frac{r^2}{w^2(z)}] \tag{13}$$

Here $exp(i2\pi fr)$ can be interpreted as a plane wave traveling in a direction that lies in the *rz*-plane, making an angle θ_i with the *z*-axis such that

$$f = \sin\frac{\theta_i}{\lambda} \tag{14}$$

According to Equations (12) and (13), the corresponding spatial spectrum at the waist $w(z)$ is

$$\tilde{E}^{(i)}(z,f) = \sqrt{\pi}E_0 w_0 \exp\left\{-\left[\pi w(z) f\right]^2\right\} \tag{15}$$

The amplitude portion of the transmission function depends sensitively on the incident angle θ_i of the various plane-wave components and polarization azimuth angle α which have been discussed in section 2.1.2, and the transmitted electric field $E^{(t)}(z,r,\alpha)$ can be expressed as

$$E^{(t)}(z,r,\alpha) = \int_{-\infty}^{+\infty} \tilde{E}^{(t)}(z,f,\alpha)\exp(i2\pi fr)df \tag{16}$$

where $\tilde{E}^{(t)}(z,f,\alpha)$ can be obtained by Equations (6) and (15). The transmitted light intensity $I^{(t)}(z,r,\alpha)$ can be written as follows

$$I^{(t)}(z,r,\alpha) = E^{(t)}(z,r,\alpha) \times E^{(t)*}(z,r,\alpha) \tag{17}$$

Finally, the normalized transmittance through the glass substrate in the z-scan can be approximately obtained by spatially integrating $I^{(t)}(z,r,\alpha)$, giving

$$T(z,\alpha) = \frac{P^{(t)}(z,\alpha)}{P^{(i)}(z)} = \frac{n_0\pi \int_0^{+\infty} I^{(t)}(z,r,\alpha)rdr}{n_0\pi \int_0^{+\infty} I^{(i)}(z,r)rdr} \tag{18}$$

where $P^{(i)}$ is the incident power, $P^{(t)}$ is the transmitted power, and n_0 is the refractive index of vacuum.

2.2. Pseudo-nonlinear Absorption Curves

This section is devoted to showing some numerical results based on the theoretical analysis outlined in the previous section. In order to investigate the effects of the incident and polarization azimuth angles on the intensity of transmitted light, a single Gaussian laser beam is assumed to go through a glass substrate sample without any nonlinear absorption in the z-scan measurement. Numerical simulations of the z-scan measurements are carried out, and the z-scan curves are observed. With a theoretical fitting process, the pseudo-nonlinear absorption curve is extracted. Two sources of the pseudo-nonlinear absorption are tested, namely the dependences on the angular semi-aperture (that is, the NA value of the converging lens), and on the polarization azimuth angle of the incident laser beam.

2.2.1. Pseudo-nonlinear Absorption Curves Induced by Changes of the Incident Angle

Figure 6 shows the calculated z-scan curves. Figure 6(a) shows the calculated result of the pseudo-nonlinear absorption curves for $\alpha = 90^\circ$ (s-polarization). The peak configuration of this z-scan is indicative of a nonlinear saturable absorption characteristic. The maximum variation in the transmittance increases with increased NA. Figure 6(b) shows the calculated result of the pseudo-nonlinear absorption curves for $\alpha = 0^\circ$ (p-polarization). The valley configuration of this z-scan is indicative of a nonlinear reverse-saturable absorption characteristic. Similar to $\alpha = 90^\circ$, the maximum variation in the transmittance increases with an increased NA. The maximum transmittance variation can reach 2.72% when $NA = 0.90$, which is a large transmittance change compared with the usual z-scan measurement. In other words, both Figures 6(a) and 6(b) show that the pseudo-nonlinear absorption becomes substantial when NA increases.

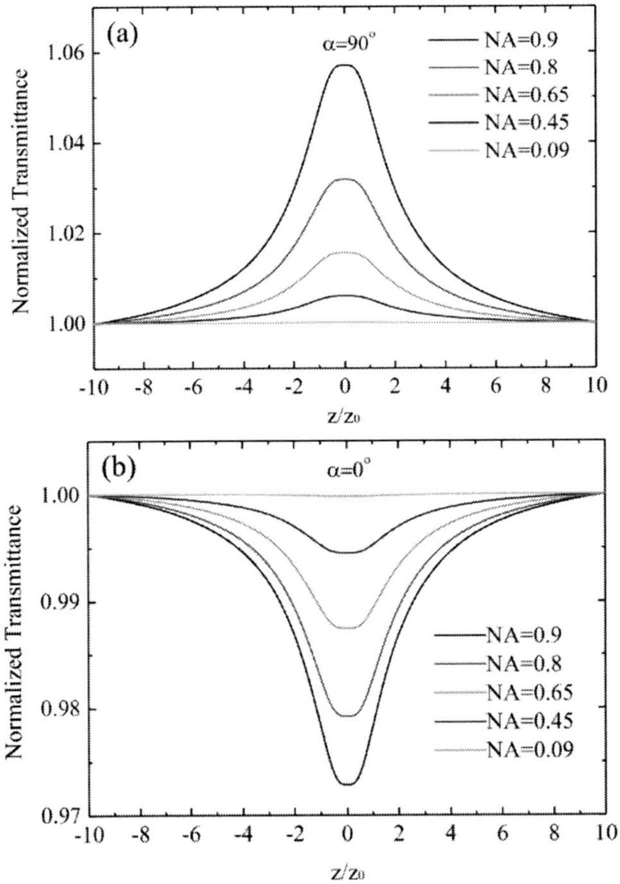

Figure 6. Calculated result of $T(z)$ for $\alpha = 0^\circ, 90^\circ$ using $\lambda = 633\ nm, n_0 = 1, n_1 = 1.5$, and $NA = 0.9, 0.8, 0.65, 0.45$, as well as 0.09: (a) s-polarization and (b) p-polarization [17].

Figure 6 also shows that when the NA is larger than 0.65, the detected result has more than 1% error in transmittance variation. The interference of the pseudo-nonlinear absorption even contains errors as much as 5.7% and 2.72% for a converging lens with a numerical aperture of 0.90 for s- and p-polarizations, respectively. Consequently, significant errors in fitting the nonlinear absorption coefficient may occur. If the incorrect data is substituted into the fitting formulae, the nonlinear coefficient would also be incorrect. Therefore, accurate experimental results can be obtained from a sensitive z-scan measurement by reducing and eliminating the influence of the change in the incident angle during the z-scan process. Figure 6 shows that the maximum variation is less than 0.02% when the laser beam is focused by a converging lens with an NA of 0.09. Thus, if a converging lens with a low NA is used in the z-scan, the error of the nonlinear absorption induced by the change in the incident angle during the z-scan process may be ignored.

2.2.2. Pseudo-nonlinear Absorption Curves Induced by Changes of the Polarization Angle

The effect of an angular semi-aperture (i.e., the NA of a converging lens) on the pseudo-nonlinear absorption curves has been discussed. According to Equation (18), the polarization azimuth angle also has an important effect on the normalized transmittance, which causes pseudo-nonlinear absorption characteristics in the z-scan measurement.

Figure 7 shows that the change of the polarization azimuth angle leads to pseudo-nonlinear absorption curves. Figures 7(a) and 7(b) show the results for NA of 0.90 and 0.45, respectively. The result for the NA of 0.09 is shown in Figure 7(c).

Figure 7(a) shows that for NA=0.90 when the polarization azimuth angles are $\alpha = 0^\circ, 20^\circ, 35^\circ, 45^\circ, 60^\circ$, and 90°, the maximum transmittance variations are $2.71\%, 1.79\%, 0.071\%, 1.35\%, 3.49\%$, and 5.71%, respectively. Figure 7(b) shows that NA=0.45, when the polarization azimuth angles are $\alpha = 0^\circ$, 20°, 35°, 45°, 60°, and 90°, the maximum transmittance variations are $0.55\%, 0.34\%, 0.02\%, 0.17\%, 0.31\%$, and 0.60%, respectively. For $NA = 0.09$, the maximum variations in the transmittance are less than 0.02%, as shown in Figure 7(c). Figure 7 also shows that the peak and valley configurations of the z-scan are indicative of a transformation from a nonlinear saturable absorption to a nonlinear reverse-saturable absorption due to different polarization azimuth angles.

Figure 7 indicates that for a fixed converging lens NA, the incident light beam with a critical polarization azimuth angle $\alpha_{critical}$ can minimize or eliminate the interference of the pseudo-nonlinear absorption on the experimental results. That is, if the converging lens is fixed in the z-scan, the NA is fixed accordingly. The polarization azimuth angle of the incident light beam can be tuned to the critical polarization azimuth angle $\alpha_{critical}$, and the polarization induced pseudo-nonlinear absorption can be eliminated. Different NA corresponds to different $\alpha_{critical}$ values. Figure 8 shows the dependence of $\alpha_{critical}$ on the NA. At $NA = 0.10$, $\alpha_{critical} = 45.2^\circ$. However, $\alpha_{critical}$ decreases with increased NA (for example, $\alpha_{critical} = 36.2^\circ$ at NA=0.90). Figure 6 to Figure 8 indicate that the pseudo-nonlinear absorption effect can be reduced or avoided by choosing the incident light beam with a critical polarization azimuth angle or using a converging lens with a low NA in a z-scan measurement.

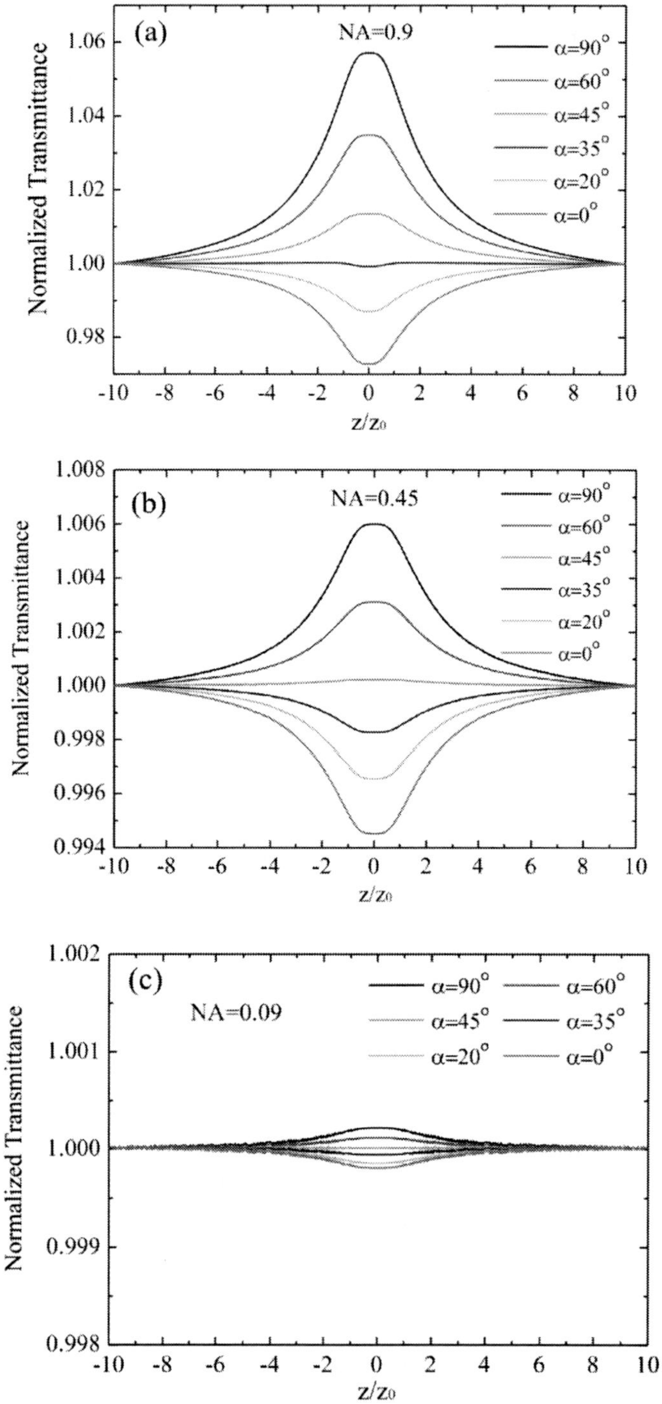

Figure 7. Calculated result of $T(z)$, using $\lambda = 633\,nm$, $n_0 = 1, n_1 = 1.5$, and (a) $NA = 0.9$, (b) $NA = 0.45$, as well as (c) $NA = 0.09$ for $\alpha = 0^\circ$ (s − polarization), 20°, 35°, 45°, 60°, and 90°(p − polarization) [17].

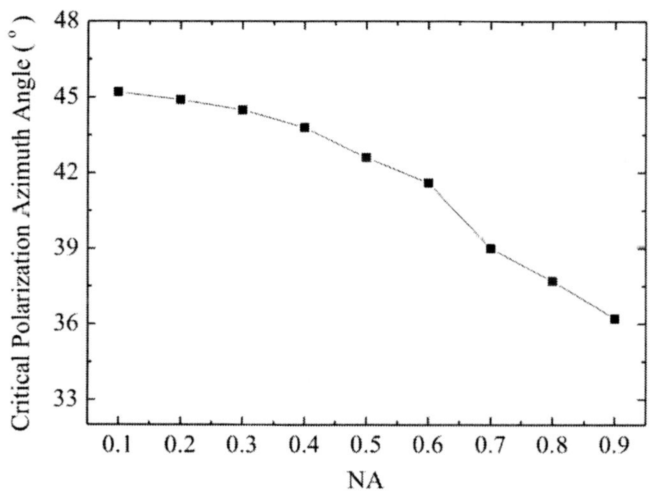

Figure 8. Calculated critical polarization azimuth angle dependence on the numerical aperture [17].

2.2.3. Elimination of Pseudo-nonlinear Absorption

In the z-scan measurement, the pseudo-nonlinear absorption has an undesirable effect on the real nonlinear absorption coefficient fitting and data processing. To obtain an approximate estimate of the pseudo-nonlinear absorption induced nonlinear absorption coefficient, Figure 9 shows the pseudo-nonlinear absorption curve (black line) and the corresponding fitting curve for an open aperture z-scan (red line) with an *NA* of 0.90. Generally, for a small nonlinear absorption, the normalized transmittance change becomes [13]:

$$\Delta T(z) \approx -\frac{q_0}{2\sqrt{2}}\frac{1}{\left(1+z^2/z_0^2\right)} \tag{19}$$

with $q_0 = \Delta\alpha L = \beta I L_{eff} << 1$, where β is the nonlinear absorption coefficient, *I* is the laser intensity, and L_{eff} is the effective sample length. In Figure 9, the calculated curve 1 gives the incident laser beam with a polarization azimuth angle α of 90° (s-polarization) and *NA* of 0.90. To fit the calculated result, $q_0 = -0.161$ is chosen to have the same peak value of 1.057 as the calculated results. The calculated curve 2 presents the incident laser beam with a polarization azimuth angle $\alpha = 0^\circ$(p-polarization) and $NA = 0.9$. $q_0 = +0.077$ is chosen to have the same valley value of 0.9728 as the calculated results. The fitting curve does not well match the calculated result because the transmittance change in the pseudo-nonlinear absorption curve is based on the incident and polarization azimuth angles of the Gaussian laser beam. Instead, in actual z-scan measurements, the transmittance change depends on the nonlinear absorption property of a sample. That is, the sample moves along the optical axis. The normalized transmittance of the pseudo-nonlinear absorption curve increases at a very low rate within the range of Rayleigh length and can basically remain invariable. The calculated curve shows a platform in the Rayleigh length range. On the contrary, for a nonlinear absorption sample, an apparent normalized transmittance change occurs in the range of the Rayleigh length, and the nonlinear absorption curve should accordingly show a sharp peak or valley in the z-scan measurement. The real value is estimated to be $-0.483 <$

$q_0 < -0.118$ for the s-polarization of calculated curve 1, and $0.061 < q_0 < 0.178$ for the p-polarization of calculated curve 2 according to the fitting curves shown in Figure 9. Generally, in a traditional z-scan measurement [2], the laser intensity is in the order of $\sim 10^{12}\ W/m^2$, and the effective thickness of the sample is $\sim 10^{-3} m$. The case of calculated curve 1 results in β being $\sim -10^{-10}\ or \sim -10^{-11}\ m/W$. The case of calculated curve 2 results in β being $\sim +10^{-10}$ or $\sim +10^{-11}\ m/W$. If a typical ZnSe material is used to measure the nonlinear absorption characteristic, the contribution from the pseudo-nonlinear absorption to the measured β is 2 or 3 times larger than the real nonlinear absorption coefficient of about $\sim 10^{-11}\ m/W$.

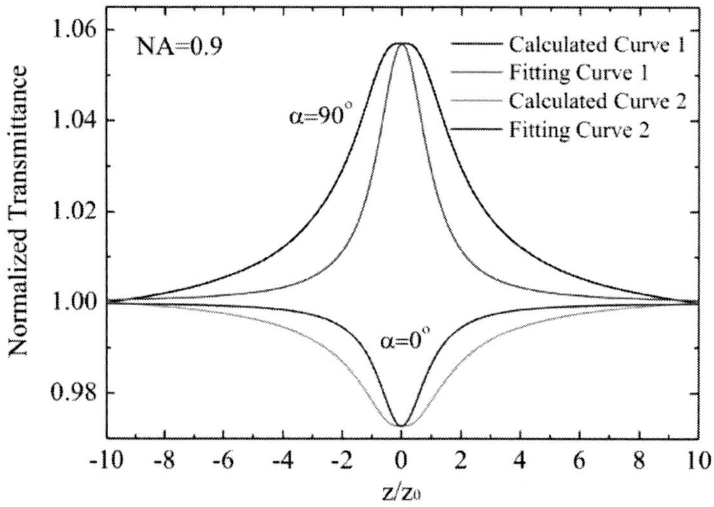

Figure 9. Calculated result of $T(z)$ for NA=0.90, and fitting using q_0 values of -0.161 for $\alpha = 90^\circ$ (curve 1) and $+0.077$ for $\alpha = 20^\circ$(curve 2) [17].

To reduce the undesirable effect of the pseudo-nonlinear absorption, a converging lens with a low NA can be adopted. Figure 10 shows the calculated and fitting results where $NA = 0.09$ is chosen. $q_0 = +5.403 \times 10^{-4}$ is chosen to have the same valley value of 0.99981 as the calculated curve 1, and $q_0 = -6.223 \times 10^{-4}$ is chosen to have the same peak value of 1.00022 as the calculated curve 2. The pseudo-nonlinear absorption induced β is about $+10^{-12}\ m/W$ to $+10^{-13}\ m/W$ for $\alpha = 0^\circ$ and $-10^{-12}\ m/W$ to $-10^{-13}\ m/W$ for $\alpha = 90^\circ$, respectively. This finding shows that the undesirable effect of the pseudo-nonlinear absorption on the real β is very small and can be ignored. The undesirable effect of the pseudo-nonlinear absorption can be also reduced by adjusting the polarization azimuth angle of the incident laser beam for a fixed NA. Figure 11 shows the calculated and fitting results where NA=0.90 and α=35°are chosen. $q_0 = +2.015 \times 10^{-3}$ is chosen to have the same valley value of 0.99929 as the calculated results. The pseudo-nonlinear absorption induced β is about $-10^{-12} m/W$, which is very low and negligible for a usual z-scan measurement curve.

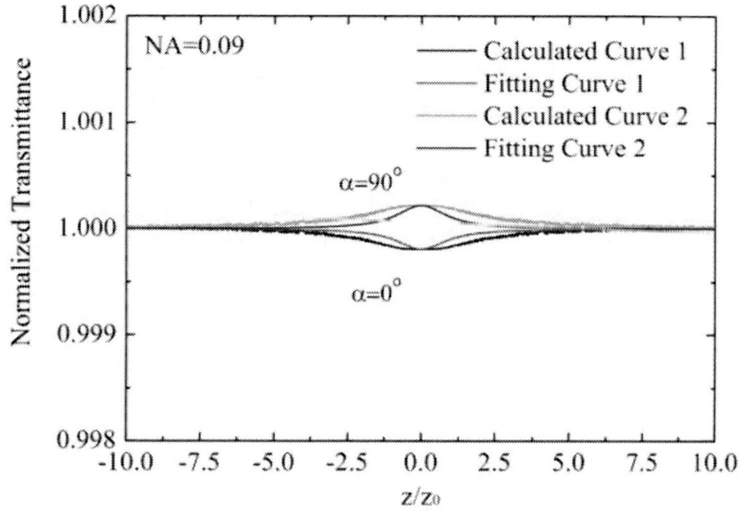

Figure 10. Calculated result of $T(z)$ for $NA = 0.45$, and fitting using curve 1: $\alpha = 0^{\circ}$, $q_0 = +6.225 \times 10^{-4}$ and curve 2: $\alpha = 90^{\circ}$, $q_0 = -6.223 \times 10^{-4}$ [17].

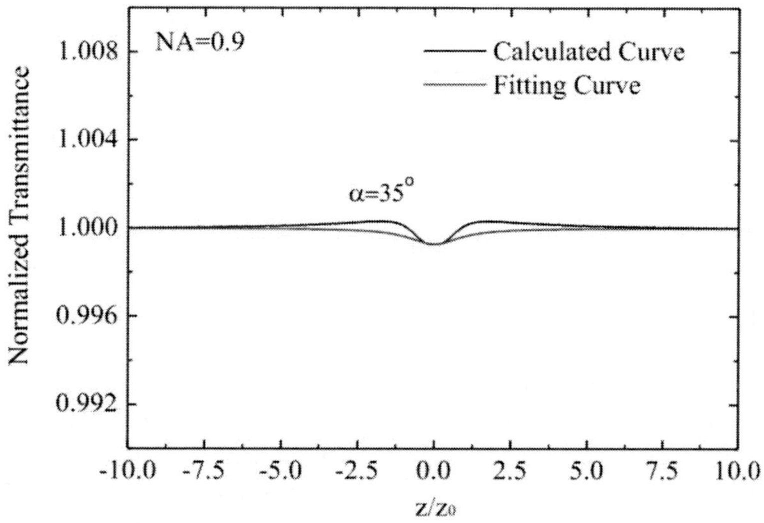

Figure 11. Calculated result of $T(z)$, for $NA = 0.90, \alpha = 35^{\circ}$, fitting using $q_0 = +2.015 \times 10^{-3}$ [17].

2.2.4. Experimental Results and Analysis

In order to confirm the theoretical analysis, a z-scan setup with an open aperture mode was established. In this setup, the cw Ar^+ laser beam with a wavelength of $488\ nm$ was used as the light source, a half-wave plate was adopted to change the laser beam polarization azimuth angle, and the angular semi-aperture was tuned by using the focusing lens with different numerical aperture. The experimental results are shown in Figure 12 where a $0.6\ mm$ thickness BK7 glass slice sample is examined. Figure 12(a) depicts the results of polarization azimuth angle $\alpha = 90^{\circ}$ where the three cases of $NA = 0.09, 0.45$ and 0.65 are presented, respectively. The experiment curves are indicative of a nonlinear saturable absorption characteristic. The effect of semi-aperture angle on the pseudo-nonlinear absorption is so small that changes of the transmittance are not significant when NA is not more than 0.09 (blue line). The maximum variation of the transmittance increases with increased NA, and can reach 0.36% and 0.57% for $NA = 0.45$ and 0.65, respectively, which effectively proves the theoretical results. Figure 12(b) shows the experimental data with $NA = 0.45$, the maximum transmittance variation are 0.36% and 0.47% for s-polarization ($\alpha = 90^{\circ}$) and p-polarization ($\alpha = 0^{\circ}$), respectively. As indicated before, in the z-scan measurement the glass slice sample moves along the optical axis, the normalized transmittance curve slowly increases especially in the range of Rayleigh length. The normalized transmittance can basically maintain invariable and the calculated curve shows a platform, and the experimental results have the same characteristics. In addition, the polarization angle can be adjusted to the corresponding critical angle to eliminate the influence of pseudo-nonlinear absorption. In experiment, we tuned the half-wave plate to the critical polarization azimuth angle $\alpha_{critical} = 43^{\circ}$, the results are presented in curve marked with blue line in Figure 12(b), one can find that the pseudo-nonlinear absorption becomes very small and can be ignored.

According to above theoretical and experimental analysis, the change of the incident and polarization angles in the z-scan can cause the pseudo-nonlinear absorption, and the significance of the effects depends on the NA of the focusing lens. When the laser is focused by a lens with a large NA, the interference of the pseudo-nonlinear absorption due to the incident angle change cannot be ignored. If a lens with a low NA is used, the error induced by the pseudo-nonlinear absorption is negligible. For an incident laser beam focused by a lens with a large numerical aperture, the polarization azimuth angle of the incident light can be adjusted to the corresponding critical angle to eliminate the influence of pseudo-nonlinear absorption.

3. Laser Instability Induced Pseudo-nonlinear Absorption

In z-scan measurement, the laser pulses irradiate on the samples repeatedly, and the light transmitted or reflected from the samples is always acquired until the measurement is completed. Thus the stability of laser power is very important to accurately get the nonlinear parameters. Generally speaking, the gas laser devices are stable compared with the solid state laser devices and semiconductor laser devices, and in the z-scan measurement the gas laser devices were generally adopted in past years. With the development of semiconductor technology, the semiconductor laser devices and semiconductor laser pump-solid state laser devices have been applied in the z-scan experiments due to the compactness and cheapness in recent years. In z-scan measurement the optical beam needs to be normal incidence on the sample surface, which causes the reflected light from sample surface goes back to the resonant cavity of laser devices. The output stability of semiconductor laser devices is very sensitive to the feedback light, thus the feedback light will lead to the instability of the laser output [19, 20, 21], and have an ill-effect on the accurate testing of the nonlinear index of the materials. In this section, we will introduce the influence of reflected light on the z-scan measurement, and give some method s to avoid the ill-effect theoretically and experimentally.

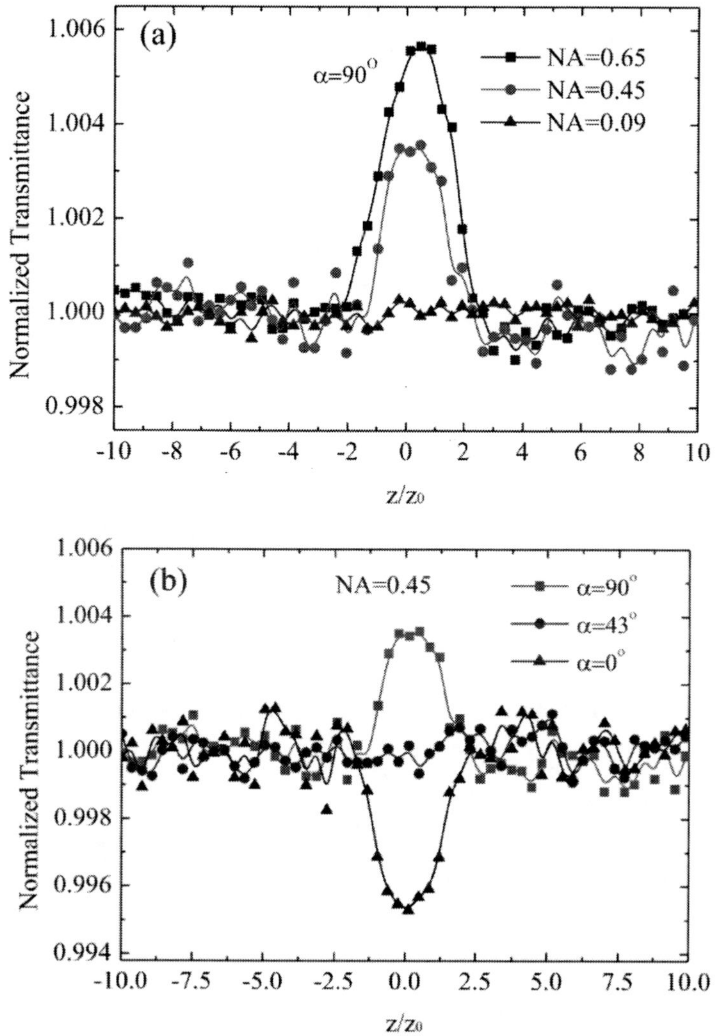

Figure 12. Normalized z-scan transmittance of BK7 glass slice sample measured using cw laser at $\lambda = 488\ nm$. (a) $\alpha = 90^\circ$, $NA = 0.09, 0.45, 0.65$ (b) $\alpha = 90^\circ, 43^\circ, 0^\circ$, $NA = 0.45$ [17].

3.1. Theoretical Basis on the Feedback Light Influence on Power Output of Semiconductor Laser Devices [22, 23]

It is well known that the feedback light has an important influence on the laser output characteristics for semiconductor laser diodes. For the laser diodes, the fabry-perot (F-P) resonator usually is formed between the two cleavage planes for the laser diode if there is no feedback light. The reflectivities of the two cleavage planes are marked as R_1 and R_2, respectively. When the light travels in the fabry-perot (F-P) resonator the light energy loss includes free-carrier induced light absorption α_{fc}, diffraction loss α_{diff}, and output end induced loss $\alpha_T = (1/2l)ln(1/R_1R_2)$, where l_1 is the length of the fabry-perot (F-P) resonator. The total light loss in the active area is

$$\alpha_{loss} = \alpha_{fc} + \alpha_{diff} + \frac{1}{2l_1}\ln\frac{1}{R_1R_2} \tag{20}$$

When the balance between gain and loss is reached the inject current density is equal to the threshold current density, which is written as

$$J_{th} = C\left(\alpha_{fc} + \alpha_{diff} + \frac{1}{2l_1}\ln\frac{1}{R_1R_2}\right) \tag{21}$$

where C is defined as the emission factor.

However, if there is feedback light into laser diode, the emission threshold current will be influenced, accordingly. Figure 13 presents the schematic of feedback light influence on laser diode, where the laser diode emits the light. The light is splitted into two beams, one enters the power detector, and the other is incident on the mirror with a tunable relectivity R_3, and the mirror reflects the light into the laser diode again, which will cause the power output of laser diode to be instable, and the power instability is monitored by the power detector.

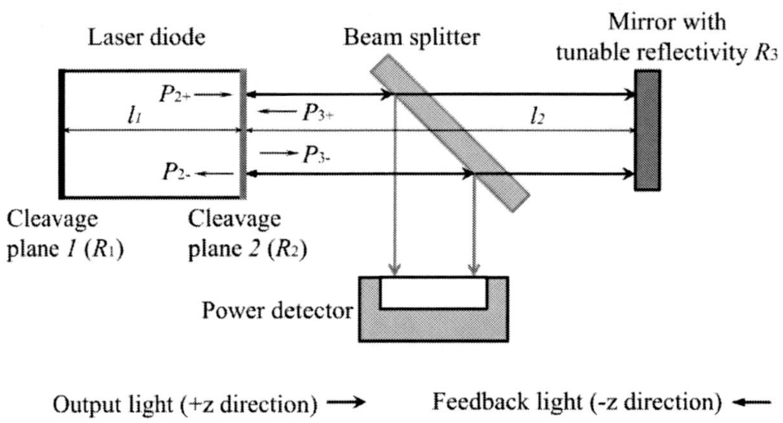

Figure 13. Schematic of feedback light influence on laser diode, l_2 ($>> \lambda$) is the outer resonator length.

The combination of cleavage plane 1 and 2 can be considered as an inner resonator, and the assembly of cleavage plane 2 with mirror can be taken as outer resonator. In order to get the influence of R_3 on the laser diode, one can consider, on the left side of cleavage plane 2, the power of light along $+z$ direction to be P_{2+} and along –z direction to be P_{2-}. On the right side of cleavage plane 2, the P_{3+} and P_{3-} are light powers along $+z$ and $+z$ deirection, respectively. One can get the following relations:

$$P_{2-} = R_2 P_{2+} + (1 - R_2) P_{3+}$$
$$P_{3-} = (1 - R_2) P_{2+} + R_2 P_{3+} \tag{22}$$

In the outer resonator, the P_{3+} can be calculated as

$$P_{3+} = R_3 P_3 \tag{23}$$

Substituting Equation (23) into Equation (22), one can obtain an effective reflectivity of cleavage plane 2 as

$$\mathrm{R}_{eff} = \frac{P_{2-}}{P_{2+}} = \frac{R_2 + R_3 - 2R_2R_3}{1 - R_2R_3} \tag{24}$$

Replacing R_2 of Equation (21) with R_{eff} one can obtain the threshold current density considering the feedback light

$$J_{th} = C\left(\alpha_{fc} + \alpha_{diff} + \frac{1}{2l_1} \ln \frac{1}{R_1} \frac{1 - R_2R_3}{R_2 + R_3 - 2R_2R_3} \right) \tag{25}$$

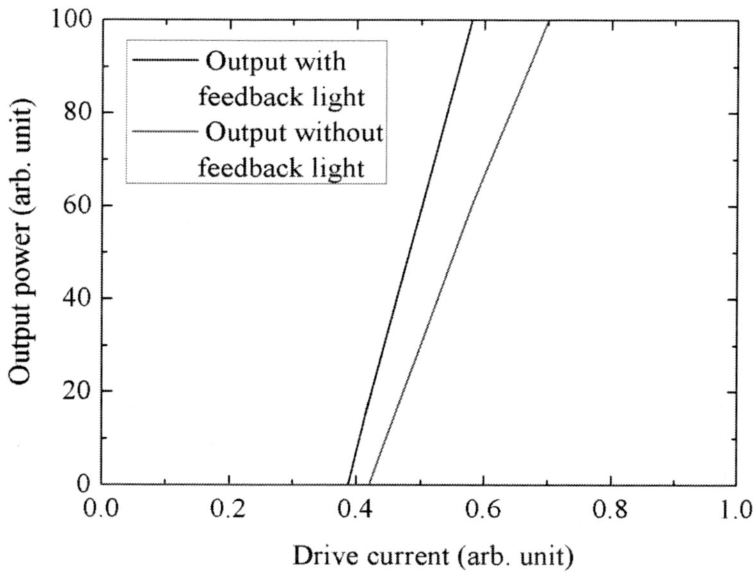

Figure 14. The comparison of output power of with feedback light with without feedback light.

Equation (25) is drawn schematically in Figure 14, it can be seen that the feebback light will cause the decrease of the threshold current. In the common application of the laser diode, the drive current keeps unchanged once the experimental parameters are adjusted to an optimum value. Thus the laser output power with feedback light becomes larger than without feedback light under the same drive current condition for low operation current. In high operation current the influence of feedback light is complex, and the readers can learn from the related textbook of semiconductor laser devices.

3.2. The Feedback Light in the Z-Scan Measurement

In the z-scan experiment, the sample moves from $+z$ to $-z$ along the optical axis of the lens, where $+z$ and $-z$ mean that the sample is situated on the left and right of the focus of lens, respectively. The sample will reflect the light, and the intensity of reflected light into laser diode changes with sample position z due to the laser induced nonlinear effects inside the sample and incident angle change. Figure 15 gives the three kind of typical sample positions. In Figure 15(a), the sample is situated on the left of the focus ($+z$ position), the partially reflected light passes through the lens and enters the laser diode, and most reflected light are scattered into environment due to limited aperture angle of lens and the limited light aperture size of the laser diode. In Figure 15(b), the sample moves to the focus of the lens, the most reflected light is collected by the lens and goes into the laser diode. The sample is placed on the right of the focus (shown in Figure 15(c)), the most reflected light is scattered into the environment, and the minor reflected light passes throught the focus and is collected by the lens, and finally enters the laser diode. Thus, the reflected light intensity entering the laser diode is the maximum when the sample is placed at the focus, and then decreases with the increase of the distance from the focus. According to Equation (25) and Figure 14, the current threshold is the minimum at the sample being placed at focus position since the R_3 reaches up to the maximum compared with the $+z$ and $-z$ positions, which causes the laser power is unstable and changes with the sample position z. The instability of laser output power needs to be eliminated in the real application.

3.3. The Elimination of the Ill-effect from Feedback Light on Z-Scan Measurement

In order to eliminate the ill-effect of laser instability on the nonlinearity measurement, the z-scan measurement is presented in Figure 16. The laser beam from the diode device passes through the $1/2$ wave plate and becomes a p-polarization light. The p-polarization light enters the polarization beam splitter and $1/4$ wave plate becomes a circular polarization light. The circular polarization light is split into two beams; one beam is collected by the noise detector, which records the laser instability and the other beam passes through the lens and is incident on the sample to induce the nonlinear effect. The sample moves through the focus from $+z$ to $-z$. The transmitted light as a function of sample position z is collected by the detector. The reflected light from the sample passes through the $1/4$ wave plate and becomes an s-polarization light. The s-polarization light enters the polarization beam splitter and the light beam direction becomes perpendicular to the incident beam. Thus the combination of $1/2$ wave plate, polarization beam splitter, and $1/4$ wave plate is actually an opto-isolator unit, and the reflected light is effectively avoided from going into the laser diode by the opto-isolator unit.

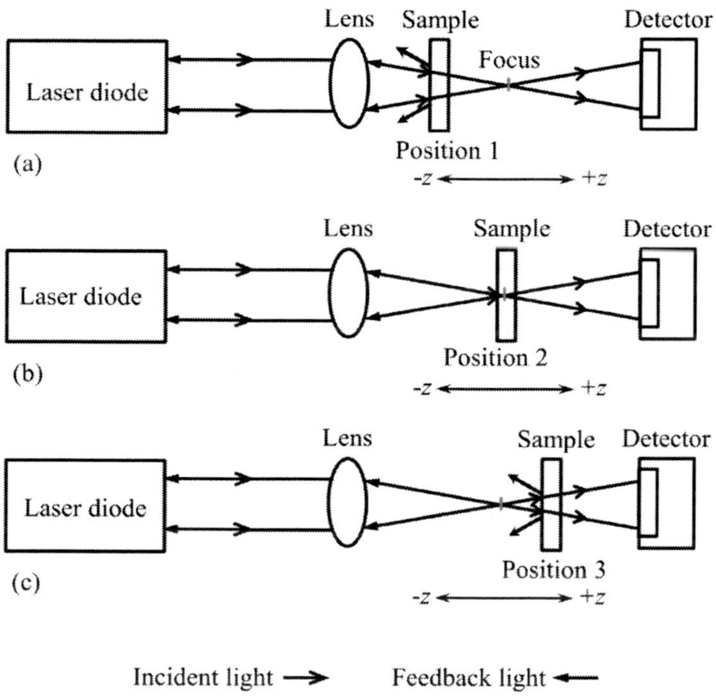

Figure 15. The reflected light change with the sample position z in the z-scan measurement, sample (a) on the left of the focus, (b) on the focus, and (c) on the right of the focus.

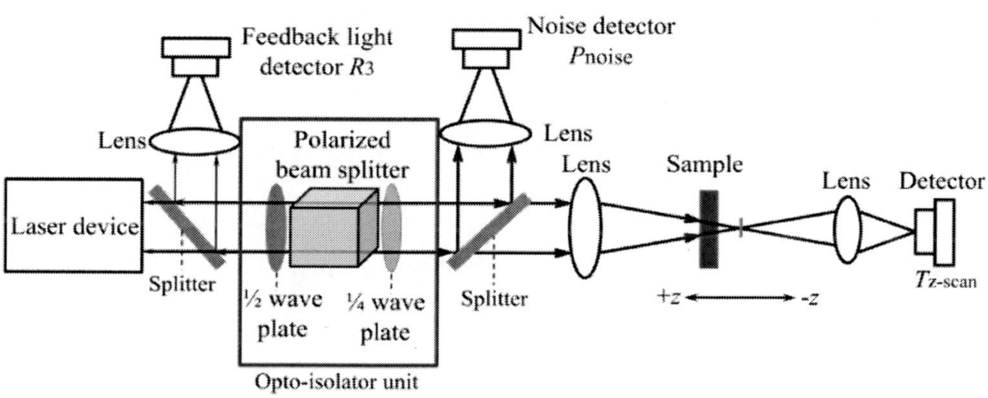

Figure 16. The light path setup for eliminating the influence of feedback light.

Theoretically, the reflected light can be thoroughly avoided into the laser diode by the opto-isolator unit. However, in real application, there can always be a leak light, and the leak light goes into the laser diode and leads to the fluctuatation of the laser power output. In order to avoid this leak light influence on the z-scan measurement, firstly in the light path the reflected light entering the laser diode is recorded by the feedback light detector, which is the R_3 of Equation (25). Secondly, the leak light induced laser power fluctuatation is gotten by the noise detector, whose data is marked as P_{noise}. Thirdly, the z-scan data is divided by the

P_{noise}. Finally, one can get more accurate z-scan measurement data induced by the optical nonlinear effect of sample.

3.4. Experimental Results and Analysis

In z-scan measurement, the feedback light from sample causes the laser to be instable, which can induce a pseudo-nonlinearity effect. The pseudo-nonlinear absorption effect can be avoided through the combination of placing an opto-isolator and systemic noise detector in the z-scan setup and advanced data processing methods. According to the light path setup shown in Figure 16, the laser diode with a wavelength of $658\ nm$ is used as the light source, and the BK7 glass is used as the sample. It is well known that there is no nonlinearity for BK7 glass at the low power laser irradiation; however, in Figure 17, the z-scan measurement without opto-isolator unit is presented, where a fallacious nonlinear saturable absorption transmittance open aperture z-scan curve (T_{z-scan}) occurs. The T_{z-scan} indicates a pseudo-nonlinearity absorption characteristic. The T_{z-scan} curve is almost the same as P_{noise} curve, where P_{noise} actually reflects the laser power changes at the z-scan measurement, and the maximum intensity is almost three times primary intensity far away from focus. Based on the section 3.2 and Equation (25), the laser power instability is due to the feedback light going into the laser diode. The feedback light change with the sample position is recorded by the R_3. Compared with the P_{noise} and T_{z-scan} one can see that the feedback light R_3 is almost the same as P_{noise} and T_{z-scan}, which indicates that the pseudo-nonlinear absorption effect is from the the feedback light R_3 induced laser power instability. The R_3 induced laser power instability is because the R_3 leads to the change of the threshold current density, which can be seen from the Equation (25) and Figure 14. The feedback light induced pseudo-nonlinearity can be reduced by the T_{z-scan}/P_{noise}, which is also shown in Figure 17. One can see that T_{z-scan}/P_{noise} curve is unchanged with the sample position in the z-scan measurement.

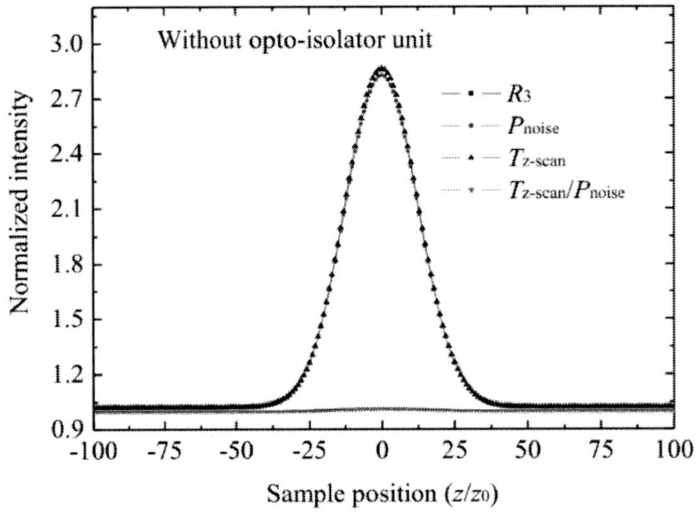

Figure 17. The feedback light influence on the z-scan measurement without opto-isolator (laser power of $1.53mW$).

The Feedback light has an ill-effect on the z-scan measurement, thus in the light path setup an opto-isolator unit is placed in between the feedback light detection unit and the noise detector unit. The opto-isolator unit can avoid the most reflected light going back to the laser diode, thus the R_3 and P_{noise} are reduced effectively, and Figure 18 gives the experimental results. Compared with the Figure 17 one can see that the normalized maximum intensities of R_3, P_{noise}, and T_{z-scan} with opto-isolator unit are decreased to about $1/2$ the intensities without opto-isolator unit, that is, the pseudo-nonlinearity is suppressed, and the T_{z-scan}/P_{noise} is also reduced to a lower value, accordingly.

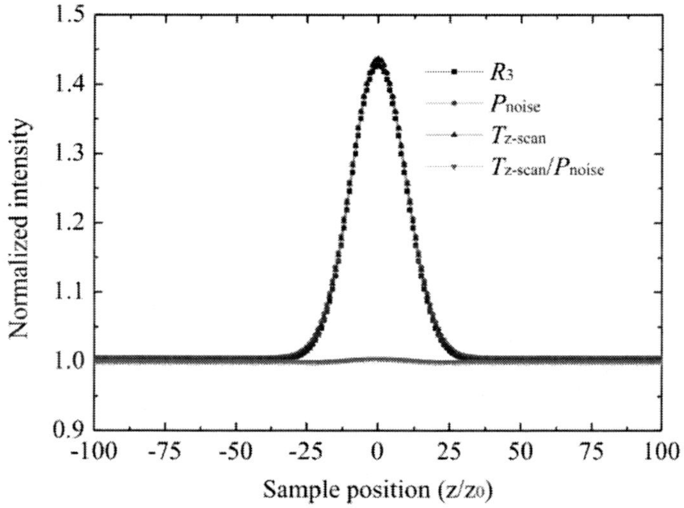

Figure 18. The feedback light influence on the z-scan measurement with opto-isolator (laser power of $1.53mW$).

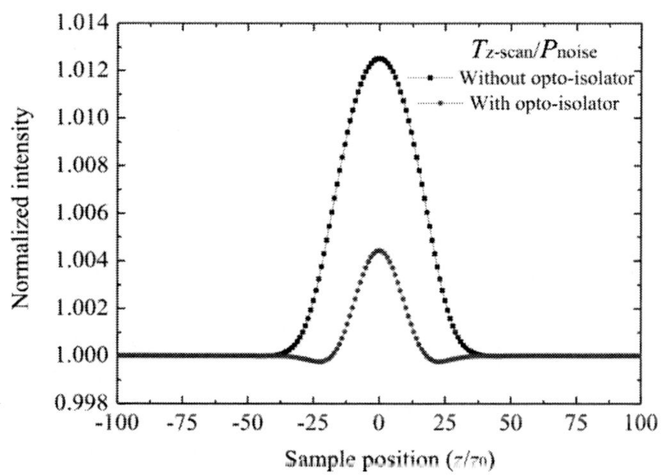

Figure 19. The comparison of with opto-isolator and without opto-isolator.

In Figure 19, the comparison of T_{z-scan}/P_{noise} with the opto-isolator and without opto-isolator is presented. The maximum T_{z-scan}/P_{noise} with the opto-isolator is further reduced down to about 1.004 compared with the without opto-isolator unit of the maximum $T_{z-scan}/$

P_{noise} of 1.012. Thus the combination of opto-isolator unit and the noise detection unit can effectively make the pseudo-nonlinearity become negligible.

The feedback light from the sample can lead to the laser diode instability, and the instable laser output will induce a pseudo-nonlinearity effect in the z-scan measurement. We designed the z-scan light path where a noise detector unit was added in the front of the converging lens, and an opto-isolator unit was placed in between the feedback light detection unit and the noise detector unit, and the most feedback light can be avoided from going back to the laser diode. The influence of leaked reflected light induced laser power instability P_{noise} was recorded by the noise detection unit. The transmittance z-scan data T_{z-scan} was divided by the P_{noise}, and finally the experimental results showed the leaked reflected light induced pseudo-nonlinearity can be reduced to about 0.4%, which is a negligible persudo-nonlienarity signal.

4. SAMPLE DAMAGE INDUCED PSEUDO-NONLINEARITY

In z-scan measurement, there are lots of photosensitive materials, such as dye thin films, photoresist thin films, and chalcogenide semiconductor phase change thin films. For example, the amorphous chalcogenide semiconductor thin films exhibit photo-induced crystallinzation when exposed to the laser with a power of less than 10 mW [24]. The absorption coefficient change $\Delta\alpha$ is usualy accompanied by a change in the refraction Δn since the absorption coefficient and refractive index are related through Kramers-Kroning relations. The photo-induced crystallinzation effect has been wide applied in the data storage due to the optical reflectivity difference between amorphous and crystalline states [25]. Therefore, in the z-scan measurement, the collected signal may be not from the nonlinear effect, but from the photo-induced ablation or structural change of materials, and the signal is called to be sample damage induced pseudo-nonlinearity. Patterson *et al* [26] showed that the z-scan data from the thin films actually arise from linear optical effects caused by the laser ablation and not from nonlinear optical effects. Fedus *et al* [27, 28, 29] adopted the z-scan trances to measure the permanent phase shift written inside thin films due to photo-induced ablation or structural transformation.

4.1. Theoretical Model on the Sample Damage Induced Light Field Profiles

Assuming a TEM_{00} Gaussian intensity profile laser beam is incident on the sample, and the laser beam electric field can be expressed by Equation (1). The sample is considered to be "thin sample", and the diffraction phenomenon is neglectible. According to Equation (1), the permanent modification inside the sample causes the spatial profile of the refractive index and absorption coefficient to be

$$n(r) = n_0 + \Delta n_0 \exp(-2r^2/w_0^2) \tag{26}$$

$$\alpha(r) = \alpha_0 + \Delta\alpha_0 \exp(-2r^2/w_0^2) \tag{27}$$

where n_0 and α_0 are the refractive index and absorption coefficient of the unirradiated area, respectively. Δn_0 and $\Delta\alpha_0$ are the photo-induced change of on-axis refractive index and absorption coefficient, respectively. If the Fresnel reflection loss can be neglected the electric field at the exit surface of the sample at position z can be expressed as

$$E_e(r,z) = E_i(r,z)\exp[-\alpha(r)L/2]\exp[-(i\Delta\phi(r)] \tag{28}$$

with

$$\Delta\phi(r) = \Delta\phi_0 \exp(-2r^2/w_0^2) \tag{29}$$

$$\Delta\phi_0 = 2\pi\Delta n_0 L/\lambda \tag{30}$$

where L is the sample thickness. The transmittance of open aperture z-scan measurement is only sensitive to the absorption, thus the normalized z-scan curve can be obtained by spatially integrating the intensity, which is written as

$$T_o(z) = \frac{n_0\pi \int_0^{+\infty} |E_e(r,z)|^2 \, rdr}{n_0\pi \int_0^{+\infty} |E_i(r,z)|^2 \, rdr} \tag{31}$$

To obtain an analytical result, Fedus and Boudebs [28] applied an incomplete gamma function theory in order to explore into the exponential of Equation (27) as was done by Samad and Vieira [30], the Equation (31) can be simplified as

$$T_o(z) = \sum_{m=0}^{\infty} \frac{(-1)^m}{m!} \frac{(\Delta\alpha_0 L)^m}{[1+m(1+z^2/z_0^2)]} \tag{32}$$

The sample damage induced pseudo-nonlinear absorption can be obtained in Equation (32) through $\Delta\alpha_0$, which is gotten by fitting the open aperture z-scan curve.

For the sample damage induced pseudo-nonlinear refraction, the exit surface electric field in far-field has to be calculated. By virtue of Huygen's principle, the electric field in far-field detection plane is a superposition of different secondary wavelets from the exit surface of sample. The complex electric field distribution in the far-field aperture plane can be calculated from the Fresnel-Kirchhoff diffraction integral equation as follows [31],

$$E_a(\rho,z) = -\frac{i}{\lambda} \int_0^{\sqrt{5}w(z)} \int_0^{2\pi} \left(\frac{1}{2} + \frac{L_1}{2L_2}\right) E_e(r,z) \frac{\exp(i2\pi L_2/\lambda)}{L_2} \, rdrd\varphi \tag{33}$$

with

$$L_2 = \sqrt{L_1^2 + (r\cos\varphi - \rho)^2 + (r\sin\varphi)^2} \tag{34}$$

where ρ the radius of any point at the detection aperture plane. The distance between the sample and aperture detection plane is L_1. The normalized transmittance for closed aperture z-scan measurement is

$$T_c(z) = \frac{\int_0^{r_a} |E_a(\rho, z)|^2 \rho d\rho}{S\int_0^{\infty} |E_a(\rho, z, \Delta\phi_0 = 0, \Delta\alpha_0 = 0)|^2 \rho d\rho} \tag{35}$$

where r_a is an aperture radius, and $S = 1 - \exp(-2r_a^2/w_a^2)$ is the aperture transmittance, and w_a is the beam radius at the aperture plane in absence of any sample damage.

4.2. Experimental Results and Analysis

In order to verify the sample damage induced nonlinear effect in z-scan measurement, here we give the open aperture z-scan testing for AgInSbTe thin films, which were wide used as the optical information storage materials. The closed aperture z-scan measurement and analysis can be found in [26, 27, 28].

The amorphous AgInSbTe thin films were deposited on the glass substrate by the magnetron-control sputtering method at the room temperature. The background pressure was approximately $1.5 \times 10^{-4} Pa$ and the sputtering pressure was about $0.8\, Pa$ of Ar environment, the sputtering power was $50W$ and the sputtering speed was about $25\, nm/min$ accordingly. Here the sputtering time was about 8 minutes, thus the thickness was about $200\, nm$. The z-scan measurement was carried out at the laser wavelength of $633\, nm$.

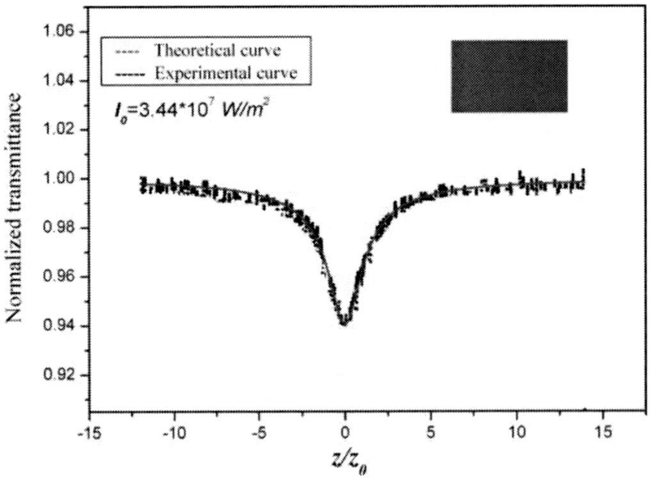

Figure 20. Experimental (black) and theoretical (red) z-scan curves of the deposited amorphous AgInSbTe thin filmand the inset is an optical picture of the amorphous thin film [32].

Figure 20 shows the typical z-scan result of the deposited amorphous AgInSbTe thin films with the light intensity $I_0 = 3.44 \times 10^7 W/m^2$, it is obviously observed that the sample is of the reverse saturable absorption; the inset is the optical picture of the deposited amorphous thin films after the z-scan measurement, and there is no evidence that the area has been structurally changed by laser irradiation, and multiple z-scan measurements at the same area keep almost the same curve. Furthermore, the z-scan curve is so symmetric that there should be no phase transition throughout the measurement process, all the evidences above confirm that this procedure is reversible, it is indicated that the z-scan curve from the light-induced intrinsic nonlinear absoprtion other than the structurally change (or damage) induced pseudo-nonlinear absorption.

However, when the light intensity increases above the crystallization threshold I_c, the deposited amorphous AgInSbTe thin films become into crystalline state, which is a white area in the central part of inset. There is an inflexion at the z-scan curve, and the curve is not symmetric (shown in Figure 21, the blue curve), which is structural transformation induced peudo-nonlinear reverse saturable absorption. While for the repeated experiment on the same area, the measured z-scan curves are symmetric (the black curve in Figure 21), which indicates an intrinsic nonlinearity of the crystalline state of AgInSbTe thin films.

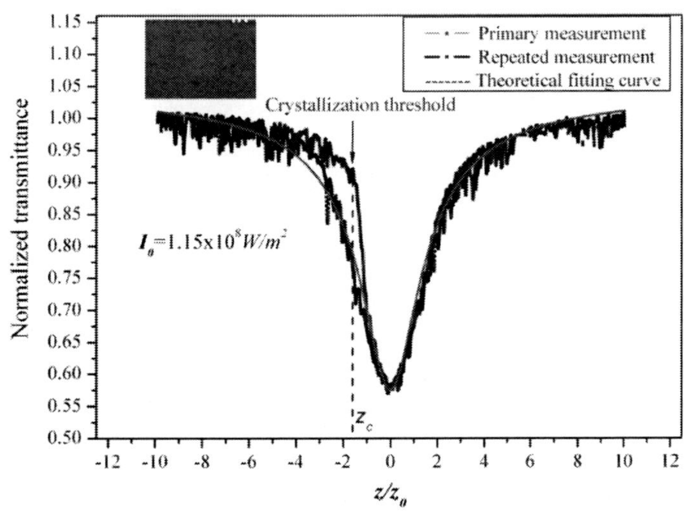

Figure 21. First time (blue) and repeated (black) experimental measurements as well as theoretical fitting (red) curve, and the inset is an optical picture of laser-irradiated area [32].

When the light intensity continues to be ascending, the thin film is molten ablation, which is shown in the inset of Figure 22, where an ablation hole is formed. By adjusting the light intensity with the variable attenuator we can get the melting threshold by analyzing the z-scan curves. In Figure 22(a), the z-scan curve still exhibits a molten ablation induced pseudo-nonlinear reverse saturable absorption characteristic with laser intensity $I_0 = 1.37 \times 10^8 \ W/m^2$. The melting ablation can be seen from the z-scan curve, a crystallization inflexion in the position z_c firstly occurs, when the sample moves to the position z_m, there is a rapid ascending in the z-scan curve, thus z_m is a melting ablation posstion. With the increase of incident laser intensity, the ablation hole becomes larger and larger, and the pseudo-nonlinear reverse saturable absorption curve shape is transformed to pseudo-nonlinear saturable absorption shape. Figure 22(b) presents the result of $I_0 = 1.46 \times 10^8 W/m^2$, one can see a large ablation hole compared with Figure 22(a), and the valley shape becomes into a peak shape. Generally speaking, the aim of the z-scan measurement is to get the intrinsic optical nonlinear parameters, and the sample damage induced pseudo-nonlinearity needs to be ruled out during the z-scan measurement, which can be realized by a real-time imaging unit in the z-scan setup. In the z-scan measurement, one can monitor in real-time in-situ the sample surface change, when the sample surface change occurs at the laser-irradiated area, the sample damage or structural transformation happens, and the incident laser intensity needs to be adjusted to a lower value. In this case the z-scan curve is only an indication of the intrinsic nonlinear characterisitcs.

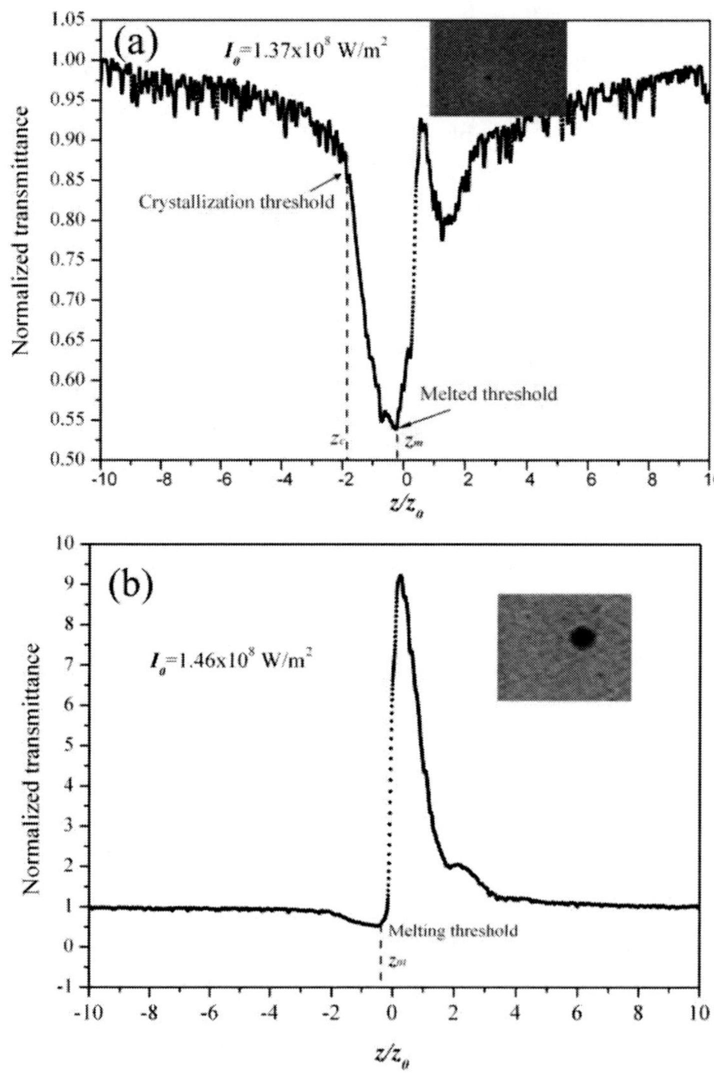

Figure 22. Z-scan curves with higher laser intensity. (a) $I_0 = 1.37 \times 10^8\ W/m^2$, (b) $I_0 = 1.46 \times 10^8\ W/m^2$, the inset is the optical pictures of the irradiated area [32].

ACKNOWLEDGMENTS

This work is partially supported by the National Natural Science Foundation of China (Grant Nos. 60977004, 51172253, and 61137002), the Instrument Developing Project of the Chinese Academy of Sciences (Grant No. YZ201140), the Science and Technology Commission of Shanghai Municipality (Grant Nos. 11JC1412700 and 11JC1413300).

REFERENCES

[1] Sheik-Bahae, M; Said, AA; van Stryland, EW. "High-sensitivity, single-beam n_2 measurements", *Opt. Lett.*, 1989, 14, 955-957.

[2] Sheik-Bahae, M; Said, AA; Wei, TH; Hagan, DJ; van Stryland, EW. "Sensitive measurement of optical nonlinearities using a single beam", *IEEE J. Quantum Electronics*, 1990, 26, 760-769.

[3] Wang, J; Sheik-Bahae, M; Said, AA; Hagan, DJ; van Stryland, EW. "Time-resolved z-scan measurements of optical nonlinearities", *J. Opt. Soc. Am. B*, 1994, 11, 1009-1017.

[4] Kozich, VP; Marcano, A; Hernandez, F; Castillo, J. "Dual-beam time-resolved z-scan in liquids to study heating due to linear and nonlinear light absorption", *Appl. Spectroscopy*, 1994, 48, 1506-1512.

[5] Petrov, PV. "Reflection z-scan technique for the study of nonlinear refraction and absorption of a single interface and thin film", *J. Opt. Soc. Am. B*, 1996, 13, 1491-1498.

[6] Rhee, B; Byun, JS; van Stryland, EW. "Z-scan using circularly symmetric beams", *J. Opt. Soc. Am. B*, 1996, 13, 2720-2723.

[7] Tsigaridas, G; Fakis, M; Polyzos, I; Persephonis, P; Giannetas, V. "Z-scan technique for elliptic Gaussian beams", *Appl. Phys. B*, 2003, 77, 71-75.

[8] Hughes, S; Burzler, JM. "Theory of z-scan measurements using Gaussian-Bessel beams", *Phys. Rev. A*, 1997, 56, R1103-R1106.

[9] Wang, J; Gu, B; Xu, YM; Wang, HT. "Enhanced sensitivity of z-scan technique by use of flat-topped beam", *Appl. Phys. B*, 2009, 95, 773-778.

[10] Zhao, W; Palffy-Muhoray, P. "Z-scan technique using top-hat beams", *Appl. Phys. Lett.*, 1993, 63, 1613-1615.

[11] Martinelli, M; Bian, S; Leite, JR; Horowicz, RJ. "Sensitivity-enhanced reflection z-scan by oblique incidence of a polarized beam", *Appl. Phys. Lett.*, 1993, 72, 1427-1429.

[12] Yan, X; Liu, Z; Zhang, X; Zhou, W; Tian, J. "Polarization dependence of z-scan measurement: theory and experiment", *Opt. Express*, 2009, 17, 6397-6406.

[13] van Stryland, EW; Sheik-Bahae, M. "Z-scan measurements of optical nonlinearities", in *Characterization Techniques and Tabulations for Organic Nonlinear Materials*, Editors: Kuzyk, M. G., Dirk, C. W. pp. 655-692, (Marcel Dekker Inc. New York, 1998).

[14] Sutherland, RL. "Effects of multiple internal sample reflections on nonlinear refractive z-scan measurements", *Appl. Opt.* 1994, 33, 5576-5584.

[15] Ren, L; Yao, B; Hou, X; Liu, L; Zhou, C. "Analyses and computations of asymmetric z-scan for large phase shift from diffraction theory", *Chin. Opt. Lett.*, 2003, 1, 111-113.

[16] Michelotti, F; Caiazza, F; Liakhou, G; Paoloni, S; Bertolotti, M. "Effects of nonlinear Fabry-Pérot resonator response on z-scan measurements", *Opt. Commun.*, 1996, 124, 103-110.

[17] Wang, R; Wei, J; Xiao, M. "Pseudo-nonlinear absorption in z-scan measurement", *J. Opt.*, 2013, 15, 025204.

[18] Riesz, RP; Simon, R. "Reflection of a Gaussian beam from a dielectric slab", *J. Opt. Soc. Am. A*, 1985, 2, 1809–1817.

[19] Masoller, C; Sicardi Schilino, AC; Cabeza, C. "Chaotic properties of the coherence collapsed state of laser diodes with optical feedback", *Opt. Commun.*, 1993, 100, 331-340.

[20] Lang, R; Kobayashi K. "External optical feedback effects on semi- conductor injection laser properties", *IEEE J. Quantum Electronics*, 1980, QE-16, 347-355.

[21] Tromborg, B; Osmundsen, JH; Olesen, H. "Stability analysis for a semiconductor laserin an external cavity", *IEEE J. Quantum Electronics*, 1984, QE-20, 1023-1032.

[22] Liu, C; Ge, JH; Chen J. "Influence of external cavity feedback on the oscillating characteristics of a semiconductor laser", *Acta Physica Sinica*, 2006, 55, 5211-5215. (in Chinese).

[23] Liu, C; Ge, JH; Chen J. "Investigation of loss and threshold characteristics in the laser diode with external feedback", *Chin. J. Lasers*, 2004, 31, 1413-1416. (in Chinese).

[24] Wei, J; Gan, F. "Theoretical explanation of different crystallization processes between as-deposited and melted-quenched amorphous $Ge_2Sb_2Te_5$ thin films", *Thin Solid Films*, 2003, 441, 292-297.

[25] Wei, J; Ruan, H; Gan, F. "Analysis of short-wavelength recording properties of AgInSbTe thin films", *Proc. of SPIE*, 2003, 5060, 163–166.

[26] Patterson, Brian M; White, W. Roc; Robbins, Todd A; Knize, Randall J. "Linear optical effects in z-scan measurements of thin films", *Appl. Opt.*, 1998, 37, 1854-1857.

[27] Fedus, K; Boudebs, G; de Araújo, Cid B; Cathelinaud, M; Charpentier, F; Nazabal, V. "Photoinduced effects in thin films of $Te_{20}As_{30}Se_{50}$ glass with nonlinear characterization", *Appl. Phys. Lett.*, 2009, 94, 061122.

[28] Fedus, K; Boudebs, G. "Determination of photo-induced changes in linear optical coefficients by the z-scan technique", *J. Opt. Soc. Am. B*, 2009, 26, 2171-2175.

[29] Boudebs, G; Fedus, K. "Linear optical characterization of transparent thin films by z-scan technique", *Appl. Opt.*, 2009, 4124-4129.

[30] Samad, RE; Vieira ND. "Analytical description of z-scan on-axis intensity based on the Huygens-Fresnel principle", *J. Opt. Soc. Am. B*, 1998, 15, 2742-2747.

[31] Wei, J; Xiao, M. "A z-scan model for optical nonlinear nanometric films", *J. Opt. A: Pure and Appl. Opt.*, 2008, 10, 115102.

[32] Liu, J; Wei, J. "Optical nonlinear absorption characteristics of AgInSbTe phase change thin films", *J. Appl. Phys.*, 2009, 106, 083112.

In: Nonlinear Performance and Characterization Methods … ISBN: 978-1-62808-093-3
Editor: Jingsong Wei

Chapter 6

THERMO-OPTICAL COEFFICIENT MEASUREMENT WITH LASER PULSE WIDTH TUNABLE Z-SCAN METHOD

Jingsong Wei*
Shanghai Institute of Optics and Fine Mechanics,
Chinese Academy of Sciences, Shanghai, China

ABSTRACT

The thermo-optical coefficients dn/dT and $d\alpha/dT$, where n, α, and T are the refractive index, absorption coefficient and temperature, respectively, are universal properties of materials, and almost all kinds of materials possess the properties. It is very important to find a simple and accurate method to determine the thermo-optical coefficients. In this chapter, an effective method on the basis of laser pulse width tunable z-scan measurement was introduced. We firstly presented the dependences of dn/dT and $d\alpha/dT$ on the laser pulse width, and then gave the experimental principle and setup of laser pulse width tunable z-scan method to measure the dn/dT and $d\alpha/dT$. The experiments were carried out to measure Sb_2Te_3 thin films, the experimental results were verified by variable-temperature spectroscopic ellipsometrymeasurements.

Keywords: Thermo-optical coefficients, z-scan method, optical nonlinearity

1. INTRODUCTION

The thermo-optic coefficients dn/dT and $d\alpha/dT$, where $n, \alpha,$ and T are the refractive index, absorption coefficient and temperature, respectively, are universal properties of materials, and almost all kinds of materials possess the properties. On one hand, the characteristic is widely used in the field of high density optical storage, optical switching, and other optical nonlinear devices [1-2]. On the other hand, for some material such as the pump

*Corresponding author e-mail: weijingsong@siom.ac.cn.

material in the laser devices [3], the thermo-optic effect would not be expected, so the dn/dT and $d\alpha/dT$ need to be measured for materials to clarify its influence on the application.

Measurements used to determine the dn/dT and $d\alpha/dT$ coefficients include ellipsometric measurement [4], interferometric method [5], thermal lensing method [6, 7], and traditional z-scan technique etc [8-10]. For these techniques, the ellipsometry is expensive, and complex operation and analysis are required.

The thermal lensing technique, interferometric method, and the pump-probe method, are based on the distortion of the wavefront of the transmitted laser beam to determine the dn/dT, but it is difficult to measure $d\alpha/dT$. The z-scan-relevant methods were based on the assumption that the refractive index change purely comes from the thermal effect, however, the refractive index change should be a mixture of many effects, this method only measures the dn/dT, but the $d\alpha/dT$ cannot be given due to the fixed laser pulse width.

Therefore, it is very important to find a simple and accurate method to determine the thermo-optical coefficients of dn/dT and $d\alpha/dT$. In this chapter, an effective method on the basis of laser pulse width tunable z-scan measurement is introduced to get the thermo-optic coefficient, and the detailed principle is described in this chapter.

2. Principle of Thermo-Optical Coefficient Measurement with Laser Pulse Width Tunable Z-Scan Method

2.1 The Dependence of the Refractive Index on Temperature [11, 12]

When a laser beam with intensity I is incident on the thin sample the intensity change along the sample thickness direction can be expressed as

$$dI / dz' = -\alpha_0 I \tag{1}$$

where α_0 is the linear absorption coefficient, and z' is the propagation length of the laser beam within the sample. The sample absorbs the laser energy and is heated. The thermally-induced change of the refractive index Δn_{th} can be written as

$$\Delta n_{th} = \frac{dn}{dT} \Delta T \tag{2}$$

where ΔT is the temperature change. For the short laser pulse heating the thermal irradiation loss can be ignored. The beam aberration resulting from free convection and forced convection is also ignored. Assuming that the thermo-physical parameters of the sample are independent on the temperature, the equation of heat conduction can be defined as

$$c\rho \frac{dT}{dt} = Q + \kappa \nabla^2 T \tag{3}$$

where c, ρ, and κ are heat capacity, density, and thermal conductivity. Q is thermal energy of unit time and unit length produced due to absorption, and is defined as

$$Q = \frac{dI}{dz'} \tag{4}$$

According to Equation (3) one can get the temperature rise at time t in the cylindrical coordinate

$$\Delta T(r,z,t) = \int_0^{\infty}\int_{-\infty}^{t} 2\pi r_1 Q(r_1,t_1,z) G(r,r_1,t-t_1) dr_1 dt_1 \tag{5}$$

where G is Green function, and is expressed as

$$G(r,r_1,t) = \frac{1}{4\pi\kappa t}\exp\left[\frac{-(r^2+r_1^2)}{4Dt}\right] I_0\left(\frac{rr_1}{2Dt}\right) \tag{6}$$

where $I_0(x) = J_0(x)$ is the zero-order Bessel function of the first kind. D is the thermal diffusivity, and is defined as

$$D = \frac{\kappa}{\rho c} \tag{7}$$

In z-scan the incident laser beam pulse with a Gaussian intensity profile is usually expressed as

$$I(r,z,t) = I(0,0,0)\frac{w_0^2}{w^2(z)}\exp\left[-\frac{2r^2}{w^2(z)}\right]A(t) = I(0,z,0)\exp\left[-\frac{2r^2}{w^2(z)}\right]A(t) \tag{8}$$

where $I(0,0,0)$ is the peak intensity. Based on Equations (1), (4), and (8). Equation (5) is written as

$$\Delta T(r,z,t) = \frac{\alpha_0 I(0,z,0)}{2\kappa} \times \int_{-\infty}^{t}\frac{A(t_1)}{t-t_1}\exp\left(\frac{-r^2}{4D(t-t_1)}\right)\int_0^{\infty} I_0\left(\frac{rr_1}{2D(t-t_1)}\right)\times\exp\left[-\frac{2r_1^2}{w^2(z)}-\frac{r_1^2}{4D(t-t_1)}\right] r_1 dr_1 dt_1 \tag{9}$$

By Laplace transformation operation on Equation (9), one can get

$$\Delta T(r,z,t) = \frac{\alpha_0 D I(0,z,0)}{\kappa}\int_{-\infty}^{t}\frac{A(t_1)\exp\left[-\frac{2r^2}{w^2(z)}\frac{1}{1+2(t-t_1)/t_c'}\right]}{1+2(t-t_1)/t_c'}dt_1 \tag{10}$$

and

$$t_c^{'} = t_c(1+\frac{z^2}{z_0^2}) \tag{11}$$

where $t_c = w_0^2/4D$ is the thermal characteristic time. Substituting Equation (10) into Equation (2) one can get thermally-induced change of the refractive index

$$\Delta n_{th}(t) = \frac{\alpha_0 D\frac{dn}{dT}I(0,z,0)}{\kappa}\int_{-\infty}^{t}\frac{A(t_1)\exp[-\frac{2r^2}{w^2(z)}\frac{1}{1+2(t-t_1)/t_c^{'}}]}{1+2(t-t_1)/t_c^{'}}dt_1 \tag{12}$$

Analyzing Equation (12) one can see that the thermally-induced change of the refractive index is not proportional to the incident laser beam intensity. Thus the Δn_{th} can not be directly gotten by common z-scan method with a fixed laser pulse width. However, in z-scan if the laser pulse width $t_w << t_c$, for 0~t_w Equation (12) can be rewritten as

$$\Delta n_{th}(t) = \frac{\alpha_0 D\frac{dn}{dT}I(0,z,0)}{\kappa}\exp(-\frac{2r^2}{w^2(z)})\int_{-\infty}^{t_w}A(t_1)dt_1 \tag{13}$$

For a temporally rectangular pulse with a pulse width of t_w, the $A(t)$ can be expressed as

$$A(t) = \begin{cases} 1 & 0 \le t \le t_w \\ 0 & t < t_w \end{cases} \tag{14}$$

Substituting Equation (14) into Equation (13) one can obtain

$$\Delta n_{th}(t) = \frac{\alpha_0 D\frac{dn}{dT}t}{\kappa}I(z,r,0) \tag{15}$$

Equation (15) can be rewritten as

$$\Delta n_{th}(t) = n_{2th}(t)I \tag{16}$$

and the thermally-induced refractive index in z-scan measurement is

$$n = n_0 + \Delta n_{nontherm} + \Delta n_{th}(t) = n_0 + [n_{2nontherm} + n_{2th}(t)]I \tag{17}$$

with

$$n_{2th}(t) = \frac{\alpha_0 D \frac{dn}{dT} t}{\kappa}$$

where $\Delta n_{nontherm}$ and $n_{2nontherm}$ are refraction index variation and nonlinear refraction coefficient induced by the non-thermal effect.

Equations (17) and (18) indicates that the thermally-induced refractive index variation is formally consistent with the nonlinear refraction expression $n = n_0 + \Delta n = n_0 + n_2 I$ in z-scan measurement, where n_0 is linear refractive index.

The thermally-induced nonlinear refraction coefficient $n_{2th}(t)$ increases with the laser pulse time, and the non-thermally-induced nonlinear refraction coefficient $n_{2nontherm}$ is independent on the laser pulse width. Thus the n_{2th} can be obtained by laser pulse width tunable z-scan method. The time-averaged $< n_{2th}(t) >$ can be obtained as follows

$$< n_{2th}(t) > = \frac{\int_{-\infty}^{+\infty} n_{2th}(t) I(0,0,t) dt}{\int_{-\infty}^{+\infty} I(0,0,t) dt} = \frac{\alpha_0 D \frac{dn}{dT}}{2\kappa} I(0,0,0)$$

Thus the defined n_{2th} as time-averaged nonlinear refraction coefficient is

$$n_{2th} = \frac{\alpha_0 D \frac{dn}{dT}}{2\kappa} \tag{18}$$

According to Equation (18) the thermo-optical coefficient dn/dT can be expressed as

$$\frac{dn}{dT} = \frac{\kappa}{D\alpha_0} \frac{n_{2th}}{t_w} \tag{19}$$

Equation (19) tells us that the thermo-optical coefficient dn/dT can be experimentally gotten through laser pulse width tunable z-scan method.

For a temporally Gaussian pulse $A(t) = exp(-t^2/\tau^2)$, similar to the rectangular laser pulse, the Equation (13) becomes

$$\Delta n_{th}(t) = \begin{cases} \frac{\sqrt{\pi}\tau\alpha_0 D(dn/dT)[1 - erf(t/\tau)]}{2\kappa} I(z,r,0), & t \le 0 \\ \frac{\sqrt{\pi}\tau\alpha_0 D(dn/dT)[1 + erf(t/\tau)]}{2\kappa} I(z,r,0), & t \ge 0 \end{cases} \tag{20}$$

with

$$n_{2th}(t)=\begin{cases}\dfrac{\sqrt{\pi}\tau\alpha_0 D(dn/dT)[1-erf(t/\tau)]}{2\kappa}, & t\le 0\\ \dfrac{\sqrt{\pi}\tau\alpha_0 D(dn/dT)[1+erf(t/\tau)]}{2\kappa}, & t\ge 0\end{cases} \tag{21}$$

The time-averaged nonlinear refraction coefficient n_{2th} is

$$n_{2th}=\frac{\alpha_0 D\dfrac{dn}{dT}}{2\kappa}\left[\sqrt{\pi}\tau+\int_{-\infty}^{+\infty} erf(\frac{t}{\tau})\exp(-\frac{t^2}{\tau^2})dt\right] \tag{22}$$

where $erf(x)$ is error function. Equation (22) is used to calculate the thermally-induced change of the refractive index dn/dT only $t_w << t_c$, and the dn/dT can be obtained by measuring the n_{2th} at different laser pulse width τ.

2.2. Dependence of Absorption Coefficient on Temperature [13]

Under the irradiaiton of laser beam, the absorption coefficient of sample is $\alpha=\alpha_0+\Delta\alpha=\alpha_0+\beta I$, where β is nonlinear absorption coefficient. The β can may be made up of two parts, one is non-thermally induced nonlinear absorption coefficient $\beta_{nontherm}$, such two-photon absorption effect, and free-carrier absorption effect etc.; the other is thermally-induced nonlinear absorption coefficient β_{th}, thus $\beta=\beta_{nontherm}+\beta_{th}$. $\beta_{nontherm}$ is independent on temperature, while β_{th} is dependent on temperature. Under the laser pulse irradiation, the sample absorbs the laser energy and is heated. The thermally-induced change of the absorption coefficient $\Delta\alpha_{th}$ can be written as

$$\Delta\alpha_{th}=\frac{d\alpha}{dT}\Delta T \tag{23}$$

If the laser pulse is rectangular as given in Equation (14), according to Equations (13)-(15), the temperature rise can be expressed as

$$\Delta T(t)=\frac{\alpha_0 Dt}{\kappa}I_0(t),\ \text{for } t<t_w \tag{24}$$

and the steady-state temperature rise can be easily obtained by the time-averaged method

$$\langle\Delta T(t)\rangle=\frac{\int_{-\infty}^{\infty}\Delta T(t)I_0(t)dt}{\int_{-\infty}^{\infty}I_0(t)dt}=\frac{\alpha_0 t_w}{2\rho c}I_0 \tag{25}$$

One can see from Equation (25) that the laser-induced temperature rise is proportional to the laser intensity, pulse width and linear absorption coefficient, while inversely proportional to the density and thermal capacity. Combining Equations (23) and (25), one can get the absorption coefficient change

$$\langle \Delta\alpha_{th} \rangle = \frac{\alpha_0 t_w I_0}{2\rho c}\frac{d\alpha}{dT} \tag{26}$$

Equation (26) shows that the thermally-induced steady-state absorption coefficient change is proportional to $d\alpha/dT$. If $d\alpha/dT < 0$, then the thermal effect induces a negative absorption coefficient change; if $d\alpha/dT > 0$, then the thermal effect induces a positive absorption coefficient change. According to Equation (26), the thermally-induced nonlinear absorption coefficient can be calculated as

$$\beta_{th} = \frac{\langle \Delta\alpha_{th} \rangle}{I_0} = \frac{\alpha_0 t_w}{2\rho c}\frac{d\alpha}{dT} \tag{27}$$

Similar to rectangular laser pulse, for the irradiation of Gaussian pulse $A(t) = exp(-t^2/\tau^2)$ the thermally-induced nonlinear absorption coefficient is

$$\beta_{th}(t) = \begin{cases} \dfrac{\sqrt{\pi}\tau\alpha_0 D)[1-erf(t/\tau)]}{2\kappa}\dfrac{d\alpha}{dT}, & t \le 0 \\ \dfrac{\sqrt{\pi}\tau\alpha_0 D[1+erf(t/\tau)]}{2\kappa}\dfrac{d\alpha}{dT}, & t \ge 0 \end{cases} \tag{28}$$

Equations (27) and (28) tell us that the thermally-induced nonlinear absorption coefficient β_{th} is proportional to the laser pulse width, and the non-thermally-induced nonlinear absorption coefficient $\beta_{notherm}$ is independent on laser pulse width. Thus the $d\alpha/dT$ can be obtained through measuring the thermally-induced nonlinear absorption coefficient β_{th} using laser pulse width tunable z-scan method.

3. Z-Scan Experimental Setup and Data Analysis

3.1. Z-Scan Principle and Setup [14, 15]

To support the theoretical derivation discussed above, we designed a laser pulse width tunable z-scan experimental setup (shown in Figure 1). In this setup, a variable attenuator is used to change the laser intensity, and an acousto-optic modulator (or high frequency TTL modulator) as well as a signal-generator is applied to tune the laser pulse width from nanosecond to continuous wave (cw). Both the nonlinear refraction and the nonlinear absorption coefficients can be obtained by close-aperture and open-aperture z-scan measurements, respectively. By changing the laser pulse width and intensity, the thermo-optical coefficients dn/dT and $d\alpha/dT$ can be gotten from the measurements, and the variable temperature ellipsometric measurements were taken to verify the results.

The laser pulse width tunable z-scan setup is shown in Figure 1. A laser beam from a laser device at the wavelength $633\,nm$ (or $405\,nm$) is modulated into pulse light by a combination of acousto-optic modulator (or high frequency TTL modulator) and a signal generator, and then converged to the sample surface by an objective lens with a numerical aperture of about 0.09. The sample moves along the z-axis by an electrically-drived stage controlled by a computer. The transmitted light is then split into two beams by a splitter prism (SP3). One beam is collected to a photo-detector (PT2) and the other is centrally collected by another photo-detector (PT3) with an aperture in the front of the photo-detector PT3 on the principal optical axis. The former is known as open-aperture mode z-scan measurement characterizing the nonlinear absorption property of the material, and the latter is the close-mode aperture z-scan measurement characterizing the property mixed by nonlinear refraction and nonlinear absorption. The signals collected by PT2 and PT3 are sent to a Boxcar (Stanford SRS250) connected to a computer (PC).

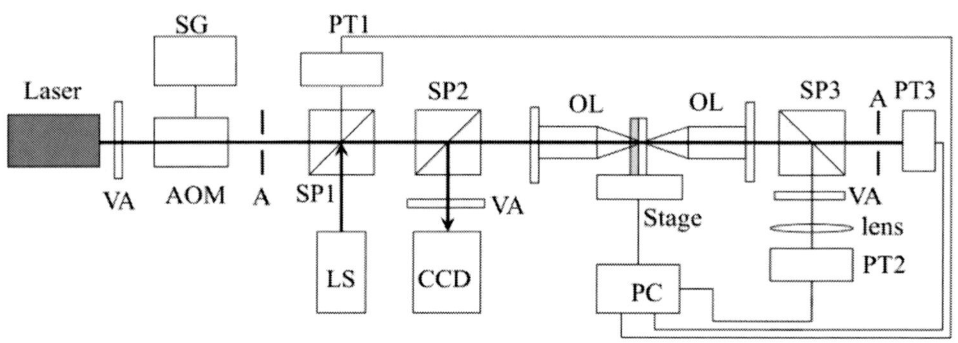

Figure 1. Schematics of the modified z-scan setup, VA: variable attenuator; AOM: acousto-optic modulator or high-frequency TTL modulator; SG: signal generator; A: aperture; SP: splitter prism; OL: objective lens; PT: photo detector; PC: computer; IS: lighting source.

If the incident light is a TEM_{00} mode Gaussian laser beam, the electric field is expressed as

$$E_i(z,r,t) = E_0(t)\frac{w_0}{w(z)}\exp(-\frac{r^2}{w^2(z)} - \frac{ikr^2}{2R(z)} - ikz)\mathrm{e}^{-i\phi(z,t)} \tag{29}$$

where $w^2(z) = w_0^2(1 + z^2/z_0^2)$ is the beam radius, $R(z) = z(1 + z^2/z_0^2)$ is the curvature radius of the wavefront at position z. $z_0 = kw_0^2/2$ is the diffraction length of the beam, $k = 2\pi/\lambda$ is the wavevector, and λ is the laser wavelength in free space. $E_0(t)$ is the incident electric field at the focus with temporal envelope of the laser pulse. The $i\phi(z,t)$ is the radial phase variations from the laser beam propagation, can be ignored under the slowly varying envelope approximation, that is, $i\phi(z,t)\sim 0$. According to the z-scan method proposed by Sheik-Bahae*et al.* [15, 16], some assumptions need to be met as follows :

(1) The sample thickness L is small enough that the beam diameter changes within the sample due to either diffraction or refraction can be neglected, the sample is regarded as "thin" so that the self-refraction effect is considered to be "external self-action".
(2) $L < z_0$ and $L << z_0/\Delta\phi(0)$, where $\Delta\phi(0)$ is nonlinear phase shift at the optical axis. This requirement is automatically met for most z-scan measurement because $\Delta\phi(0)$ is usually small.

According to the assumptions above, the intensity $I(= E_i \times E_i^*)$, where E_i^* is the conjugate of the electric field, and phase $\Delta\phi$ of the light beam within the sample is governed in the slowly varying envelope approximation through the following equations,

$$\frac{d\Delta\phi}{dz'} = \Delta nk \qquad \frac{dI}{dz'} = -\alpha I \tag{30}$$

with

$$n = n_0 + n_2 I \quad \Delta n = n_2 I \tag{31}$$

$$\alpha = \alpha_0 + \beta I \quad \Delta\alpha = \beta I \tag{32}$$

$$I_i(z,r,t) = E_i E_i^* = I_0(t)\frac{w_0^2}{w^2(z)}\exp(-\frac{2r^2}{w^2(z)}) \tag{33}$$

where $I_0(t)$ is the laser intensity at the beam waist position. If the Fresnel reflection losses can be ignored, the I_i can be considered to be irradiance within the sample.

The intensity profile and phase shift of the light beam at the exit surface of the sample is

$$I_e(z,r,t) = \frac{I_i(z,r,t)e^{-\alpha L}}{1+q(z,r,t)} \tag{34}$$

with

$$\Delta\phi(z,r,t) = \frac{k\gamma}{\beta}\ln[1+q(z,r,t)] \tag{35}$$

where

$$q(z,r,t) = \beta I_i(z,r,t)L_{eff} \tag{36}$$

The complex electric field at the exit surface of the sample is

$$E_e(z,r,t) = E_i(z,r,t)e^{-\frac{\alpha L}{2}}[1+q(z,r,t)]^{(\frac{ik\gamma}{\beta}-\frac{1}{2})} \tag{37}$$

The transmitted laser power through the sample can be recorded by placing an aperture with a radius of r_a in the far-field. The normalized aperture size to the light beam size can be repressed as

$$S = 1-\exp(-\frac{2r_a^2}{w_a^2}) \tag{38}$$

where w_a is the light beam radius at the aperture plane. For open-aperture z-scan measurement $S = 1$ is used to get the nonlinear absorption coefficient β, and for close-aperture z-scan measurement $S < 1$ is used to get the nonlinear refraction coefficient n_2.

3.2. Nonlinear Absorption Measurement and Data Analysis

For the nonlinear absorption measurement, the $S = 1$, and the total transmitted laser power needs to be obtained by spatially integrating Equation (34) at z over r as follows,

$$P(z,t) = P_i(t)e^{-\alpha L}\frac{\ln[1+q_0(z,t)]}{q_0(z,t)} \tag{39}$$

where

$$q_0(z,t) = \frac{\beta I_0(t)L_{eff}}{1+z^2/z_0^2} \tag{40}$$

where $L_{eff} = (1-e^{-\alpha L})/\alpha$ is an effective sample length, and $P_i(t)$ is an instantaneous incident laser power within the sample, and is defined as

$$P_i(t) = \frac{\pi}{2}w_0^2 I_0(t) \tag{41}$$

For a temporally Gaussian pulse the Equation (40) can be time integrated to give the normalized energy transmittance

$$T(z,S=1) = \frac{1}{\sqrt{\pi}q_0(z,0)}\int_{-\infty}^{+\infty}\ln[1+q_0(z,0)e^{-\tau^2}]d\tau \tag{42}$$

If the $|q_0| < 1$, Equation (42) is expanded into a summation form more suitable for numerical evaluation in terms of the peak irradiance

$$T(z,S=1)=\sum_{m=0}^{\infty}\frac{[-q_0(z,0)]^m}{(m+1)^{3/2}} \tag{43}$$

For usual data analysis, the first two terms are only considered to be fit the z-scan curve. That is

$$T(x)=1-\frac{2x^2}{(x^2+9)(x^2+1)}q_0 \tag{44}$$

Once the open-aperture ($S = 1$) z-scan is performed, the nonlinear absorption coefficient β can be determined. The nonlinear refraction coefficient n_2 can also be extracted from the closed-aperture ($S < 1$) z-scan curve based on the known β value.

3.3. Nonlinear Refraction Measurement and Data Analysis

Equation (37) combines the nonlinear absorption and nonlinear refraction. If the nonlinear absorption is very small and in the limit of nonlinear absorption $\beta \rightarrow 0$, Equation (37) can be simplified as

$$E_e(r,z,t)=E_i(r,z,t)e^{-\alpha L/2}e^{i\Delta\phi(r,z,t)} \tag{45}$$

with

$$\Delta\phi(z,r,t)=\Delta\phi_0(z,t)\exp(-\frac{2r^2}{w^2(z)}) \tag{46}$$

$$\Delta\phi_0(z,t)=\frac{\Delta\phi_0(t)}{1+z^2/z_0^2} \tag{47}$$

where the on-axis phase shift at the beam waist position $\Delta\phi_0(t)$ is defined as

$$\Delta\phi_0(t)=k\Delta n_0(t)L_{eff} \tag{48}$$

with

$$\Delta n_0(t)=n_2 I_0(t)$$

Substituting Equations (29) and (33) into Equation (45) one can get the complex electric field as follows

$$E_e(z,r,t) = E_0(t)\frac{w_0}{w(z)}e^{-\alpha L/2}\exp[-\frac{r^2}{w^2(z)} - \frac{ikr^2}{2R(z)} - ikz]\exp[-i\Delta\phi_0(z,t)\exp(-\frac{2r^2}{w^2(z)})] \quad (49)$$

In Equation (49), the first exponential term represents the beam attenuation due to sample absorption; the second exponential term is the electric field amplitude distribution and the intrinsic phase of the laser beam wavefront; the last exponential term gives nonlinear phase following a Gaussian profile.

In z-scan setup, the light intensity is detected in far-field, thus the complex electric field in far-field has to be calculated. By virtue of Huygen's principle, the complex electric field in far-field detection plane is the superposition of different secondary wavelets from the exit surface of sample. From Figure 2, the complex electric field distribution in the far-field aperture plane can be calculated from the Fresnel-Kirchhoff diffraction integral equation as follows,

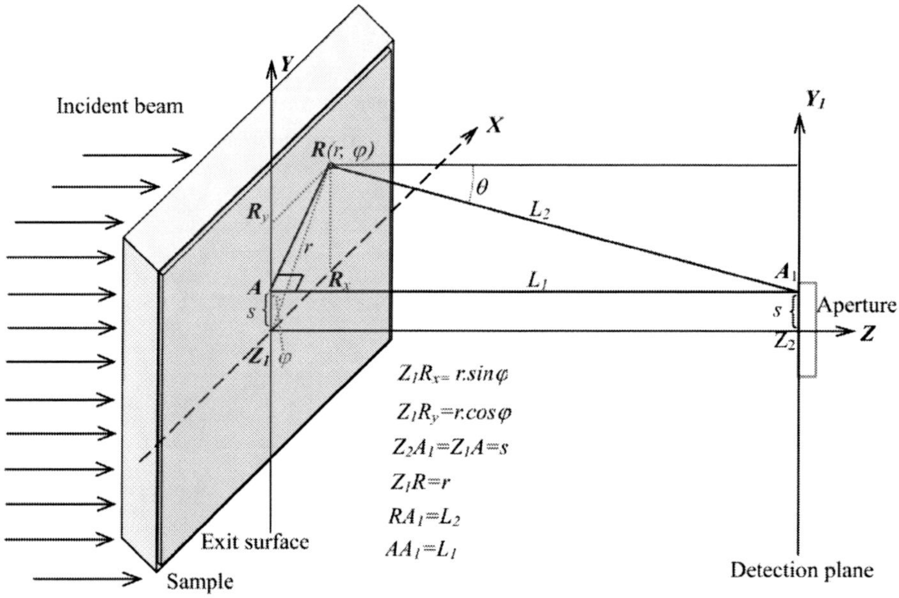

Figure 2. The geometrical relations in the Fresnel-Kirchhoff diffraction integral.

$$E_{aperture}(z,s,t) = \frac{-i}{\lambda}\int_0^{\infty}\int_0^{2\pi} F(\theta_0+\theta).E_e(z+L,r,t)\frac{\exp(i\frac{2\pi}{\lambda}L_2)}{L_2}rdrd\varphi \quad (50)$$

with

$$L_2 = \sqrt{L_1^2 + (r\cos\varphi - s)^2 + (r\sin\varphi)^2} \tag{51}$$

$$\theta = \arctan\frac{\sqrt{(r\sin\varphi)^2 + (r\cos\varphi - s)^2}}{L_1} \tag{52}$$

$$F(\theta_0 + \theta) = \frac{\cos\theta_0 + \cos\theta}{2} \tag{52}$$

where s the radius of the point A_1 from the center Z_2 of the detection aperture plane, Z_1 and Z_2 are the sample and detection plane positions, respectively. The distance between the sample and aperture detection plane is L_1. The origin of axial ordinate is defined at the waist of Gaussian laser beam. $F(\theta_0 + \theta)$ is the inclination factor, θ_0 is the angle formed by the incident light and the secondary wavelet (for normal incident light, $\theta_0 = 0$), and θ is the angle formed by the point to be calculated on the aperture plane and the secondary wavelet, $\cos\theta = L_1/L_2$. For normal incident light on the sample, thus

$$F(\theta_0 + \theta) = \frac{1+\cos\theta}{2} = \frac{1}{2} + \frac{L_1}{2L_2} \tag{53}$$

According to the characteristics of the TEM$_{00}$ Gaussian laser beam, when $r = \sqrt{5}w(z)$, $exp\{-[r^2/w^2(z)]\}$ and $\Delta\phi(r)$ tends to zero. Therefore, the integral limits can be defined as $r = 0 \to \sqrt{5}w(z)$, and $\varphi = 0 \to 2\pi$. The Equation (50) can be rewritten as

$$E_{aperture}(z,s,t) = \frac{-i}{\lambda}\int_0^{\sqrt{5}w(z)}\int_0^{2\pi}\left(\frac{1}{2} + \frac{L_1}{2L_2}\right)E_e(z+L,r,t)\frac{\exp(i\frac{2\pi}{\lambda}L_2)}{L_2}rdrd\varphi \tag{54}$$

For a circular aperture of radius r_a, the total light power is

$$P(z,t) = 2\pi\int_0^{r_a}\left|E_{aperture}(z,s,t)\right|^2 sds \tag{55}$$

By scanning z, (changing the position of sample), the z-scan curve can be obtained from Equation (55). Considering the pulse temporal variation, the normalized transmittance curve $T(z)$ can be calculated as

$$T(z) = \frac{\int_{-\infty}^{+\infty}P(z,t)dt}{S\int_{-\infty}^{+\infty}P_i(t)dt} \tag{56}$$

Equation (56) gives the common z-scan data analysis and fitting. One can get the nonlinear refraction coefficient by the searching for the n_2 value to matching the measuring the z-scan curve, which needs to complex mathematical operation and computing. Here it should be noted that the pure nonlinear refraction z-scan is difficult to obtain, the measured closed-aperture z-scan curve is a coupling of nonlinear absorption and nonlinear refraction, and the pure nonlinear refraction curve is usually extracted from the normalized closed-aperture z-scan curve being divided by the normalized open aperture z-scan curve.

Case 1: Small Shift ($|\Delta\phi_0|\leq\pi$)

However, for some small phase shift $|\Delta\phi_0|\leq\pi$ and $S\sim 0$, the fitting can be done as follows. Generally speaking, in z-scan measurement the far-field condition $L_1 >> z_0$ is easily met, and the geometry-independent normalized transmittance is

$$T(x) = 1 - \frac{4x\Delta\phi_0}{(x^2+9)(x^2+1)} \tag{57}$$

where $x = z/z_0$, the positions of peak and valley of the z-scan curve can be calculated by solving $dT(x,\Delta\phi_0)/dz = 0$ of Equation (57) to be $x_{p,v}\sim\pm 0.858$, and the peak-valley separation is $\Delta Z_{p-v}\sim 1.7z_0$. The peak-valley transmittance change is

$$\Delta T_{p-v} = \frac{8|x_{p,v}|}{(x_{p,v}^2+9)(x_{p,v}^2+1)}\Delta\phi_0 = 0.406|\Delta\phi_0| \tag{58}$$

For larger aperture, the following relationship can be used to fit the z-scan data

$$\Delta T_{p-v} \simeq 0.406(1-S)^{0.25}|\Delta\phi_0| \text{ for } |\Delta\phi_0| \leq \pi \tag{59}$$

Case 2: Thick Sample

In above analysis, the sample is thought to be "thin", and the beam diameter will not be changed within the sample. $L < z_0$ and $L << z_0/\Delta\phi_0$ need to be met. However, some thick samples need to use other methods to fit the measured z-scan data. Tian *et al* [17, 18] proposed a method that the thick sample is considered to be the stack of lots of "thin" sample. For small phase variation the normalized closed-aperture transmittance z-scan curve can be expressed as

$$T(x) = 1 - \frac{4\Delta\Phi_0(t)x}{(x^2+9)(x^2+1)} - O[\Delta\Phi_0^2(t)] \tag{60}$$

The former two terms are consistent with the Equation (57). The last term is second order small and can be usually ignored. For the stacking of lots of "thin" sample, the normalized transmittance is

$$T(x)=\prod_{i=1}^{n}[1+\frac{4\Delta\Phi_{0i}(t)x_i}{(x_i^2+9)(x_i^2+1)}]\cong \exp[\sum_{i=1}^{n}\frac{4\Delta\Phi_{0i}(t)x_i}{(x_i^2+9)(x_i^2+1)}] \tag{61}$$

where i is the ith thin sample. $x_i = z_i/z_0$, $\Delta\Phi_{0i}(t) = k\Delta n_0(t)L_i = \Delta\Phi_R(t)\Delta x_i$, with $\Delta\Phi_R(t) = k\Delta n_0(t)z_0$, z_i is the position of the ith thin sample, L_i is the thickness of ith thin sample. Letting $L_i \to 0$ and representing the position of the thick sample by the position of the front surface of the thick sample, one can get

$$T(x,t)=\left\{\frac{[(x+l)^2+1]\times(x^2+9)}{[(x+l)^2+9]\times(x^2+1)}\right\}^{\Delta\varphi_R(t)/4} \tag{62}$$

where $l = L/z_0$. The exponential factor $\Delta\Phi_R(t)/4$ of Equation (62) should be corrected to be $C_\Phi[\Delta\Phi_R(t), x, l]$ due to the coupling effect among "thin" samples. By a series of mathematical approximation operation $C_\Phi[\Delta\Phi_R(t), x, l]$ is written as

$$C_\Phi(t)\cong\Delta\Phi_R(t)+\tanh(l/3)\times\left\lfloor[\Delta\Phi_R(t)]^2/4+[\Delta\Phi_R(t)]^3/16\right\rfloor \tag{63}$$

By the mathematical operation of $dT/dx = 0$ on Equation (62), the distance between peak and valley is

$$x_{p-v}=\pm\{\frac{(l^2/2-10)+[(l^2+10)^2+108]^{1/2}}{6}\}^{1/2} \tag{64}$$

The transmittance difference of peak to valley is

$$\Delta T_{p-v}=\left|2\sinh\{C[\Delta\Phi_R(t),l]\times f(x_{p-v},l)\}\right| \tag{65}$$

with

$$f(x_{p-v},l)=\frac{[(|x_{p-v}|+l)^2+9]\times(|x_{p-v}|^2+1)}{[(|x_{p-v}|+l)^2+1]\times(|x_{p-v}|^2+9)} \tag{66}$$

Equations (62)-(66) can be used to fit the large nonlinear phase variation. Correspondingly, the open-aperture nonlinear absorption curve is

$$T(x)=\frac{1}{1+1/2C_q(t)[\tan^{-1}(x+l)-\tan^{-1}(x)]} \tag{67}$$

with

$$C_q(t) \cong Q_R(t) \times (1 + \tanh(l/2) \times \{\frac{3\Delta\Phi_R(t)}{10} + \frac{[\Delta\Phi_R(t)]^2}{8}\}) \qquad (68)$$

In the section, we introduced the data fitting of z-scan curve for different cases. Generally speaking, the open-aperture z-scan measurement is fitted, and the nonlinear absorption coefficient β is firstly determined.

The pure nonlinear refraction curve is then usually extracted from the normalized closed-aperture z-scan curve being divided by the normalized open aperture z-scan curve due to the measured closed-aperture z-scan curve being a coupling of nonlinear absorption and nonlinear refraction. Finally, the nonlinear refraction coefficient n_2 is obtained by fitting the pure nonlinear refraction curve.

In order to obtain the thermally-induced nonlinear refraction and absorption coefficients, the z-scan measurements with tunable laser pulse width are carried out, because the non-thermally-induced nonlinear refraction and absorption coefficients, based on Equations (18), (21), (27), and (28), are unchanged with the laser pulse width, and only thermally-induced nonlinear refraction and absorption coefficients are changed with the laser pulse width.

4. Experimental Results and Analysis

4.1. Measurement of Dependence of Refractive Index on Temperature [14]

In order to verify the effectiveness and correctness of derived equations and experimental setup, the measurement of the thermo-optical coefficient dn/dT of the crystalline Sb_2Te_3 thin films was carried out. The Sb_2Te_3thin films were deposited on the glass substrate by the radio frequency magnetron-control sputtering method at the room temperature.

The background pressure was approximately $1.5 \times 10^{-4}\ Pa$ and the sputtering pressure was about $0.8\ Pa$ of Ar environment. The thin film thickness was about $20\ nm$. The deposited Sb_2Te_3 thin films are amorphous state, so the sample was further crystallized at $200°C$ for about 1 hour with a vacuum degree of $1.2 \times 10^{-3}\ Pa$.

To get the valu*e* dn/dT, efforts were taken by changing the width of the rectangular laser pulse by in the z-scan setup shown in Figure 1.

The dependence of the nonlinear refraction coefficient n_2 at laser wavelength of $633\ nm$ on the pulse width t_w is illustrated in Figure 3, where the circles are the experimental data, and the solid line is a fitting curve. As is shown in Figure 3, the effective nonlinear refraction coefficient n_2 increases linearly with the laser pulse width t_w, which indicates that for the Sb_2Te_3 thin films the refractive index is sensitive to the temperature. According to the theoretical derivation of Equation (18), the thermo-optic coefficient dn/dT of crystalline Sb_2Te_3 materials can be obtained from the slope of the fitting curve and the thermal-physical parameters listed in Table 1, and $dn/dT = 1.25 \times 10^{-3}/\mathrm{K}$ is gotten by the laser pulse width tunable z-scan measurements.

In order to further verify the accuracy of the laser pulse width tunable z-scan measurements, the variable−temperature ellipsometric measurements at the $632.8\ nm$ wavelength were carried. In the measurements, the ellipsometry (Sopra, France) can heat the sample from room temperature to 200°C (473 K), and in the process of heating Ar gas was applied to protect the sample from oxidation during the measurements.

Figure 4 presents the dependence of the refractive index on the temperature. One can find that the refractive index increases linearly with the temperature, and the slope of this curve is about $0.62 \times 10^{-3}/K$, that is, the thermo-optical coefficient obtained by variable temperature ellipsometric measurement is $dn/dT = 0.62 \times 10^{-3}/K$, which is approximately consistent with the laser pulse width tunable z-scan measurements in the order of the magnitude.

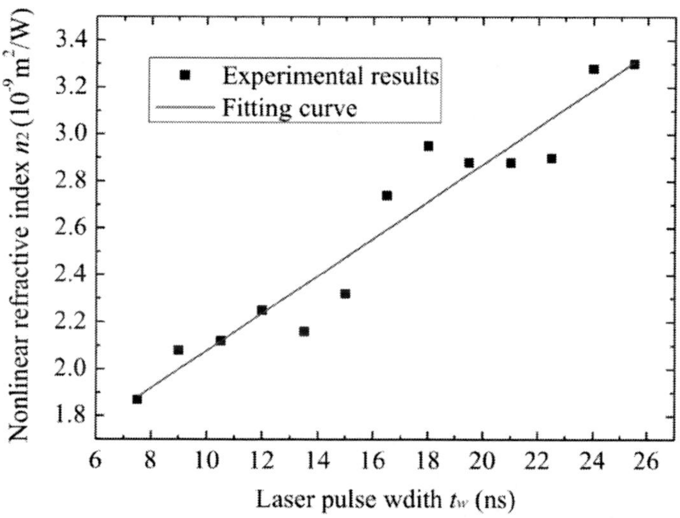

Figure 3. Dependence of nonlinear refraction coefficient n_2 of Sb_2Te_3 versus the pulse width t_w as measured from the z-scan measurements (laser wavelength of 633 nm). The circles are the experimental results and the solid line is the linear fitting results.

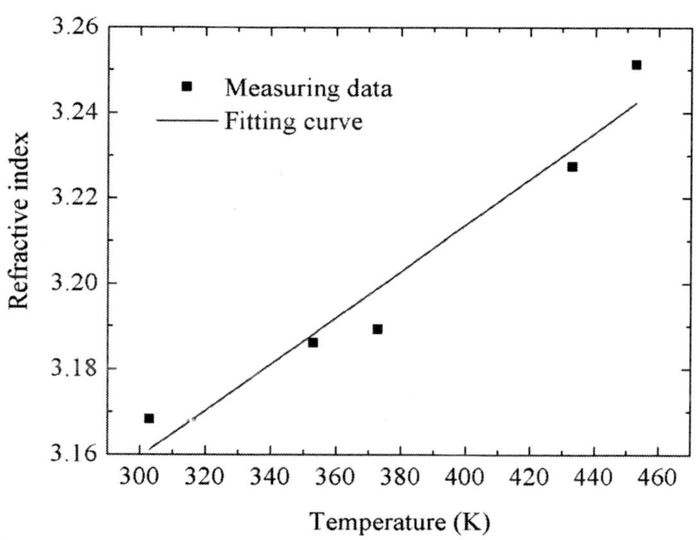

Figure 4. Dependence of the refractive index of Sb_2Te_3 thin films on temperature at 632.8 nm light wavelength from the ellipsometric measurement.

The small difference may be because the laser pulse is not perfectly rectangular pulse, which has an ill-influence on the measurement accuracy. The other reasons are from the thermal-physical parameters. The thermal-physical parameters listed in Table 1 are usually changed with the temperature, however, in our calculation, and the thermal-physical parameters are independent on temperature.

Table 1. Thermo-physical parameters of Sb_2Te_3 used in the numerical calculation. The linear absorption coefficient $\alpha_0$is obtained from our experiment, while the density $\rho$and heat capacity$c$are taken from thereferences [19, 20]

Materials	α_0 at 405 nm	α_0 at 633 nm	ρ (10^3Kg/m^3)	c (J/Kg · K)
Sb_2Te_3	7.78×10^7/m	3.65×10^7/m	6.50	205.49

In summary, in order to get the thermo-optical coefficient dn/dT of thin film materials, the laser pulse width tunable z-scan method was proposed. In this method, the continuous laser beam was modulated into pulse light by the combination of acousto-optic modulator (or high frequency TTL modulator) and signal generator. The laser pulse width can be changed from nanosecond to continuous waves. The Sb_2Te_3 crystalline thin films are taken as an example, and the thermo-optical coefficient dn/dT measured by the laser pulse width tunable z-scan method is about $1.25 \times 10^{-4}/K$, which is basically in agreement with the variable-temperature ellipsometric measurement result $0.62 \times 10^{-3}/K$. The theoretical derivation and experimental results indicate that the laser pulse width tunable z-scan method is very effective and simple for determining the thermo-optical coefficient of thin film materials.

4.2. Measurement of Dependence of Absorption Coefficient on Temperature [21]

The thermo-optical coefficient of $d\alpha/dT$ is also measured by the laser pulse width tunable z-scan method where the high frequency TTL modulated GaN semiconductor laser device with a wavelength of 405 nm was used. Figure 5 shows typical open-aperture (OA) z-scan results with a 50 ns rectangular laser pulse width, where the numerical aperture of focusing lens is 0.09.

The curves show that the thin films exhibit saturable absorption characteristic, indicating that the transmitted light intensity increases as the sample moves close to the focal plane. Moreover, the observed saturable absorption is highly repeatable, and no change in surface morphology of the sample is observed after the irradiation. Therefore, the nonlinearity of Sb_2Te_3 thin films can be safely attributed to their inherent mechanism, and the effect of morphological change is excluded. Data fitting was performed to quantitatively estimate the nonlinearity in Figure 5(a). The nonlinear absorption coefficients β of Sb_2Te_3 thin films are determined as $-1.36 \times 10^{-2}\ m/W$. The z-scan measurement was also performed through changing the laser pulse width from 50 to 120 ns to get the effect of pulse width on $\beta$and the fitting values for $\beta$are plotted in Figure 5(b).

One can see that β shows a quick decreasing trend with pulse width. Based on Equation (27), the thermal effect is more prominent as pulse width becomes longer due to the large linear absorption coefficient and poor thermal conductivity. According to Equation (27) and Table 1, the $d\alpha/dT$ can be calculated to be $\sim -2 \times 10^4/(m \cdot K)$.

Variable temperature spectroscopic ellipsometry measurements were performed in the $300\ K$ to $480K$ temperature range under vacuum to further confirm the thermal origin of β. The model used to fit the experimental ellipsometry is composed of three contributions: (1) a Cauchy term, which describes the polarization in the high energy range; (2) a Drude term, which accounts for the contribution from free carriers; (3) a Lorentz oscillator term to depict the onset of optical transitions.

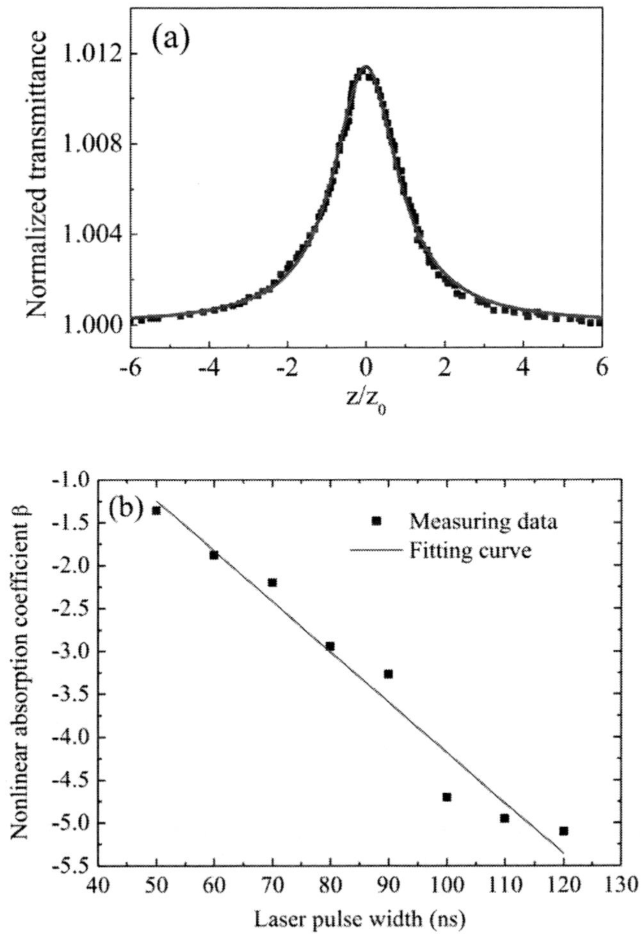

Figure 5. The experimental results of optical nonlinear absorption coefficients at $405\ nm$ laser wavelength, (a) OA z-scan results at laser pulse width of $50\ ns$, (b) dependence of β on laser pulse width.

Figure 6 shows the plots of the temperature dependence of the absorption coefficient at $405nm$ light wavelength. The absorption coefficients decrease within creasing temperature. Using the linear fitting, the $d\alpha/dT$ can be determined as $-0.737 \times 10^4/(m \cdot K)$. The measured $d\alpha/dT$ is agreement with the laser pulse width tunable z-scan method in the order of the magnitude. The small difference in the coefficient may be because the laser pulse is not perfectly rectangular pulse, which has an ill-influence on the measurement accuracy. The other reasons are from the thermal-physical parameters. The thermal-physical parameters listed in Table 1 are usually changed with the temperature, however, in our calculation, and the thermal-physical parameters are independent on temperature.

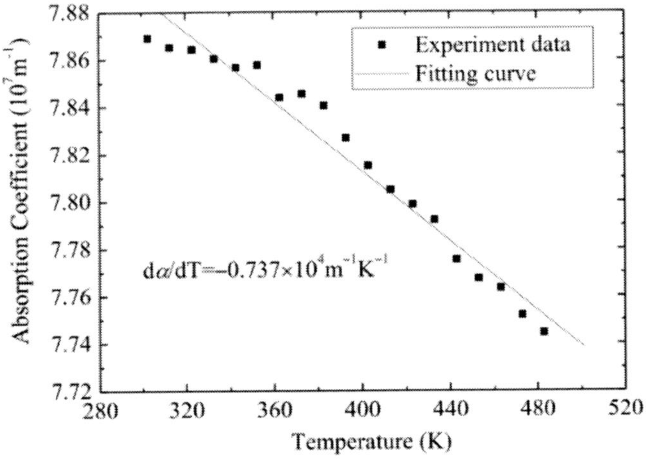

Figure 6. Dependence of absorption coefficient on temperature.

Conclusion

The thermo-optical coefficients dn/dT and $d\alpha/dT$ are very important for the interaction of laser pulse and materials. In this chapter, we presented the dependences of dn/dT and $d\alpha/dT$ on the laser pulse width, and gave the experimental setup of laser pulse width tunable z-scan method to measure the dn/dT and $d\alpha/dT$.

The data analysis and processing of the z-scan method were also introduced by considering the nonlinear absorption and nonlinear refraction. The laser pulse width tunable z-scan experiments were carried out to measure the dn/dT and $d\alpha/dT$ of Sb_2Te_3 thin films, the experimental results were verified by variable-temperature spectroscopic ellipsometry measurements.

The laser pulse width tunable z-scan method is simple and practical for characterizing the thermo-optical coefficients dn/dT and $d\alpha/dT$.

Acknowledgments

The work is partially supported by National Natural Science Foundation of China (Nos. 60977004, 51172253, and 61137002), the Instrument Developing Project of the Chinese Academy of Sciences (Grant No. YZ201140), and Science and Technology Commission of Shanghai Municipality (11JC1412700, and 11JC1413300).

REFERENCES

[1] Wei, J; Liu, S; Geng, Y; Wang, Y; Li, X; Wu, Y; Dun, A. "Nano-optical information storage induced by the nonlinear saturable absorption effect", *Nanoscale*, 2011, 3, 3233-3237.

[2] Wei, J. "Super-resolution optical effects of nanoscale nonlinear thin film structure and ultrahigh density information storage", in *Lecture Notes in Nanoscale Science and Technology—Toward Functional Nanomaterials*, Wang, ZM. (Ed.), (Springer Publisher 2009).

[3] Muknopadhyay, PK; Nautiyal, A; Gupta, PK; Ranganathan, K; George, J; Sharma, SK; Nathan, TPS. "Experimental determination of the thermo-optic coefficient (*dn /dT*) and the effective stimulated emission cross-section (σ_e) of an a-axis cut 1.-at.% doped Nd:GdVO4 crystal at 1.06 μm wavelength", *Appl. Phys. B*, 2003, 77, 81-87.

[4] Zhao, GY; Ishikawa,H; Yu, G; Egawa, T; Watanabe, J; Soga, T; Jimbo, T; Umeno, M. "Thermo-optical nonlinearity of GaN grown by metalorganic chemical-vapor deposition", *Appl. Phys. Lett.*, 1998, 73, 22-24.

[5] Nicola, D; Finizio, A; Pierattini, G; Carbonara, G. "Interferometric method for concurrent measurement of the thermo-optic coefficients of quartz retarders", *Pure Appl. Opt.*, 1994, 3, 209-213.

[6] Pilla, V; Munin, E; Gesualdi, MRR. "Measurement of the thermo-optic coefficient in liquids by laser-induced conical diffraction and thermal lens techniques", J. *Opt. A: Pure Appl. Opt.*, 2009, 11, 105201.

[7] Wei, J; Gan, F. "Time-resolved thermal lens effect of Sb thin films induced by structural transformation near melting temperature", *Opt. Commun.*, 2003, 219, 261-269.

[8] Falconieri, M; Salvetti, G. "Simultaneous measurement of pure-optical and thermo-optical nonlinearities induced by high-repetition-rate, femtosecond laser pulses: application to CS_2", *Appl. Phys. B*, 1999, 69, 133-136.

[9] Sinha, S; Sasikumar, S; Ray, AK; Dasgupta, K. "Z-scan measurement of optical nonlinearity in solid-state dye doped media", *Appl. Phys. B*, 2007, 87, 145-149.

[10] Tian, J; Zhang, C; Zhang, G; Li, W. "Position dispersion and optical limiting resulting from thermally induced nonlinearities in Chinese tea liquid", *Appl. Opt.*, 1993, 32, 6628-6632.

[11] Jia, ZH; Li, Qu; Chen, YX; Chen, YL. "Linear absorption induced thermal effect in z-scan measurements", A*cta Optica Sinica*, 1996, 16(5): 635～639 (in Chinese).

[12] Jia, ZH; Zhou, J; Fu, GJ; Li, Q. "Study on the thermal effects in nonlinear medium by z-scan technique", *Acta Photonica Sinica*, 1997, 26, 56-60 (in Chinese).

[13] Liu, S. *Thermally-induced nonlinear absorption of phase change materials stimulated by nanosecond laser pulse*, A Doctor's Dissertation of University of Chinese Academy of Sciences, 2012.

[14] Liu, J. *Optical nonlinearity in the super resolution structures*, A Master's Dissertation of Graduate University of Chinese Academy of Sciences, 2010.

[15] Sheik-Bahae, M; Said, AA; Wei, TH; Hagan, DJ; Van Stryland, EW. "Sensitive measurement of optical nonlinearities using a single beam", *IEEE J. Quantum Electron.*, 1990, 26, 760-769.

[16] Sheik-Bahae, M; Said, AA; Wei, T; Van Stryland, EW. "High-sensitivity, single-beam n_2 measurements", *Opt. Lett.*, 1989, 14, 955-957.

[17] Tian, JG; Zang, WP; Zhang, CZ; Zhang, GY. "Analysis of beam propagation in thick nonlinear media", *Appl. Opt.*, 1995, 34, 4331-4336.

[18] Zang, WP; Tian, JG; Liu, ZB; Zhou, WY; Song, F; Zhang, CP. "Analytic solutions to z-scan characteristics ofthick media with nonlinearrefraction and nonlinear absorption", *J. Opt. Soc. Am. B*, 2004, 21, 63-66.

[19] Anderson, TL; Krause, HB. "Refinement of the Sb_2Te_3 and Sb_2Te_2Se structures and their relationship to nonstoichiometric $Sb_2Te_{3-y}Se_y$ compounds", *Acta Cryst.*, 1974, B30, 1307-1310.

[20] Jette, ER; Foote, F. "Precision Determination of Lattice Constants", *J. Chem. Phys.*, 1935, 3, 605-616.

[21] Liu, S; Wei' J; Gan, F. "Nonlinear absorption of Sb-based phase change materials due to the weakening of the resonant bond" *Appl. Phys. Lett.*, 2012, 100,111903.

In: Nonlinear Performance and Characterization Methods … ISBN: 978-1-62808-093-3
Editor: Jingsong Wei

Chapter 7

Z-SCAN TECHNIQUE CONSIDERING THE FRESNEL REFLECTION LOSS

Jingsong Wei*
Shanghai Institute of Optics and Fine Mechanics,
Chinese Academy of Sciences, Shanghai, China

ABSTRACT

The Fresnel reflection loss is usually neglected in the conventional z-scan measurement. However, the Fresnel reflection loss is not a constant, but is a function of the sample position z, which can lead to the modification of the incident electric field within the sample. Thus the Fresnel reflection loss needs to be considered for more accurate characterization of the nonlinear properties including nonlinear refraction and absorption, especially for nanofilms and strong absorption samples. In this chapter, we firstly analyze the production of Fresnel reflection loss, and then the theoretical model of characterization of nonlinear refraction and absorption for the "thin" sample was derived by considering the sample surface (or interface) reflection, and the experimental results were given. For the very "thin" sample, called as nanofilm sample, the Fresnel reflection loss is mainly from the internal multi-interference effect, the theoretical calculation formula were given, and the numerical simulations for different cases were depicted.

Keywords: Z-scan, Fresnel reflection, nonlinear refraction, nonlinear absorption

INTRODUCTION

With the development of nanotechnology, the optical nonlinear materials have been extensively applied in optical information storage, optic-electronic devices, optical limiting devices, and so on. The optical nonlinearity is an intrinsic characteristic of the materials. The nonlinearity originates from lots of mechanisms, such as electronic polarization, magnetostriction, thermal effect, surface plasmon resonance, band filling, excitonic screening,

* E-mail: weijingsong@siom.ac.cn.

excited states absorption, multi-photon absorption, thermal effect, and so on. The accurate measurement for these effects induced the coefficients of nonlinear refraction and nonlinear absorption is important to the characterization of optical properties of materials. Since the optical nonlinearity was discovered, many methods and techniques have been developed for characterizing and measuring nonlinear refraction and nonlinear absorption. The measurement techniques include degenerate four-wave mixing, beam distortion measurements, nonlinear interferometry, ellipse rotation, etc. Among them, the single beam z-scan technique proposed by Sheik-Bahae [1, 2] has been widely used due to its simplicity and high sensitivity. In the z-scan method, a Gaussian intensity laser beam irradiates the sample and induced the nonlinear refraction and nonlinear absorption. The refraction and absorption are proportional to the laser intensity, thus presents Gaussian profile. The nonlinear refraction will cause the wavefront phase to be changed when the laser beam passes through the sample. That is, the sample within the laser beam irradiation area functions as a phase aperture. The aperture can change the laser beam intensity distribution in the far-field. By the sample being scanned along the optical axis (marked as z-direction) in the focal region of the single focused laser beam, the intensity through the nonlinear sample is recorded in far-field. The intensity changes as a function of sample position. Since the z-scan method was proposed, a variety of improvements have also been proposed in recent years, such as the time-resolved z-scan method [3, 4] and the reflection z-scan technique [5]. As for the incident beams, a number of different beam profiles have been so far tested, such as the circularly symmetric beam [6], the elliptic Gaussian [7], the Gaussian–Bessel [8], the flat-topped [9], the top-hat [10], and the polarization beams [11, 12], etc. In addition, the sensitivity and accuracy of the measurements are also improved with various data processing and fitting methods [13–17]. In this chapter, we mainly introduce the basic theoretical model of the z-scan technique considering the Fresnel reflection loss.

1. The Fresnel Reflection Loss in the Z-scan Measurement

Let us look back to a basic z-scan technique, a collimated laser beam with a Gaussian intensity profile is focused through a lens, and the sample is scanned along the z-axis (optical axis) through the focus. The nonlinear refractive index is obtained by measuring the transmittance of a nonlinear sample through close-aperture ($S < 1$, where S is defined as the ratio of the aperture size to spot size at the aperture plane position,) in the far field as a function of the sample position z with respect to the focus ($z = 0$). The sensitivity to nonlinear refraction is determined by the S value and the removal of the aperture completely eliminates the effect. However, in this case, the z-scan will be sensitive to nonlinear absorption. Nonlinear absorption coefficients can be extracted from open-aperture ($S = 1$) measurement, where the z-scan traces are symmetric with respect to the focus ($z = 0$).

In the z-scan experiment, we use a thin slab of material in which the thickness is smaller than the Rayleigh length of a Gaussian beam scanned near the beam's focus. However, in the previous study, when optical nonlinear refraction and absorption coefficients are small enough we can neglected the Fresnel reflection losses. However, the Fresnel reflection loss induced by the nonlinear refraction and absorption is significant and cannot be neglected in the z-scan measurement for some samples, such as the nanofilms or strong absorption materials. That is, if there is significant nonlinear absorption and refraction, the effective incident light intensity I_{eff} within the sample at different z-scan position is not a constant, but is a function of sample position z, and the incident light intensity within the sample should be expressed as $I_{eff}(z)$. For a typical z-scan configuration, a collimated Gaussian laser beam with electric field intensity E is used as the light source. The light beam is split into two parts by a beam splitter, one is focused to detector 1 to monitor the beam power stability, and the other is focused by lens 2 to function as nonlinear z-scan measurement light beam, as is shown in Figure 1. The sample is z-scanned along the optical axis near the focus point. The light beam through the sample is fully collected by detector 2 for nonlinear absorption measurement, and nonlinear refraction is measured by placing an aperture at the front of detector 2.

However, there is large surface reflection for certain kinds of samples, and the Fresnel surface reflection loss is different at different position z at the z-scan measurement due to the laser beam induced nonlinear absorption and nonlinear refraction within the sample. For example, when the sample z-scans to position z_1 the incident laser beam intensity is I_1. The laser beam radius and the electric field intensity are marked as r_1 and E_1, respectively. The refractive index $n_1 = n_0 + \gamma I_1$, and $\alpha_1 = \alpha_0 + \beta I_1$ accordingly. n_0 and α_0 are the linear refractive index and absorption coefficients, respectively. γ and β are the nonlinear refraction coefficient and nonlinear absorption coefficient, respectively. The Fresnel reflection loss is marked as R_1. However, when the sample z-scans to position z_2 the incident laser intensity becomes I_2 because the laser beam radius is r_2, and $r_1 \neq r_2$ leads to $I_1 \neq I_2$. At position z_2, the laser beam induced refractive index and absorption coefficient are $n_2 = n_0 + \gamma I_2$, and $\alpha_2 = \alpha_0 + \beta I_2$, respectively. The Fresnel reflection loss R_2 at position z_2 is different with the R_1 at position z_1 that is, $R_2 \neq R_1$ because $n_2 \neq n_1$ and $\alpha_2 \neq \alpha_1$. Therefore, the Fresnel reflection loss at the z-scan measurement cannot be ignored because the reflection loss is not a constant, but is a function of position z, that is $R(z)$.

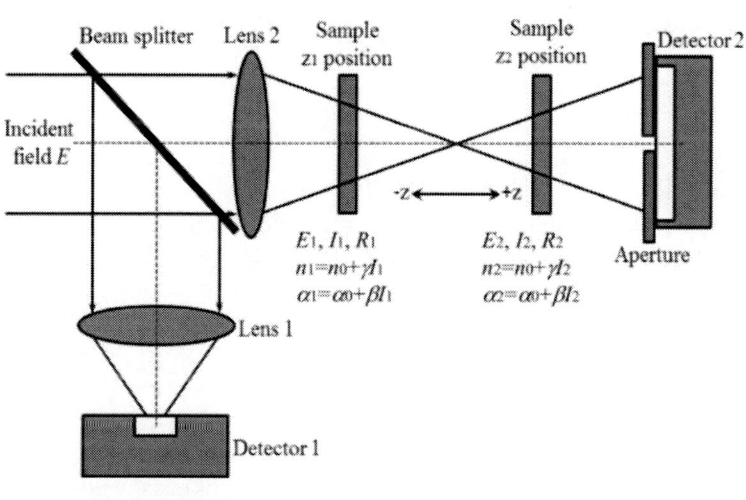

Figure 1. The conventional configuration of z-scan setup (E_1–electric field, I_1–intensity, R_1–reflection, n_1–refractive index, α_1–absorption coefficient at position z_1; E_2–electric field, I_2–intensity, R_2–reflection, n_2–refractive index, α_2–absorption coefficient at position z_2).

2. Thin Samples

2.1. The Theoretical Model for Characterization of Nonlinear Refraction and Absorption[18]

Let us analyze the Fresnel reflection in the z-scan measurement in detail. When the sample moves to position z the effective laser intensity inside the sample should be a function of z and is set to be $I_{eff}(z)$. Both the induced refractive index n and absorption coefficient α are also the functions of position z, and can be expressed as

$$n(z) = n_0 + \gamma I_{eff}(z) \tag{1}$$

$$\alpha(z) = \alpha_0 + \beta I_{eff}(z) \tag{2}$$

In z-scan technique, assuming an incident TEM_{00} Gaussian laser beam with a beam waist radius w_0 traveling to the $+z$ direction, the coordinate zero is at the beam waist (w_0), and the incident plane of nonlinear sample is at position z. The incident complex electric field at position z can be approximately written as

$$E_i(r,z) = E_0 \frac{w_0}{w(z)} \exp\left\{ -\frac{r^2}{w^2(z)} - \frac{ikr^2}{2R_0(z)} - ikz \right\} \tag{3}$$

where E_0 and w_0 are the on-axis electric amplitude at beam waist position and beam waist radius, respectively. $R_0(z) = z[1 + (z_0/z)^2]$ is the radius of wavefront at position z. $z_0 = \pi w_0^2/\lambda$ is the diffraction length of beam and $k = 2\pi/\lambda$ wave number. λ is laser wavelength. The radius of Gaussian beam as a function of coordinate z can be formulated as

$$w(z) = w_0 \sqrt{1 + (z/z_0)^2} \tag{4}$$

If the sample thickness is small enough that changes in the beam diameter within the sample due to either diffraction or nonlinear refraction can be neglected, the sample is considered to be "thin" sample, in which case the self-refraction process is referred to as "external self-action". For linear diffraction neglection, the sample thickness L needs to meet the following condition,

$$L << z_0, \tag{5}$$

While for nonlinear refraction neglection, the sample thickness needs to meet the another condition as follows,

$$L << z_0/\Delta\phi(0), \tag{6}$$

where $\Delta\phi(0)$ is nonlinear refraction induced phase variation at axial direction. In z-scan technique, the Fresnel reflection loss can be considered to be from two respects. One is induced by the internal multi-interference effect for the thin sample with a thickness of $L < 4.65/\alpha$, which we call the "very thin" samples such as the nanofilms. This case will be analyzed in the next section. The other is from the surface (or interface) reflection, we call the samples as "thin sample". The internal multi-interference effect is negligible if the sample thickness is more than $L > 4.65/\alpha$ based on the textbook of Born and Wolf [19]. For example, for the film sample with an absorption coefficient $\alpha = 4 \times 10^7 m^{-1}$, the film thickness must meet the condition $L \geq 120\ nm$ if the internal multi-interference effect needs to be ignored. The reflection loss from sample surface at position z can be calculated as

$$R(z) = \frac{[n(z)-1]^2 + [a(z)/2k]^2}{[n(z)+1]^2 + [a(z)/2k]^2} \tag{7}$$

At position z the effective laser intensity $I_{eff}(z)$ is key to characterize the induced $n(z)$ and $\alpha(z)$. For the incident Gaussian laser beam, the intensity at the sample surface at position z can be written as

$$I_i(r,z) = E_i E_i^* = I_0 \frac{w_0^2}{w^2(z)} \cdot \exp(-\frac{2r^2}{w^2(z)}) \tag{8}$$

where E_i^* is the conjugate of E_i, and I_0 is the intensity at beam waist. The effective laser beam intensity $I_{eff}(z,r) = [1-R(z)] \times I_i(z,r)$ inside the sample at position z can be calculated as

$$I_{eff}(z,r) = [1-R(z,r)] \times \left\{ I_0 \frac{w_0^2}{w^2(z)} \cdot \exp(-\frac{2r^2}{w^2(z)}) \right\} \tag{9}$$

Based on the Equations (7)-(9), one can see that it is difficult to extract the $I_{eff}(z)$ as an explicit function, to solve the $R(z)$ at position z, we can calculate approximately $R(z)$ using $I_{eff-}(z)$ instead of the $I_{eff}(z)$, where $I_{eff-}(z)$ is the effective intensity inside the sample placed at the position z_{-}, and z_{-} defined as the previous position close to position z. Thus,

$$I_{eff}(z,r) \simeq [1 - R_{-}(z_{-},r)] \times \left\{ I_0 \frac{w_0^2}{w^2(z)} \cdot \exp(-\frac{2r^2}{w^2(z)}) \right\} \tag{10}$$

with

$$R_{-}(z_{-},r) = \frac{[n_0 + \gamma I_{eff-}(z,r) - 1]^2 + \left[\frac{\alpha_0 + \beta I_{eff-}(z,r)}{2k} \right]^2}{[n_0 + \gamma I_{eff-}(z,r) + 1]^2 + \left[\frac{\alpha_0 + \beta I_{eff-}(z,r)}{2k} \right]^2} \tag{11}$$

For the thin samples with a thickness $z_0 >> L > 4.65/\alpha$, and the slowly varying envelope approximation (SVEA) can be adopted. We mainly study the transmitted light intensity, so all other phase changes are neglected, and the effective laser intensity inside the sample $I_{eff}(z)$ will decays exponentially along the optical axis z-direction as

$$\frac{dI_{eff}(z,r)}{dz'} = -\alpha I_{eff}(z,r) \tag{12}$$

where z' is the propagation depth in the sample. The intensity at the exit surface of the sample can be calculated as

$$I_e(z,r) = \frac{I_{eff}(z,r)\exp(-\alpha_0 L)}{1+\beta I_{eff}(z,r)} \tag{13}$$

In order to get the nonlinear absorption coefficient β, we can spatially integrate Equation (13) to calculate the total transmitted light intensity (i.e. open-aperture z-scan $S = 1$). So the diffraction in the free-space need not be made. The normalized open aperture mode transmitted z-scan curve is

$$T(z, S = 1) = \int_0^{+\infty} \frac{I_{eff}(z,r)exp(-\alpha_0 L)}{1+\beta L_{eff} I_{eff}(z,r)} .rdr/(\tfrac{1}{4} I_0 w_0^2) \tag{14}$$

The normalized open-aperture mode reflected z-scan curve is

$$R(z, S = 1) = \int_0^{\infty} R(z) I_i(z,r) .rdr/(\tfrac{1}{4} I_0 w_0^2) \tag{15}$$

The z-scan measurement process and data analyzing are as follows:

1) The open-aperture reflected z-scan measurement is performed to obtain the $R(z)$ curve of experimentally. The measured $R(z)$ curve is theoretically determined by both the nonlinear refraction coefficient γ and nonlinear absorption coefficient β.
2) According to the measured $R(z)$ curve, the effective laser intensity I_{eff} ($= [1- R(z)]I_i$) inside the sample is calculated by Equation (9).
3) The open-aperture transmitted z-scan measurement is then carried out to get $T(z)$ curve experimentally.
4) The measured $T(z)$ is fitted by Equation (14) and I_{eff}, the nonlinear absorption coefficient β can be firstly obtained.
5) The measured open-aperture reflected z-scan curve is fitted by the Equation (15) and given β value; the nonlinear refraction coefficient γ is finally obtained.

2.2. Experimental Measuring System

In order to confirm the theoretical model, an experimental measuring system with open-aperture mode reflected and transmitted z-scan measurement was established. Figure 2 describes the schematic. In this measuring system, a laser beam with a wavelength of 405 nm was used as the incident light source, and the incident light is divided into two parts. One small beam is focused into detector 1 to monitoring the power stability. The other beam is converged to the sample by the lens 2. The sample is placed at the stepper motor to scan along the optical axis z-direction. The light passes through the sample to induce the nonlinear refraction and absorption, which is called as transmitted z-scam measurement. The transmitted light is divided into two parts by a beam splitter, one is totally collected as the open-aperture mode transmitted z-scan measurement, and the other is centrally collected as the close-aperture mode transmitted z-scan measurement (in our data fitting, the close-aperture mode transmitted z-scan measurement was not used). The reflected light from the sample surface is also collected to be used as open-aperture mode reflected z-scan measurement. Finally, all collected signal is put into computer to conduct the data fitting process.

2.3. The Experimental Results and Data Fitting

We take the In doped Sb thin film as an example, where the In doped Sb thin film was used as active layer in the nonlinear optical information storage. The experimental results of both open-aperture transmitted and reflected z-scan measurement are shown in Figure 3, where a 30 nm thickness In doped Sb deposited on glass substrate was examined.

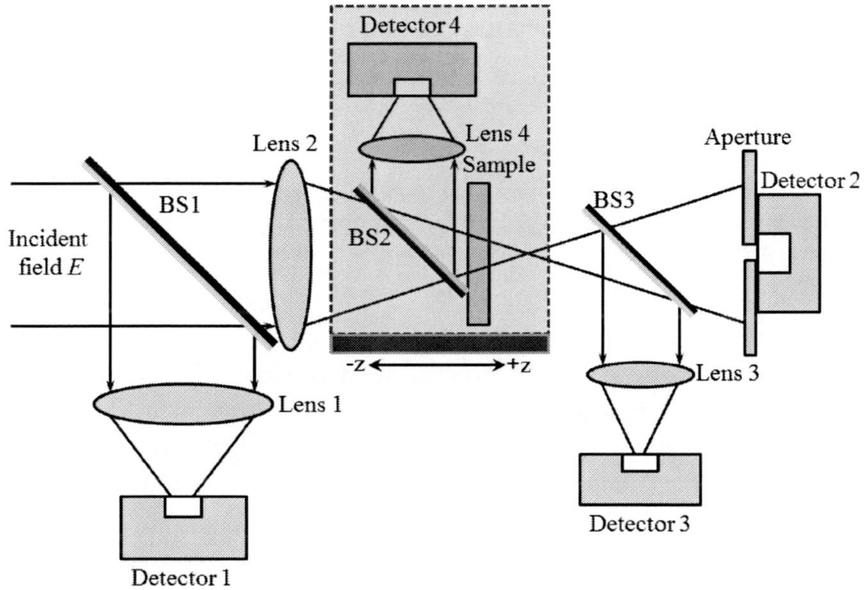

Figure 2. Experimental measuring system. The reflected light is collected by detector 4. (To avoid the reflected spot size changes drastically with the z-scan movement, the reflected light collection modular unit, which includes BS2, Lens 4, and detector 4, synchronously conducts z-scan movement with the sample, where BS is beam splitter).

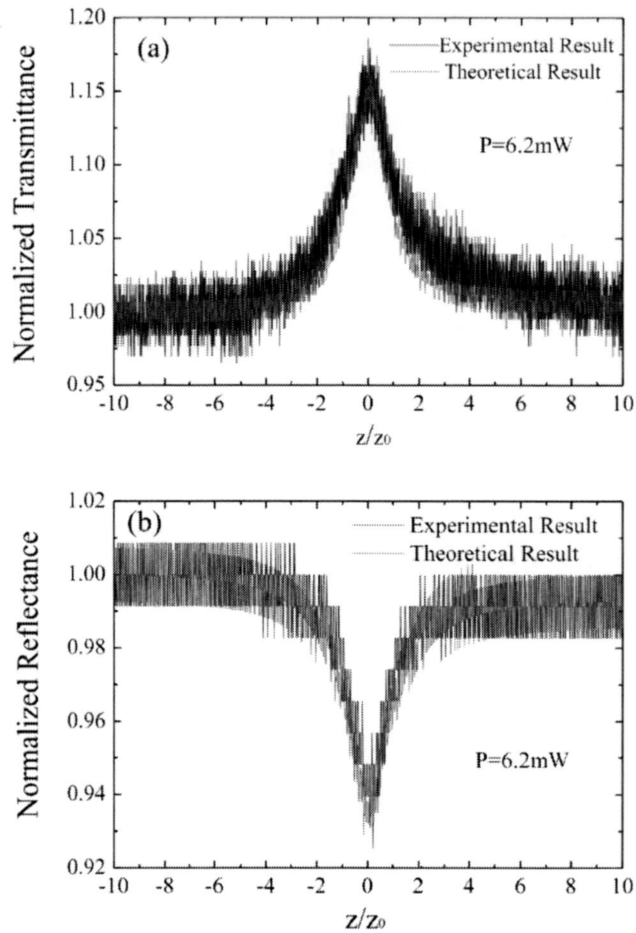

Figure 3. The experimental results of nonlinear z-scan measurement (a) Normalized open aperture mode transmitted z-scan curves. $\beta = -5.3 \times 10^{-2} m/W$ is obtained by theoretical fitting. (b) Normalized open aperture mode reflected z-scan curves. $\gamma = -1.0 \times 10^{-9}\, m^2/W$ is obtained by theoretical fitting.

Figure 3(a) depicts the open aperture mode transmitted z-scan data at the total irradiation power of 6.2 mW, and Figure 3(b) gives the open aperture mode reflected z-scan date. Firstly, $R(z)$ is obtained from the Figure 3(b); secondly, the effective laser intensity inside the sample is calculated according to $R(z)$ and Equation (9). Thirdly, the Figure 3(a) is fitted with the red curve based on $I_e \quad (z, r)$ and Equation (13), and the nonlinear absorption coefficient $\beta = -5.3 \times 10^{-2} m/W$. Finally, the Figure 3(b) is fitted with the red curve based on the given β value and Equation (15), and the nonlinear refraction coefficient $\gamma = -1.0 \times 10^{-9}\, m^2/W$. One can find from Figure 3 that the fitting curves are very consistent with the measurement curves, which indicates that the theoretical model and experimental measuring system are reasonable and feasible in the real application.

In this subsection, we gave an open aperture mode reflected and transmitted z-scan technique for measuring both the nonlinear refractive index and nonlinear absorption coefficient of thin films with strong absorption. We took the In doped Sb thin film as an example, and the fitting curves were very consistent with the measured data, which indicated that the theoretical model and experimental measuring system were reasonable and feasible in the real application.

3. NANOFILM SAMPLES

3.1. Theoretical Model [20]

In section 2 we analyzed and gave the z-scan measurement for the "thin" sample with a thickness $L \geq 4.65/\alpha$, where the internal multi-interference effect is negligible, and the Fresnel reflection loss is mainly from the sample surface and interface. However, for the very thin sample with a thickness $L < 4.65/\alpha$ (called as nanofilm), the Fresnel reflection loss is determined by the internal multi-interference effect. The exiting electric field should be written as

$$E_e(z,r) = E_i(z,r)\tilde{t}(z,r)\exp[-i(\Delta\phi(z,r))] \tag{16}$$

where $\tilde{t}(z,r)$ is the complex transmission coefficient of the incident electric field due to internal multi-interference, and $\Delta\phi(z,r)$ is the phase variation of the lens-like phase aperture induced by the nonlinear refraction. $\Delta\phi(z,r)$ can be expressed as

$$\Delta\phi(z,r) = \frac{4\gamma P_0}{\lambda\alpha w^2(z)}(1-e^{-\alpha L})\exp[-\frac{2r^2}{w^2(z)}] \tag{17}$$

with P_0 denoting the laser irradiation power. In general, for z-scan samples, the thin films are directly deposited on a substrate. Therefore a z-scan sample can be modeled as a two-interface layered system. The transmission for a two-interface layered system in normal incidence is

$$\tilde{t}(z,r) = \frac{t_{12}(z,r)t_{23}(z,r)\exp[i\frac{2\pi}{\lambda}\tilde{n}(z,r)L]}{1+r_{12}(z,r)r_{23}(z,r)\exp[i\frac{4\pi}{\lambda}\tilde{n}(z,r)L]} \tag{18}$$

where

$$\begin{aligned} r_{12}(z,r) &= \frac{n_1-\tilde{n}(z,r)}{n_1+\tilde{n}(z,r)} \quad & r_{23}(z,r) &= \frac{\tilde{n}(z,r)-n_3}{\tilde{n}(z,r)+n_3} \\ t_{12}(z,r) &= \frac{2n_1}{n_1+\tilde{n}(z,r)} \quad & t_{23}(z,r) &= \frac{2\tilde{n}(z,r)}{\tilde{n}(z,r)+n_3} \end{aligned} \tag{19}$$

with

$$\tilde{n}(z,r) = n(z,r) + iK(z,r) \tag{20}$$

$$K(z,r) = \frac{\lambda}{4\pi}\alpha(z,r) \tag{21}$$

where n_1 and n_2 are marked as the refractive index of air medium and substrate, respectively.

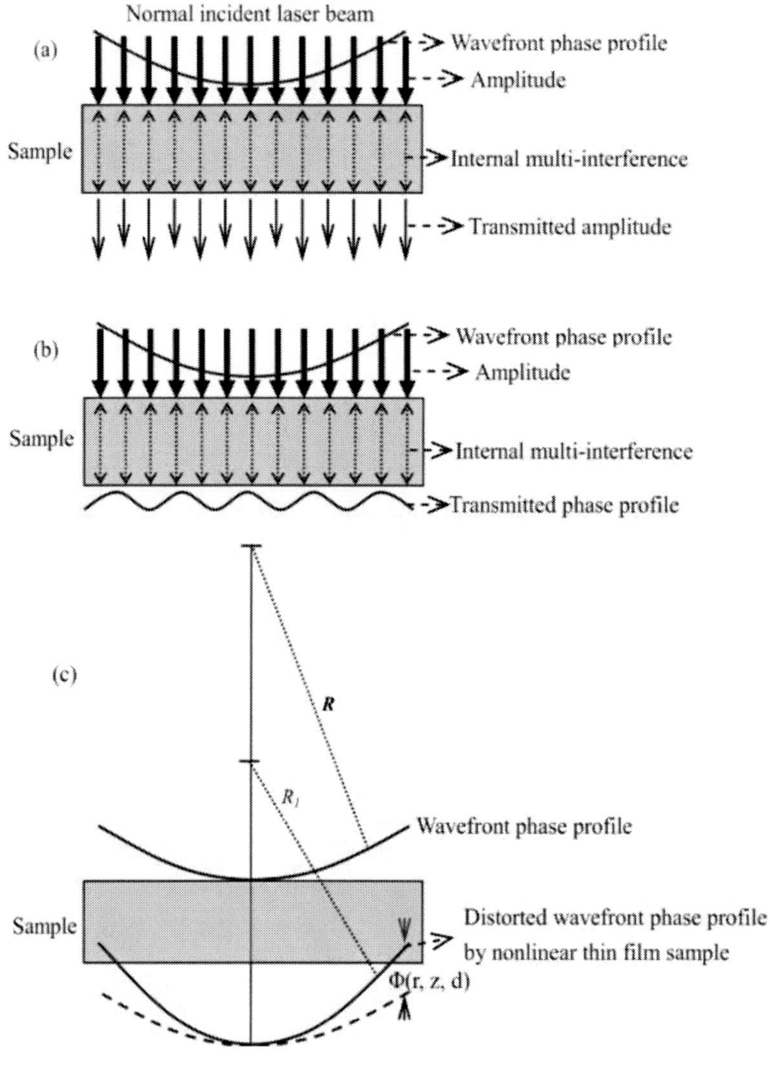

Figure 4. Schematic showing that the electric field is modified by the nonlinear nanofilm samples in (a) electric field amplitude, (b) phase of complex transmission due to internal multi-interference, and (c) phase shift of the lens-like phase aperture.

The inclusion of the complex refractive index, the local nonlinear absorption and the multi-interference in Equation (16) enables us to calculate the exiting field distribution rigorously for nanofilm samples. One can see from Equation (16) that the complex transmission (including amplitude and phase) occurs in the exiting electric field. The schematic of the exiting electric field is depicted in Figure 4, which was previously used in [21, 22] to illustrate changes both in the amplitude and in the phase of the exiting electric field. Figures 4(a) and 4(b) illustrate how the internal multi-interference causes both intensity and phase variations in the exiting electric field, and consequently brings about a distortion in the exiting electric field. Figure 4(c) shows the phase shift of the lens-like phase aperture that is induced by the Gaussian-like intensity profile of the incident light beam, which is also considered by general z-scan theories. In this case, the local refractive index as well as the local nonlinear absorption coefficient becomes a function of the light intensity. Since the intensity varies with the sample position, the nonlinear refraction and absorption causes a variant transmittance.

Finally, the transmitted light is received by a detection system at a distance L_1 away from the sample. At each point of the entrance pupil, the field is a superposition of the irradiated field from each point of the exiting field distribution on the sample surface. Assuming a circular detector pupil, one calculates the field at a point ρ from the center on the detection plane with the Fresnel–Kirchhoff integral

$$E(z,\rho) = -\frac{i}{\lambda}\int_0^{\infty}\int_0^{2\pi}\left(\frac{1}{2}+\frac{L_1}{2D}\right)E_e(z,r)\frac{\exp(i\frac{2\pi}{\lambda}D)}{D}rdrd\varphi \tag{22}$$

with

$$D = \sqrt{L_1^2 + (r\cos\varphi - \rho)^2 + r^2\sin^2\varphi} \tag{23}$$

For a circular detector of radius R_a, the total received light power is

$$P = 2\pi\int_0^{R_a}\left|E(\rho)\right|^2\rho d\rho \tag{24}$$

3.2. Numerical Simulation

3.2.1. The Numerical Testing for Strong Nonlinear Refraction

In the following, we perform some numerical tests with the formalism outlined above. We have repeatedly mentioned that taking into account the multi-interference and the nonlinear absorption would bring about significant changes in the transmitted light distribution. Our primary task is therefore to prove our assertions. In the first simulation we carried out, we considered a relatively large nonlinear refractive coefficient $\gamma = 5\times 10^{-10}\ m^2/W$ based on the experimental results in the literature [23], the parameters for the simulation were chosen as follows: $P = 500\ mW$, $w_0 = 5\ \mu m$, $\lambda = 532\ nm$, $\alpha_0 = 4\times 10^7\ m^{-1}$, $\beta = 1\times 10^{-7}\ m/W$, $n_{20} = 2$, $L_1 = 0.5\ m$, $n_1 = 1$, $n_3 = 1.5$ and $R_a = 0.2\ mm$. The numerical results of the simulation are presented in Figures 5 and 6.

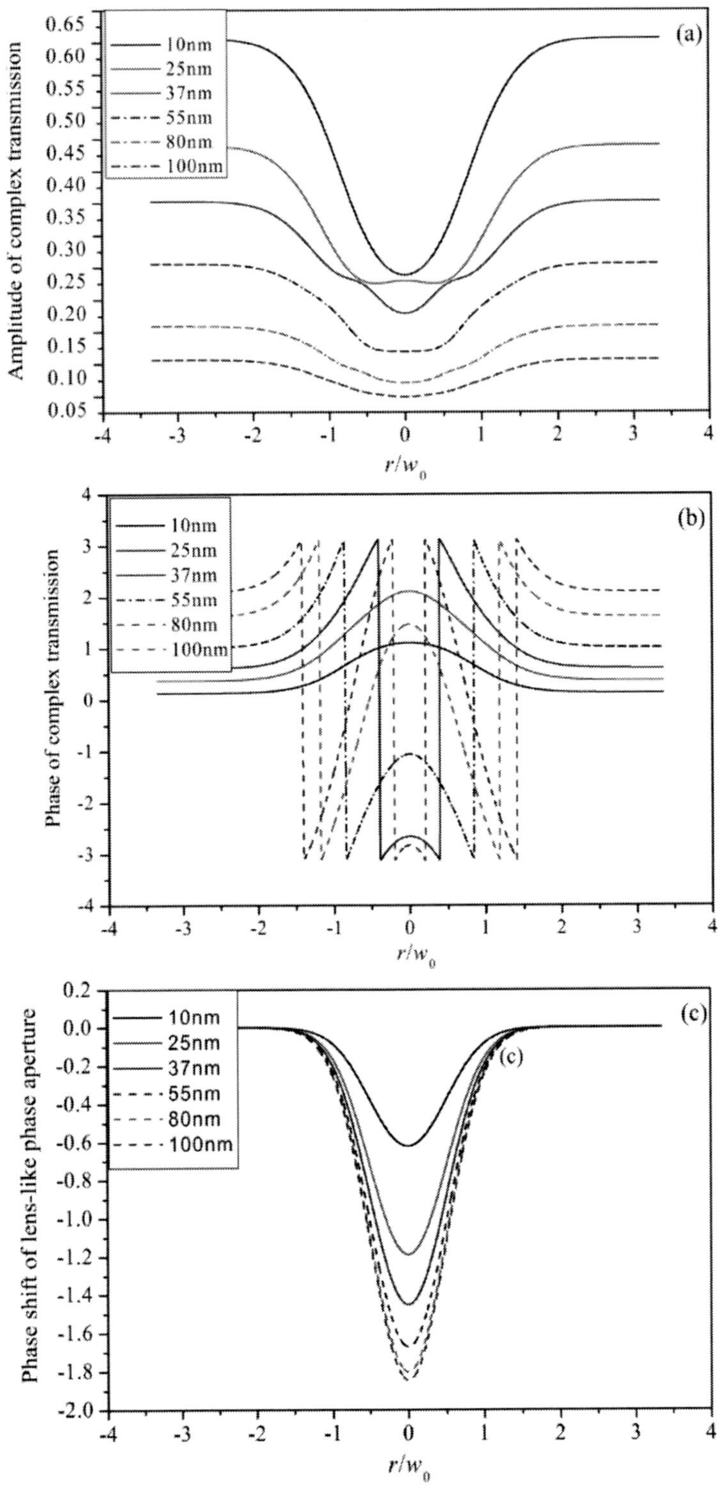

Figure 5. Complex transmission and lens-like phase shift in radial direction for film thickness from 10 to 100 nm: (a) electric field amplitude, (b) phase of complex transmission due to internal multi-interference, and (c) phase shift of the lens-like phase aperture.

Figure 5 presents three groups of curves with respect to the focal point at the smallest beam size w_0. In Figure 5(a), the amplitude of the transmission $\tilde{t}(0, r)$ is computed for a set of film thicknesses. One takes note of the following from the curves in Figure 5(a). The transmission is not a constant for thin films and it changes with the radial coordinate. The relative variation becomes increasingly significant when the thickness decreases. In addition there is a notable shape modification. In Figures 5(b) and (c) phase shifts are presented. Figure 5(c) contains the phase shift of the lens-like phase aperture, which is usually considered by conventional z-scan theories, whereas Figure 5(b) contains the extra phase changes due to the internal multi-interference. One notices that the curves in Figure 5(c) are consistent with previous results in the literature, whereas the curves in Figure 5(b) are dramatically modified by the strong extra phase changes induced by the internal multi-interference. In light of the numerical tests in Figure 5, one expects that including the effects of multi-interference and nonlinear absorption will change the exiting electric field significantly for thin films.

Note that the phase shift contribution curve in Figure 6(a) actually represents the fitting curve in previous z-scan calculations in which the transmission is usually assumed to be constant. Therefore, the curves in Figure 6(a) provide a direct comparison between the present model calculation and the conventional z-scan calculations. One realizes that the fitting curve from the model calculation deviates significantly from the curve of conventional z-scan theory, both in magnitude and in shape, and the reason for the deviation stems from the inclusion of the internal multi-interference and nonlinear absorption into the complex transmission. In order to have a closer look at the fitting curves and to make an analysis on the stability of the numerical process, we amplified the fitting curves in Figure 6(a) by a factor of 10, i.e. we reduced the discrete step in the numerical process from $dz = 0.05$ to $dz = 0.005$. The amplified curves are presented in Figure 6(b). One can see that the numerical process appears to be stable near the focal area.

Therefore, provided enough discrete points are taken in the area, the numerical integrations in Equations (22) and (24) are well convergent and stable. Furthermore, within the complex transmission curve (the dash–dotted line in Figure 6(a)) there are contributions from the amplitude variations and the phase changes.

In order to further identify these two contributions, we carried out additional numerical calculations for the same fitting curve. From the curves in Figure 6(c), it is clear that both terms contribute significantly to the fitting curve.

In light of the numerical results in Figure 6, one concludes that including the effects of multi-interference and nonlinear absorption does indeed change the exiting electric field significantly for thin films.

In passing, we make note of the origins of the small oscillations associated with the curves in Figure 6. We have found that the oscillations stem from three sources in the formalism: (1) the $exp[-i\pi r^2/\lambda R_0(z)]$ term in the incident Gaussian wave in Equation (3); (2) the phase aperture $\Delta\phi(z, r)$ in Equations (16) and (17); (3) the axial dependence of the beam size $w(z) = w_0\sqrt{1 + z^2/z_0^2}$. In other words, the oscillations are inherited from the incident Gaussian beam due to diffraction.

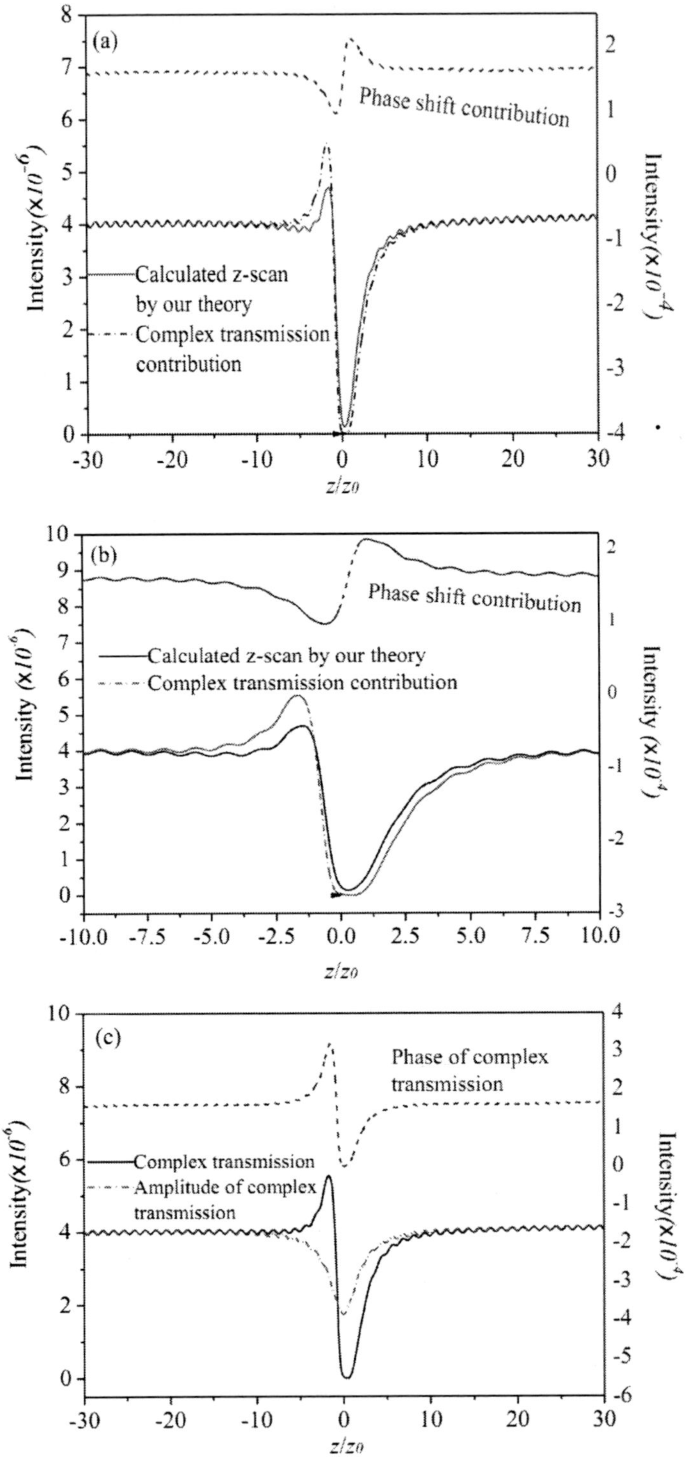

Figure 6. Numerical results from the first simulation for $L = 80\ nm$: (a) the fitting curve as well as the contribution from the complex transmission and the lens-like phase shift (conventional z-scan) contribution, (b) 10 times amplified view of the fitting curve in (a), (c) the complex transmission curve as well as the contributions from the amplitude and phase of the complex transmission.

The complete z-scan fitting curve was then obtained, and the results are presented in Figure 6(a) with a comparison to the calculations without the two effects included, i.e. results from the conventional z-scan theory. In Figure 6(a), the solid line represents the fitting curve obtained with the present model calculation. The two principal contributions to the fitting curve are from the phase shift (dotted line) and the complex transmission (dash–dotted line). In order to have a better view and comparison, the phase shift contribution curve is plotted with the intensity scale $\times 10^{-4}$, i.e. the right-hand side y-axis in Figure 6(a), while the complex transmission contribution is plotted with the intensity scale $\times 10^{-6}$, i.e. the left-hand side y-axis in Figure 6(a). In the following figures, similar arrangement for the two different intensity scales will be adopted.

3.2.2. The Numerical Testing for Weak Nonlinear Refraction

In the second simulation we carried out, we considered another set of parameters closely related to the experimental results in [24, 25]. The parameters for simulation were chosen as follows: $\gamma = 2.5 \times 10^{-15}\, m^2/W$, $P = 400\, mW$, $w_0 = 5\, \mu m$, $\lambda = 527\, nm$, $\alpha_0 = 10^3\, m^{-1}$, $\beta = 0.25 \times 10^{-7}\, m/W$, $n_{20} = 1.35$, $L_1 = 0.5\, m$, $n_1 = 1$, $n_3 = 1.48$ and $R_a = 0.2\, mm$. One notices especially that the nonlinear refractive coefficient is much weaker than that in the first simulation, and the linear refractive index is almost the same for the film and for the substrate. Note also that in the present simulation the thickness of the film is much smaller than in [24, 25], which is intended to show the importance of the additional effects when the films are in nanometric scales. The numerical results of the second simulation are presented in Figures 7 and 8, which are similarly arranged as in the first simulation in Figures 5 and 6.

In Figure 7, three groups of curves are presented. In Figure 7(a), the amplitude of the transmission $t(r, 0, d)$ is computed for a set of film thicknesses.

Figure 7(c) contains the phase shift of the lens-like phase aperture, which is usually considered by conventional z-scan theories, whereas Figure 7(b) contains the extra phase shift due to the internal multi-interference. Since the nonlinear refractive coefficient is reduced five orders of magnitude from the previous simulation, the changes in Figure 7 are reduced accordingly. Although similar conclusions can be made from Figure 7 with careful observations, the detailed difference between the present model calculation and the conventional z-scan calculations can be seen more clearly from the complete z-scan fitting curves as presented in Figure 8.

The complete z-scan fitting curve was then obtained, and the results are presented in Figure 8(a) with a comparison to the calculations without the two effects included, i.e. results from the conventional z-scan theory. In Figure 8(a), the solid line represents the fitting curve obtained with the present model calculation.

The two principal contributions to the fitting curve are from the phase shift (dotted line) and the complex transmission (dash–dotted line). In order to have a better view and comparison, the phase shift contribution curve is plotted with the intensity scale marked on right side y-axis in Figure 8(a), while the complex transmission contribution is plotted with the intensity scale marked on the left side y-axis in Figure 8(a).

One realizes from the curves in Figure 8(a) that the fitting curve (solid line) from the present calculation appears significantly different from the curve (represented by the phase shift contribution curve) calculated with the conventional z-scan theory, both in magnitude and in shape.

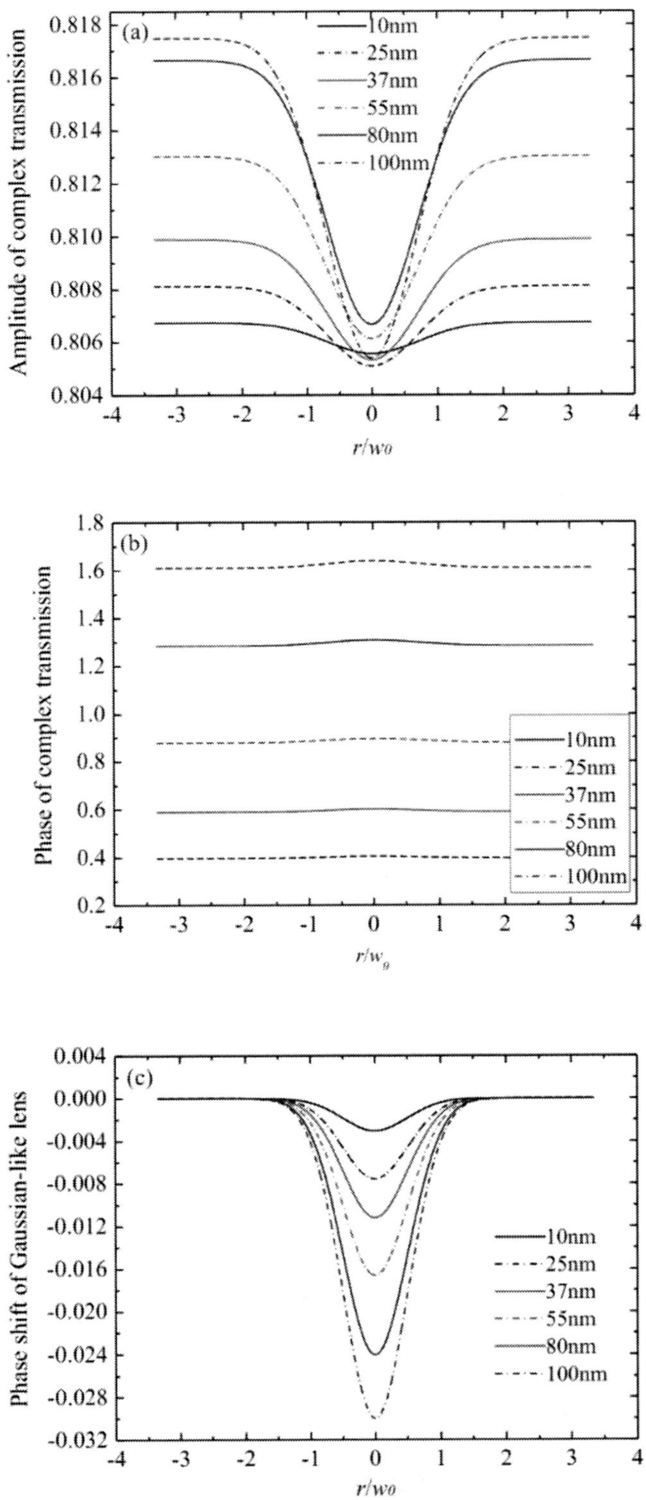

Figure 7. Complex transmission and phase shift in the radial direction for film thickness from 10 to 100 nm: (a) electric field amplitude, (b) phase of complex transmission due to internal multi-interference, and (c) phase shift of the lens-like phase aperture.

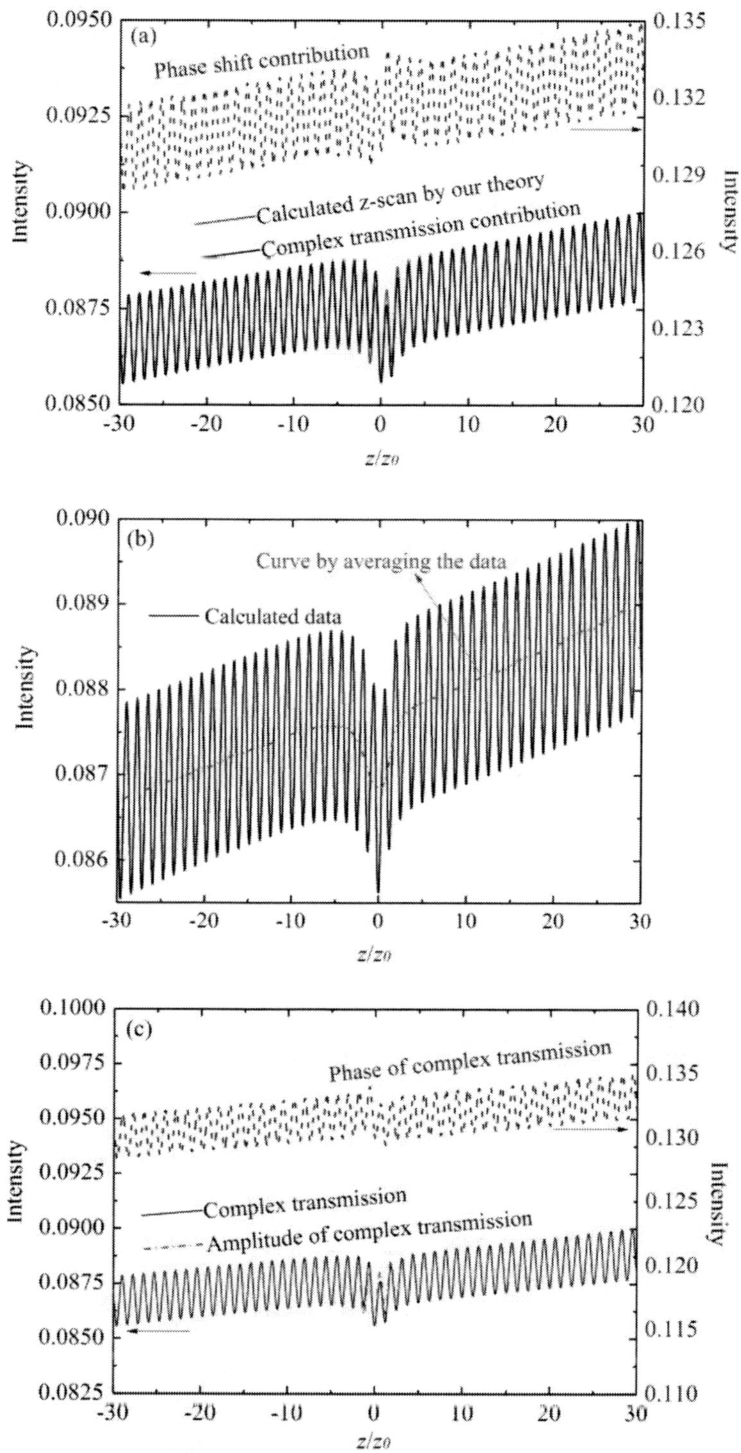

Figure 8. Numerical results from the second simulation for $L = 80\ nm$: (a) the fitting curve as well as the contributions from the complex transmission and the lens-like phase shift (conventional z-scan) contribution, (b) the averaged fitting curve, (c) the complex transmission curve as well as the contributions from the amplitude and phase of the complex transmission.

It is worth mentioning that the fitting curve becomes notably different from the curves in the literature [24, 25] for two main reasons: the film thickness used in the present simulation is less than 100 nm, much smaller than the thickness adapted in [24, 25], and the present simulation includes the internal multi-interference and the nonlinear absorption in the complex transmission.

In addition, the contribution from the complex transmission (dash–dotted line) looks mostly coincident with the fitting curve (solid line). Since the overall variation is reduced, the oscillation stemming from the incident Gaussian beam becomes relatively large. The strong oscillations are also directly related to the use of a film thickness in the nanometric scale. For relatively thick films, the oscillations should be reduced. In order to have a better view on the changes near the focal area, we have taken an average of the data, and the numerical results are presented in Figure 8(b), which is the fitting curve to be adapted for the z-scan measurement.

Furthermore, for the complex transmission curve (the dash–dotted line in Figure 8(a)), there are contributions from the amplitude variations and the phase changes. In order to further identify these two contributions, we carried out additional numerical calculations for the same fitting curve. From the curves in Figure 8(c), it is clear that both terms contribute significantly to the fitting curve.

In general, similar conclusions can be made based on the numerical results from the second simulation. A notable difference is that the variations are relatively weak and the oscillations due to the diffraction of the incident laser beam may be too strong for the fitting curve to be seen clearly. In this case, it is suggested that a data average would remove the oscillations and the fitting curve could be more clearly seen, which becomes more necessary especially for cases of weak nonlinear refractive coefficient.

3.2.3. Comparison of Strong and Weak Nonlinear Absorptions

Finally, the cases of strong and weak nonlinear absorption coefficients are compared for the complex transmission in Figure 9 and for the z-scan curves in Figure 10, respectively. In the calculations, the common parameters were chosen as $P = 500\,mW$, $w_0 = 5\,\mu m$, $\lambda = 532\,nm$, $a_0 = 4 \times 10^7\,m^{-1}$, $n_{20} = 2$, $L_1 = 0.5\,m$, $n_1 = 1$, $n_3 = 1.5$, $R_a = 0.2\,mm$. Three different nonlinear absorption coefficients were considered, as indicated below.

In Figure 9, the complex transmission (including the amplitude and phase of complex transmission) as well as the Gaussian-lens phase shift is plotted for three different nonlinear absorption coefficients: $\beta = 1 \times 10^{-3}\,m/W$, $\beta = 1 \times 10^{-7}\,m/W$, $\beta = 0$. One sees clearly that while there is no difference between $\beta = 1 \times 10^{-7}\,m/W$ and $\beta = 0$ (weak and zero nonlinear absorption coefficient), a significant difference occurs when $\beta = 1 \times 10^{-3}\,m/W$ (strong nonlinear absorption coefficient), particularly for the amplitude of complex transmission and Gaussian-like lens phase shift.

In order to further analyze the influence of the complex transmission (internal multi-interference effect) on the z-scan curve, the fitting z-scan curve comparisons between $\beta = 1 \times 10^{-3}\,m/W$ and $\beta = 0$ are presented in Figure 10, where the fitting z-scan curves by our theory and the contributions from complex transmission and Gaussian-lens phase shift are given in Figures 10(a)–10(c), respectively.

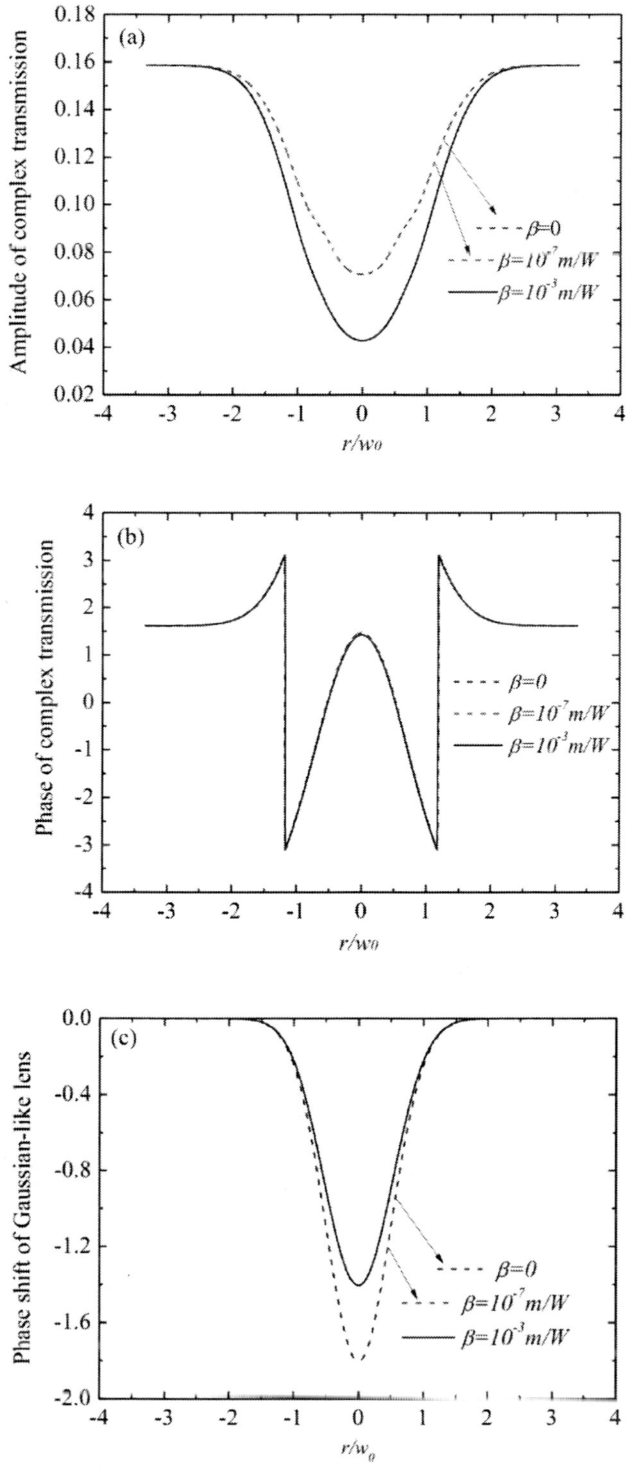

Figure 9. The comparison of complex transmission and phase shift in the radial direction for different nonlinear absorption coefficients ($\beta = 1 \times 10^{-3}\ m/W$, $\beta = 1 \times 10^{-7}\ m/W$, $\beta = 0$); the film thickness is fixed at $L = 80\ nm$: (a) the amplitude of the complex transmission, (b) the phase of the complex transmission, and (c) the phase shift of the Gaussian-like phase aperture.

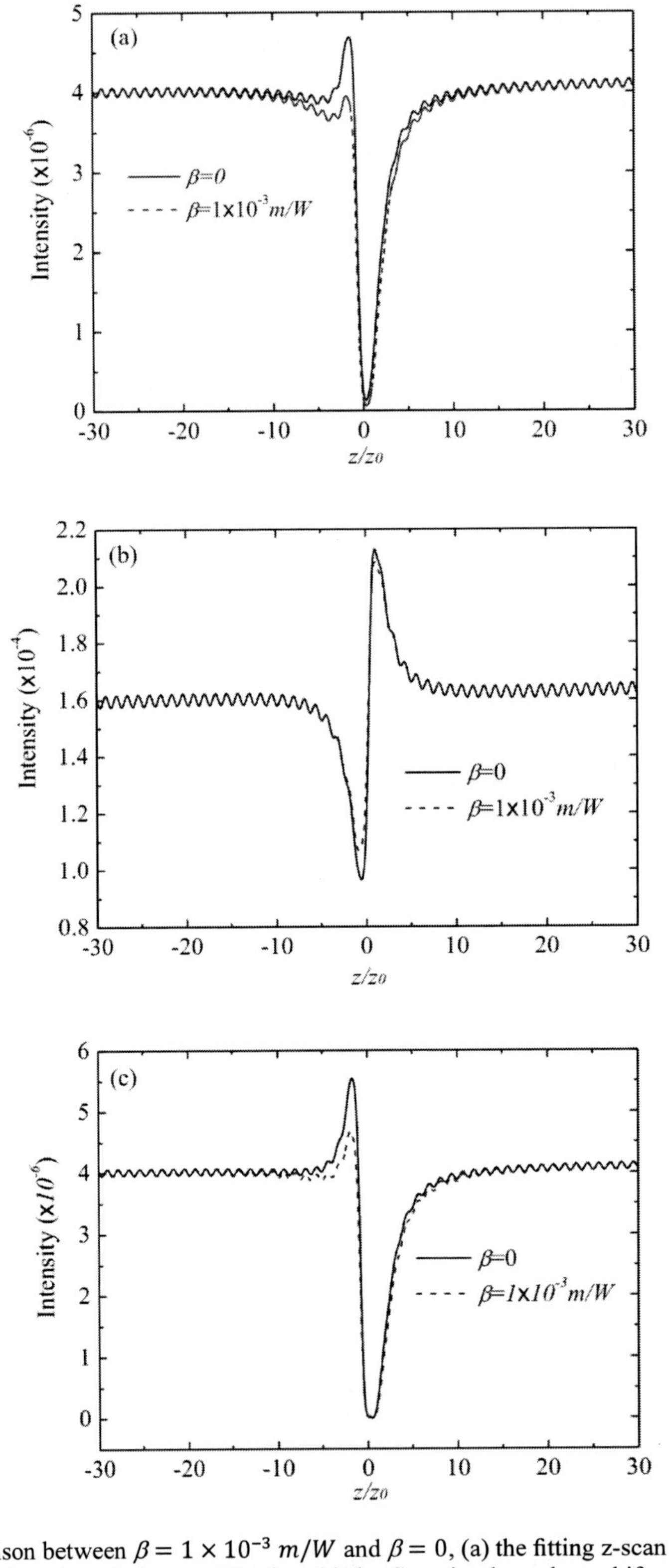

Figure 10. Comparison between $\beta = 1 \times 10^{-3}\ m/W$ and $\beta = 0$, (a) the fitting z-scan curves by our theory, (b) the complex transmission contribution, (c) the Gaussian-lens phase shift contribution.

One can find from the comparisons that there is an obvious difference between $\beta = 1 \times 10^{-3}\ m/W$ and $\beta = 0$, which indicates that, when the nonlinear absorption effect is strong, the z-scan curve is significantly modified.

In this subsection, a rigorous z-scan model calculation for nanofilm samples has been proposed. The theory takes into account both the losses due to internal multi-interference and the nonlinear absorption inside the thin film. Conventional z-scan theories ignore these two effects, and thus are restricted to relatively thick films.

Especially, for films thinner than about $100\ nm$, the two effects become dominant, and thus cannot be ignored. The proposed model calculation has been numerically tested both for strong and weak nonlinear refractive coefficients.

In the simulations, parameters were chosen based on experimental results in the literature. In both cases significant deviations are demonstrated from the conventional z-scan calculations. Finally, the stability of the numerical process has been analyzed, and it is shown that the numerical integrations near the focal area are stable as far as sufficient discrete steps are taken in the numerical computation.

CONCLUSION

The Fresnel reflection loss has an important influence on the characterization of the optical nonlinear refraction and absorption in the z-scan technique. We analyzed the production of Fresnel reflection loss, and then the theoretical model of characterization of nonlinear refraction and absorption for the "thin" sample was derived by considering the sample surface (or interface) reflection, and the experimental results were also given. For the very "thin" sample, called as nanofilm sample, the Fresnel reflection loss is mainly from the internal multi-interference effect, the theoretical calculation formula were given, and the numerical simulations for different cases were depicted.

ACKNOWLEDGMENTS

The work is partially supported by National Natural Science Foundation of China (Nos. 60977004, 51172253, and 61137002), the Instrument Developing Project of the Chinese Academy of Sciences (Grant No. YZ201140), and Science and Technology Commission of Shanghai Municipality (11JC1412700, and 11JC1413300).

REFERENCES

[1] Sheik-Bahae, M; Said, AA; Wei, TH; Hagan, DJ; Van Stryland, EW. "Sensitive measurement of optical nonlinearities using a single beam", *IEEE J. Quantum Electron.*, 1990, 26, 760–769.

[2] Sheik-Bahae, M; Said, AA; Van Stryland, EW. "High sensitivity single beam n_2 measurement", *Opt. Lett.*, 1989, 14, 955–958.

[3] Wang, J; Sheik-Bahae, M; Said, AA; Hagan, DJ; Van Stryland, EW. "Time-resolved z-scan measurements of optical nonlinearities", *J. Opt. Soc. Am. B*, 1994, 11, 1009-1017.

[4] Kawazoe, T; Kawaguchi, H; Inoue, J; Haba, O; Ueda, M. "Measurement of nonlinear refractive index by time-resolved z-scan technique", *Opt. Commun.*, 1999, 160, 125-129.

[5] Petrov, DV. "Reflection z-scan technique for the study of nonlinear refraction and absorption of a single interface and thin film", *J. Opt. Soc. Am. B*, 1996, 13, 1491-1498.

[6] Rhee, BK; Byun, JS; VanStryland, EW. "Z-scan using circularly symmetric beams", *J. Opt. Soc. Am. B*, 1996, 13, 2720-2723.

[7] Tsigaridas, G; Fakis, M; Polyzos, I; Persephonis P; Giannetas, V. "Z-scan technique for elliptic Gaussian beams", *Appl. Phys. B*, 2003, 77, 71-75.

[8] Hughes, S; Burzler, JM. "Theory of z-scan measurements using Gaussian-Bessel beams", *Phys. Rev. A*, 1997, 56, R1103-R1106.

[9] Wang, J; Gu, B; Xu, YM; Wang, HT. "Enhanced sensitivity of z-scan technique by use of flat-topped beam", *Appl. Phys. B*, 2009, 95, 773-778.

[10] Zhao, W; Palffymuhoray, P. "Z-scan technique using top-hat beams", *Appl. Phys. Lett.*, 1993, 63, 1613-1615.

[11] Yan, XQ; Liu, ZB; Zhang, XL; Zhou, WY; Tian, JG. "Polarization dependence of z-scan measurement: theory and experiment", *Opt. Express*, 2009, 17, 6397-6406.

[12] Liu, ZB; Yan, XQ; Tian, JG; Zhou, WY; Zang, WP "Nonlinear ellipse rotation modified Z-scan measurements of third-order nonlinear susceptibility tensor", *Opt. Express*, 2007, 15, 13351-13359.

[13] Chapple, PB; Staromlynska, J; McDuff, RG. "Z-scan studies in the thin and thick sample limits", *J. Opt. Soc. Am. B*, 1994, 11, 975–982.

[14] Samad, RE; Vieira, ND. "Analytical description of z-scan on-axis intensity based on the Huygens–Fresnel principle", *J. Opt. Soc. Am. B*, 1998, 15, 2742–2747.

[15] Yao, B; Ren, L; Hou, X. "Z-scan theory based on a diffraction model", *J. Opt. Soc. Am. B*, 2003, 20, 1290–1294.

[16] Tsigaridas, G; Polyzos, I; Persephonis, P; Giannetas, V. "A novel approach for analyzing open z-scan experiments", *Opt. Commun.*, 2006, 266, 284-289.

[17] Mincu, N. "Improvements of the z-scan method used in the study of optical nonlinearities", *J. Optoelectron. Adv. Mater.*, 1999, 1, 25-33.

[18] Wang, R; Wei, J. "Accurate determination of optical nonlinear absorption coefficients by the combination of reflection and transmission open-aperture z-scan measurements", *Opt. Commun.*, 2013, 297, 121-124.

[19] Born, M; Wolf, E. *Principle of Optics* (Cambridge University Press, 1999).

[20] Wei, J; Xiao, M. "A z-scan model for optical nonlinear nanometric films", *J. Opt. A: Pure Appl. Opt.*, 2008, 10, 115102.

[21] Wei, J; Xiao, M. "Super-resolution scanning laser microscopy with a third-order optical nonlinear thin film", *Appl. Phys. B*, 2008, 91 337–341.

[22] Wei, J; Xiao, M. "Laser tunable Toraldo super-resolution with a uniform nonlinear pupil filter", *Appl. Opt.*, 2008, 47, 3689–3693.

[23] Choi, Y; Park, I; Kim, MR; Jhe, W. "Direct observations of self-focusing near the diffraction limit in polycrystalline silicon film", *Appl. Phys. Lett.*, 2001, 78 856–858.

[24] Cattaruzza, E; Battaglin, G; Gonella, F; Polloni, R; Scremin, BF; Mattei, G; Mazzoldi, P; Sada, C. "Au–Cu nanoparticle in silica glass as composite material for photonic application", *Appl. Surf. Sci.*, 2007, 254, 1017–1021.

[25] Haglund Jr, RF; Mogul, HC; Weeks, RA; Zuhr, RA. "Changes in the refractive index of fused silica due to implantation of transition-metal ions", *J. Non-Cryst. Solids*, 1991, 130, 326–331.

In: Nonlinear Performance and Characterization Methods ... ISBN: 978-1-62808-093-3
Editor: Jingsong Wei

Chapter 8

TRAPPED MODES AND SYMMETRY BREAKING OF PHOTONIC AND ATOMIC WAVES IN ROTATING DOUBLE-WELL POTENTIALS

Yongyao Li[1,2], Wei Pang[3] and Boris A. Malomed[1*]

[1]Department of Physical Electronics, School of Electrical Engineering, Faculty of Engineering, Tel Aviv University, Tel Aviv, Israel

[2]Department of Applied Physics, South China Agricultural University, Guangzhou, China

[3]Department of Experiment Teaching, Guangdong University of Technology, Guangzhou, China

ABSTRACT

We study modes trapped in a rotating ring carrying the self-focusing (SF) or self-defocusing (SDF) cubic nonlinearity and a double-well potential $\cos^2\theta$, where θ is the angular coordinate. The model, based on the nonlinear Schrödinger (NLS) equation in the rotating reference frame, describes the light propagation in a twisted pipe waveguide, as well as in other optical settings, and also a Bose-Einstein condensate (BEC) trapped in a torus and dragged by the rotating potential. In the SF and SDF regimes, five and four trapped modes of different symmetries are found, respectively. The shapes and stability of the modes, and the transitions between them are studied in the first rotational Brillouin zone. In the SF regime, two symmetry-breaking transitions are found, of subcritical and supercritical types. In the SDF regime, an antisymmetry-breaking transition occurs. Ground states are identified in both the SF and SDF systems.

1. INTRODUCTION

The concept of the spontaneous symmetry breaking (SSB) in nonlinear systems was introduced in Ref. [1]. Its significance has been later recognized in various physical settings,

* Corresponding author e-mail: malomed@post.tau.ac.il.

including numerous ones originating in nonlinear optics [2-13], Bose-Einstein condensates (BECs) [14-33], and degenerate fermionic gases [34]. A general analysis of the SSB phenomenology was developed too [35-39], which is closely related to the theory of bifurcations in nonlinear systems [40].

Fundamental manifestations of the SSB occur in nonlinear systems based on symmetric double-well potentials DWPs) or dual-core configurations. A paradigmatic example of the latter in nonlinear optics is the twin-core nonlinear fiber, which may serve as a basis for the power-controlled optical switching [2-8]. DWP settings in optics were analyzed theoretically and implemented experimentally in photorefractive crystals [13]. In the realm of matter waves, main effects predicted in DWPs are Josephson oscillations [20-22], the asymmetric self-trapping of localized modes [23-24], and similar effects in binary mixtures [25-28]. Both the Josephson and self-trapping regimes were implemented in the atomic condensate with contact repulsive interactions [29-30]. The SSB was also analyzed in one- and two-dimensional (1D and 2D) models of BEC trapped in dual-core configurations [31-33].

Another dynamical setting which has produced a number of interesting effects in media with the intrinsic nonlinearity, especially in BEC, is provided by rotating potentials. It is well known that stirring the self-repulsive condensate typically leads to the formation of vortex lattices [42], although it was found experimentally [43] and demonstrated theoretically [44-51] that giant vortices, rather than lattices, may also be formed under special conditions (when the centrifugal force nearly compensates the trapping harmonic-oscillator potential). On the other hand, the rotation of self-attractive condensates gives rise to several varieties of stable localized modes, such as vortices, "crescents" (mixed-vorticity states), and the so-called center-of-mass modes (quasi-solitons) [52-60]. Further development of this topic was achieved by the consideration of rotating lattice potentials, which can be implemented experimentally as an optical lattice induced in BEC by a broad laser beam transmitted through a revolving sieve [61-62], or using twisted photonic-crystal fibers in optics [63]. In these systems, quantum BEC states and vortex lattices have been studied [64-67], as well as solitons and solitary vortices depinning from the lattice when its rotation velocity exceeds a critical value [68-69].

A specific implementation of the latter settings is provided by the quasi-1D lattice, or a single quasi-1D potential well, revolving about its center in the 2D geometry [70]. In particular, the rotation makes it possible to create fundamental and vortical soliton in the self-repulsive medium, where, obviously, nonrotating quasi-1D potentials cannot maintain bright solitons [71]. Furthermore, the rotation of a DWP gives rise to azimuthal Bloch bands [72, 73].

As mentioned above, the static DWP and its limit form reducing to dual-core systems are fundamental settings for the onset of the SSB [2-33]. A natural problem, which is the subject of this chapter, is the SSB and related phenomenology, i.e., the existence and stability of symmetric, antisymmetric, and asymmetric modes, in rotating DWPs (recently, a revolving DWP configuration was considered in a different context in Refs. [75-76], as a stirrer generating vortex lattices). To analyze basic features of the phenomenology, we here concentrate on the one dimensional DWP placed onto a rotating ring. As shown in Figure 1, in optics this setting may be realized as a hollow pipe waveguide twisted with pitch $2\pi/\omega$, while the azimuthal modulation of the refractive index, that gives rise to the effective potential $V(\theta)$, is written into the material of the pipe. Alternatively, a helical potential structure can be created in a straight sheath waveguide by means of optical-induction techniques, using pump waves with the ordinary polarization in a photorefractive material (while the probe wave is to be launched in the extraordinary polarization [77-86]), or the method of the electromagnetically-induced transparency (EIT) [89-91], including its version recently proposed for supporting spatial solitons [92-95]. In the latter case, one can make the pipe out of Y_2SiO_5 crystal doped by Pr^{3+} (Pr: YSO) ions [96-97]. In either case of the use of the photorefractive material or EIT, the helical structure may be induced by a superposition of a pair of co-propagating vortical pump waves, with equal amplitudes, a small mismatch of the propagation constants $k_{1,2}$ ($\Delta k \equiv k_1 - k_2 \leq k_1$), and opposite vorticities ($\pm S$), which will give rise to an effective potential profile,

$$V(\theta, z) \sim r^S \cos(\Delta k \cdot z + 2S\theta), \tag{1}$$

where z is the propagation distance, while r and θ are the polar coordinates in the transverse plane. In terms of the BEC, a similar setting may be based on ring-shaped (toroidal) traps, which have been created in experiments [98-102] and investigated in various contexts theoretically [103-117]. In that case, the rotating periodic potential can be added to the toroidal trap [61-62], which is equivalent to the consideration of the rotating ring [73, 118].

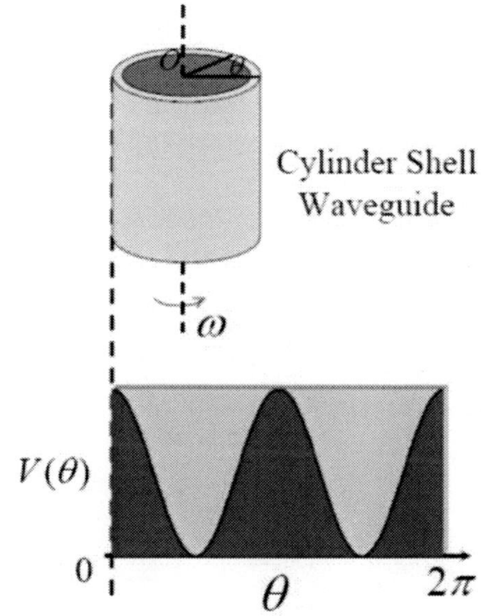

Figure 1. (Color online) The pipe waveguide with the intrinsic potential, twisted at rate ω.

In this chapter, we introduce basic types of trapped modes and their SSB phenomenology in the 1D rotating ring, in both cases of the self-focusing and self-defocusing (SF and SDF) cubic nonlinearities. In Sec. II we formulate the model and present analytical results, which predict a boundary between the symmetric and asymmetric modes, the analysis being possible for the small-amplitude potential and the rotation rate close to $\omega = 1/2$. Numerical results are reported in a systematic form, and are compared to the analytical predictions, in Secs. III and IV for the SF and SDF nonlinearities, respectively. The chapter is concluded by Sec. IV.

2. The Model and Analytical Considerations

As said above, we consider the limit of a thin helical shell, which implies a fixed value of the radius in Equation (1), $r = r_0$, that we normalize to be $r_0 = 1$. Taking the harmonic periodic potential in the form of Equation (1) with $s = 1, V(\theta, z) = 2Acos^2(\theta - \omega z)$, the corresponding scaled nonlinear Schrödinger equation is

$$i\frac{\partial}{\partial z}\Psi = \left[-\frac{1}{2}\frac{\partial^2}{\partial \theta^2} + V(\theta, z) - \sigma|\Psi|^2\right]\Psi, \tag{2}$$

where $\sigma = +1$ and -1 refer to SF and SDF nonlinearities, respectively. Then, Equation (2) is rewritten in the helical coordinate system, with $\theta' \equiv \theta - \omega z$:

$$i\frac{\partial}{\partial z}\psi = \left[-\frac{1}{2}\frac{\partial^2}{\partial \theta'^2} + i\omega\frac{\partial}{\partial \theta'} + 2A\cos^2(\theta') - \sigma|\psi|^2\right]\psi, \tag{3}$$

where the solution domain is defined at $-\pi \leq \theta' \leq \pi$ For the narrow toroidal BEC trap with the rotating potential, the respective Gross-Pitaevskii equation, written in the co-rotating reference frame (cf. Refs. [72, 73]), differs by replacing the propagation distance z, with time t [74].

Stationary modes with real propagation constant $-\mu$ (in terms of the BEC, μ is the chemical potential) are sought after as $\Psi(\theta', z) = exp(-i\mu z)\phi(\theta')$, with complex function $\phi(\theta')$ obeying equation

$$\mu\phi = \left[-\frac{1}{2}\frac{d^2}{d\theta'^2} + i\omega\frac{d}{d\theta'} + 2A\cos^2(\theta') - \sigma|\phi|^2\right]\phi. \tag{4}$$

Equation (3) conserves the total power (norm) of the field and its Hamiltonian (energy),

$$P = \int_{-\pi}^{+\pi}|\psi(\theta')|^2 d\theta', \tag{5}$$

$$H = \int_{-\pi}^{+\pi}\left[\frac{1}{2}\left|\frac{\partial\psi}{\partial\theta'}\right|^2 + \frac{i}{2}\omega\left(\psi^*\frac{\partial\psi}{\partial\theta'} - \psi\frac{\partial\psi^*}{\partial\theta'}\right) + V(\theta')|\psi|^2 - \frac{\sigma}{2}|\psi|^4\right]d\theta', \tag{6}$$

and the asterisk stands for the complex conjugate.

The periodic boundary conditions, $V(\theta' + 2\pi) = V(\theta')$ and $\Psi(\theta' + 2\pi) = \Psi(\theta')$, make Equation (3) invariant with respect to the boost transformation, which allows one to change the rotation speed from ω to $\omega - N$ with arbitrary integer N:

$$\Psi(\theta', z; \omega - N) = \Psi(\theta', z; \omega)exp\left[-iN\theta' + i\left(\frac{1}{2}N^2 - N\omega\right)z\right], \tag{7}$$

hence the speed may be restricted to interval $0 \le \omega < 1$. Furthermore, Equation (3) admits an additional invariance, relating solutions for opposite signs of the rotation speed: $\Psi(\theta', z; \omega) = \Psi^*(\theta', -z; -\omega)$. If combined with shift $\omega \to \omega + 1$, the latter transformation implies that the solutions for the rotation speeds ω and $1 - \omega$ (with $0 < \omega < 1$) are mutually tantamount, therefore the rotation speed may be eventually restricted to interval

$$0 \le \omega \le 1/2, \tag{8}$$

which plays the role similar to that of the fist Brillouin zone in solid-state media [119].

An analytical approach for small-amplitude modes can be developed if the amplitude of the potential in Equation (3) is small too, $|A| \ll 1$, and ω is close to the right edge of zone (8), $\delta \equiv 1/2 - \omega \ll 1/2$. In the simplest approximation, the corresponding stationary solutions are looked for as a combination of the zeroth and first angular harmonics,

$$\phi(\theta') = a_0 + a_1 \exp(i\theta'), \tag{9}$$

where a_0 is fixed to be real, while amplitude a_1 may be complex. Indeed, setting $\omega = 1/2$, $A = 0$, and dropping the cubic term, it is obvious that Equation (9) is an exact solution to Equation (4). Then, under the aforementioned conditions, the substitution of ansatz (9) into Equation (4) yields, in the first approximation (taking care of the balance of the zeroth and first harmonics in the equation), the following algebraic equations:

$$\begin{aligned} \mu a_0 &= A a_0 - \sigma a_0^3 - 2\sigma |a_1|^2 a_0, \\ \mu a_1 &= \delta a_1 - \sigma |a_1|^2 a_1 - 2\sigma a_0^2 a_1 + A a_1. \end{aligned} \tag{10}$$

First, setting $a_1 = 0$ or $a_0 = 0$ in ansatz (9) corresponds, respectively, to the CW (continuous-wave) mode, and the uniform vortical one, with $|\Psi(\theta')| = const$:

$$a_1 = 0,\ a_0^2 = \sigma(A - \mu), \tag{11}$$

$$a_0 = 0,\ |a_1|^2 = \sigma(A - \mu + \delta). \tag{12}$$

(recall that $\sigma^2 = 1$). In terms of full Equation (3) with $A \neq 0$, these two solutions extend into non-uniform ones [with $|\Psi(\theta')| \neq const$], which remain (and are categorized as) symmetric with respect to point $\theta' = 0$. The respective vortical mode, obtained as the extension of solution (12), is investigated in a numerical form in the next section, under the name of FHS (fundamental-harmonic symmetric) mode.

Solutions to Equations (10) with $a_0 a_1 \neq 0$ give rise to the species of asymmetric modes (categorized as FHA, i.e., fundamental-harmonic asymmetric mode, in the next section), with

$$a_0^2 = \frac{\sigma}{3}(A - \mu + 2\delta),\ |a_1|^2 = \frac{\sigma}{3}(A - \mu - \delta) \quad (13)$$

According to Equations (5) and (9), the propagation constant is related to total power (5) of soliton (13), $P = 2\pi(a_0^2 + |a_1|^2)$, as follows:

$$\mu = A + \frac{\delta}{2} - \frac{3\sigma}{4\pi}P \quad (14)$$

In particular, in the case of the SF nonlinearity, with $\sigma = \pm 1$, Equation (14) demonstrates that the asymmetric mode meets the Vakhitov-Kolokolov (VK) criterion, $d\mu/dP < 0$, which is a necessary stability condition for modes supported by the SF terms [120-122]. On the other hand, in the case of the SDF sign of the nonlinearity, $\sigma = -1$, the asymmetric mode satisfies the "anti-VK" criterion, $d\mu/dP > 0$, which, as argued in Ref. [123], may also play the role of a necessary stability condition, in the respective setting.

Further, Equation (13) predicts a transition between the symmetric (FHS) and asymmetric (FHA) vortical modes at $a_0^2 = 0$, i.e., at $\mu = A + 2\delta$. Then, Equation (14) yields the location of this boundary in terms of the total power, $P_{min} = 2\pi\sigma\delta$. In view of the above-mentioned equivalence between the modes pertaining to rotation speeds $\omega = 1/2 \pm \delta$, the latter result predicts the coexistence of the FHS and FHA modes at

$$P \geq P_{\min} = 2\pi|\delta|. \quad (15)$$

Comparison of this prediction with numerical findings is presented below.

3. NUMERICAL RESULTS FOR THE SELF-FOCUSING NONLINEARITY ($\sigma = \pm 1$)

3.1. Symmetric, Asymmetric, and Antisymmetric Modes

Solutions to stationary Equation (4) were constructed by means of numerical code "PCSOM" elaborated in Refs. [124-125]. In the SF case, the use of different input waveforms makes it possible to identify five distinct species of stationary modes, as listed in Table I.

The symmetry, asymmetry and antisymmetry of the modes designated in Table 1 is realized with respect to point $\theta' = 0$, while the solution is considered, as defined above, in the region of $-\pi \leq \theta' \leq +\pi$. Further, the fundamental or second harmonic ("FH" or "2H", respectively) in the nomenclature adopted in the table refers to a dominant term in the Fourier decomposition of the stationary solution (which is made obvious by their shapes, see below). In particular, the stationary patterns of the FHA and 2HA types are generated by the same input in Table 1, $1 + \sin\theta'$, but in non-overlapping regions of the parameter space, (ω, A, P), as shown below. Note also that all the inputs displayed in the table are real functions, but the numerical modes found for $\omega \neq 0$ have a complex structure.

Table I. Different species of stationary modes in the case of the SF nonlinearity ($\sigma = \pm 1$), labeled by input waveforms which generate them

Inputs	Types of Modes
$\cos\theta'$	Fundamental-harmonic symmetric (FHS)
$1 + \sin\theta'$	Fundamental-harmonic asymmetric (FHA)
$\sin^2\theta'$	Second harmonic symmetric (2HS)
$1 + \sin\theta'$	Second-harmonic asymmetric (2HA)
$\sin\theta'$	Anti-symmetric (AnS)
$b + \sin\theta', 0 < b \leq 1$	Broken-antisymmetry (BAnS)

The stability of the stationary solutions has been identified through the numerical computation of eigenvalues for infinitesimal perturbation modes. To this end, the perturbed solution was taken in the usual form,

$$\psi = e^{-i\mu z}[\phi(\theta') + u(\theta')e^{i\lambda z} + v^*(\theta')e^{-i\lambda^* z}] \tag{16}$$

where $u(\theta')$ and $v(\theta')$ are perturbation eigenmodes, and λ is the corresponding eigenfrequency. The substitution of expression (16) into Equation (3) and linearization leads to the eigenvalue problem in the following form:

$$\begin{pmatrix} \mu - \hat{h} - i\omega\dfrac{\partial}{\partial\theta'} & \sigma\phi^2 \\ -\sigma(\phi^*)^2 & -\mu + \hat{h} - i\omega\dfrac{\partial}{\partial\theta'} \end{pmatrix} \begin{pmatrix} u \\ v \end{pmatrix} = \lambda \begin{pmatrix} u \\ v \end{pmatrix}, \tag{17}$$

where $\hat{h} = -(1/2)(\partial^2/\partial\theta'^2) + 2A\cos^2(\theta') - 2\sigma|\phi|^2$ is the respective single-particle Hamiltonian. The underlying solution ϕ is stable if all the eigenvalues are real.

In the case of the SF nonlinearity, the numerical solutions reveal the existence of all the species of the modes indicated in Table 1, except for BAnS, which exist under the SDF nonlinearity, see below. Typical examples of the five species of the stationary modes which are supported by the SDF cubic term are displayed in Figures 2, 3 and 4. The analysis demonstrates that all these examples are stable. Moreover, the symmetric and asymmetric modes dominated by the fundamental harmonic, FHS and FHA, demonstrate mutual bistability in Figure 2, as these stable modes coexist at common values of the parameters. A distinctive feature of the FHS mode, which is evident in Figure 2, is that maxima and minima of its intensity coincide with local maxima and the minima of the harmonic potential, while its FHA counterpart features an intensity maximum in one potential well, and a minimum in the other.

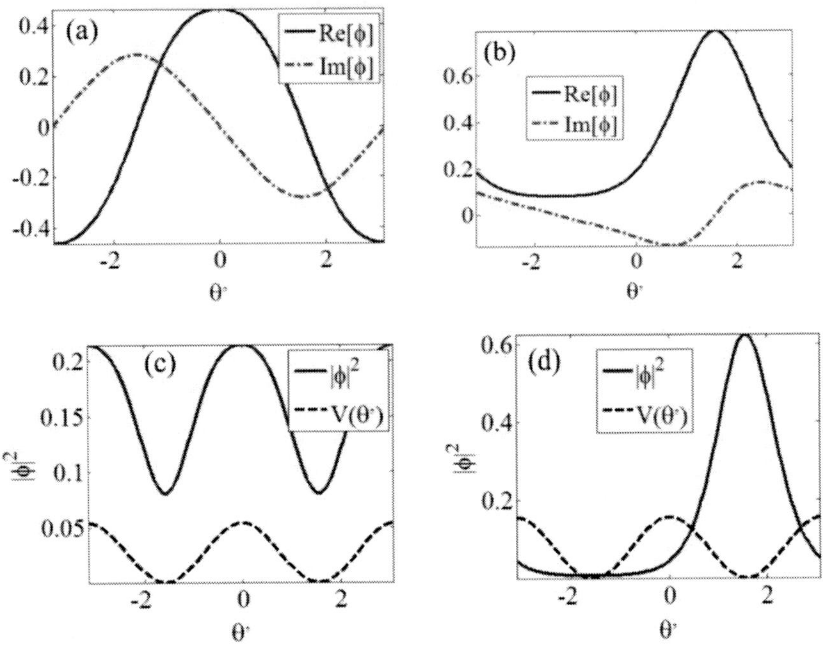

Figure 2. (Color online) Examples of stable FHS and FHA (fundamental-harmonic symmetric and asymmetric) modes, in the case of the SF nonlinearity, corresponding to the same parameter set, $(\omega, A, P) = (0.35, 0.5, 1)$. (a) and (b) Real and imaginary parts of the FHS and FHA stationary solutions, respectively. (c) and (d) Intensity profiles of the same solutions. The dashed curves in (c) and (d), and in similar panels displayed below, depict the potential, $V(\theta') = 2Acos^2(\theta')$.

This structure of the FHS suggests that it may correspond to a maximum, rather than minimum, of the energy (see below), but, nevertheless, this mode has a region of stability against small perturbations.

On the other hand, Figures 3 and 4 demonstrate that both the 2HS and AnS modes have two symmetric intensity peaks trapped in the two potential wells (while local minima of the intensity coincide, quite naturally, with potential humps), hence these mode have a chance to realize a minimum of the energy (in the best case, it may be the system's ground state).

The 2HA mode features a similar property in Figure 3, but with unequal density peaks trapped in the two potential minima.

In the limit of the uniform ring (without the potential, $A = 0$) the FHS and AnS modes go over into the above mentioned uniform vortical state (12), the 2HS pattern degenerates into the CW state (11), while the FHA mode takes the form of a 2π-periodic cnoidal-wave solution of the nonlinear Schrödinger equation (the 2HA pattern does not exist at all in the limit of $A = 0$).

The evolution of the shape of the modes with the increase of A is shown in Figure 5 [The evolution figure of AnS mode, which is not shown here, is similar to the 2HS type displayed in Figure 5(c)].

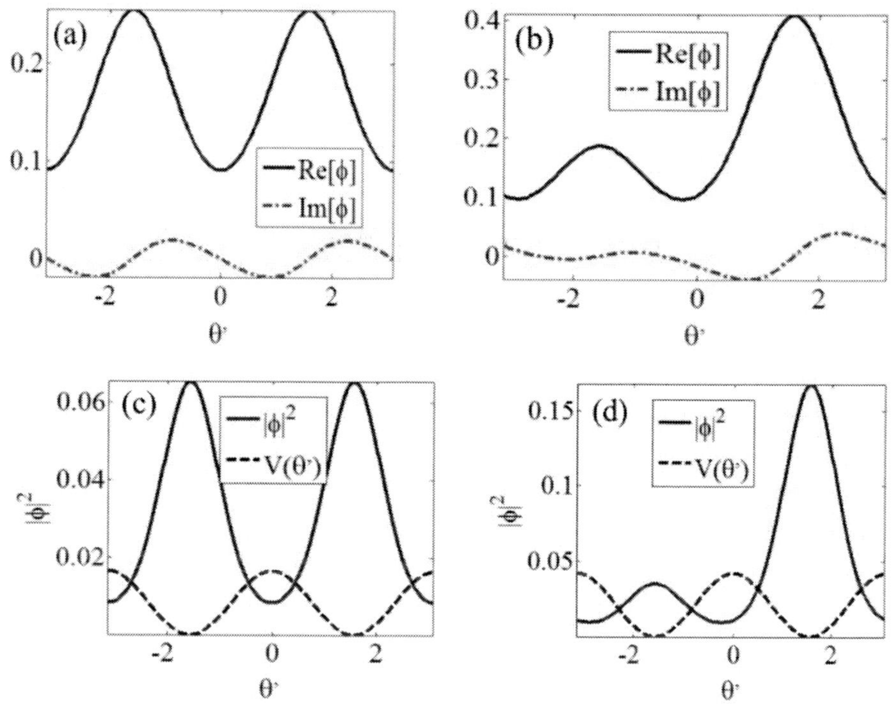

Figure 3. (Color online) Examples of stable modes of the 2HS and 2HA (second-harmonic symmetric and asymmetric) types, in the case of the SF nonlinearity, found at sets of the parameter values $(\omega, A, P) = (0.25, 1, 0.2)$ and $(0.25, 1, 0.3)$, respectively. (a) and (b) Real and imaginary parts of the 2HS and 2HA stationary solutions, respectively. (c) and (d) Intensity profiles of the same solutions. Unlike the fundamental-harmonic modes shown in Figure 2, these ones, based on the second harmonic, feature no bistability (coexistence of the stable modes).

Existence and Stability Diagrams for the Different Modes

Results of the systematic numerical analysis are summarized in Figure 6, in the form of diagrams for the existence of the stable FHS and FHA modes in the plane of (P, ω) at several fixed values of amplitude A of the harmonic potential. In the blank areas, only unstable modes of the FHS type are found [in direct simulations they feature an oscillatory instability, which is accounted for by a quartet of complex eigenvalues generated by Equation (17)], while the FHA modes are completely stable in their existence region.

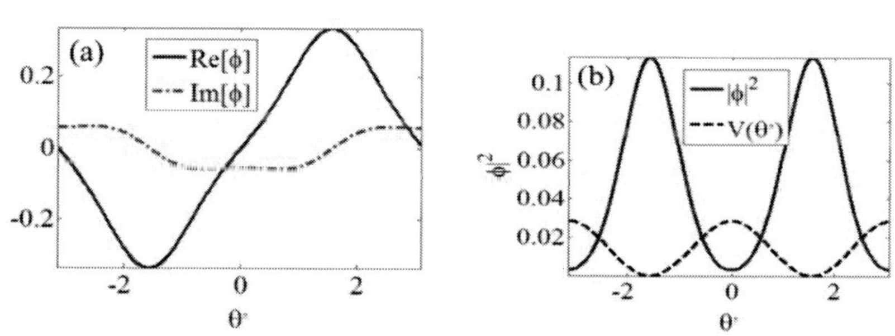

Figure 4. (Color online) An example of a stable mode of the AnS (antisymmetric) type, at $(\omega, A, P) = (0.25, 1, 0.3)$: (a) real and imaginary parts of the stationary solution; (b) its intensity profile.

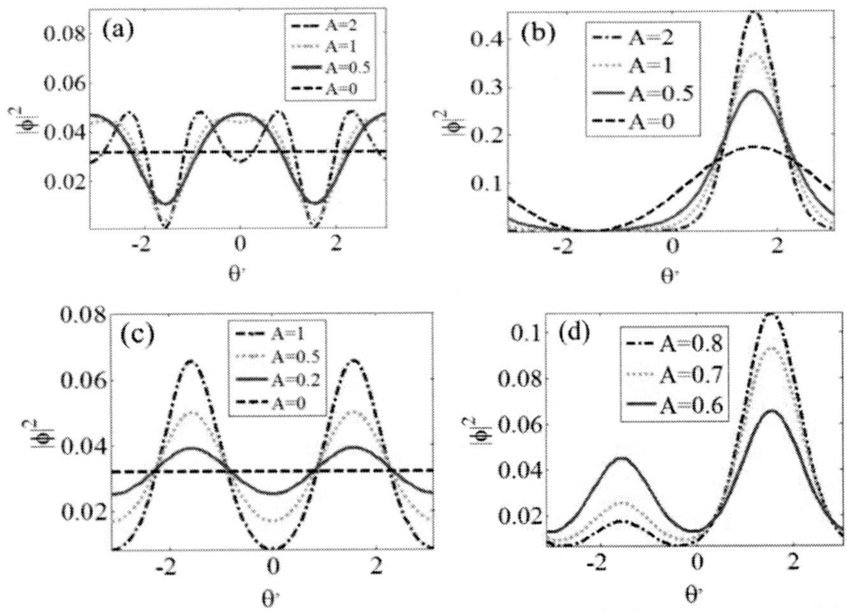

Figure 5. (Color online) (a) The FHS mode, in the case of the SF nonlinearity with $(\omega, P) = (0.25, 0.2)$, at different values of the potential's strength, A. (b) The same for the FHA mode with $(\omega, P) = (0.5, 0.5)$. (c) The same for the 2HS mode with $(\omega, P) = (0.25, 0.2)$. (d) The same for the 2HA mode with $(\omega, P) = (0.4, 0.2)$. All the modes shown in this figure are stable.

The dashed lines in Figures 6(a) and (b) demonstrate that Equation (15) correctly predicts the bistability boundary at small values of $\delta \equiv 1/2 - \omega$ and P, provided that A is small enough too [recall it is the condition under which Equation (15) was derived]. Further, the fact that vertical cuts of the panels in Figure 6, that can be drawn through $\omega = const$, go, with the increase of the total power (P), from the region of the monostability of the FHS mode through the FHS/FHA bistability area into the monostability region of the FHA mode (at least, for ω sufficiently close to $1/2$), clearly suggests that the symmetry-breaking bifurcation of the subcritical type, which features the bistability [2-8, 40], occurs along this route.

This conclusion is confirmed by a direct consideration of the numerical data (not shown here in detail).

As concerns the AnS solitons, in the case of the SF nonlinearity they do not undergo any bifurcation, and are stable in the entire region of their existence, which is shown in the (ω, P) plane for fixed values of the lattice's strength, A, in Figure 7 [as mentioned above, at A=0 the AnS mode coincides with the FHS one, hence the respective existence area is the same as the red (bottom) one in Figure 6(a)]. From here we conclude that the stability region of the AnS mode quickly expands with the increase of A, and the FHS/FHA bistability areas in panels (b) and (c) of Figure 6 actually support the tristability, as the AnS mode is stable, too.

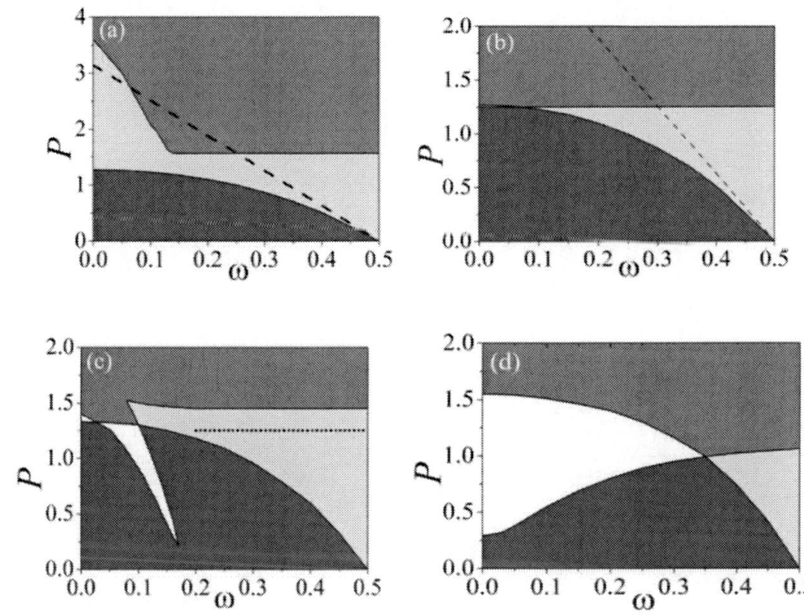

Figure 6. (Color online) Stability diagrams for the symmetric and asymmetric modes dominated by the fundamental harmonic in the case of the SF nonlinearity, in the plane of the rotation speed (ω) and total power (P), at several fixed values of the potential's strength: (a) $A = 0$, (b) $A = 0.2$, (c) $A = 0.5$, and (d) $A = 1$. The green (top) and red (bottom) areas designate regions where solely the FHS or FHA mode is stable, respectively. The bistability (coexistence of these stable modes) occurs in the yellow (middle) region. In the blank area, no stable modes of these types are found (in fact, an unstable FHS mode exists in this area). In panels (a) and (b), the dashed lines represent the analytical prediction (15), $P = 2\pi|1/2 - \omega|$, for the existence boundary of the FHA mode (the prediction is relevant at small values of $|1/2 - \omega|$, P, and A). The horizontal dotted line in (c) is the cross section of the bistability region, along which energies of the FHS and FHA modes is compared in Figure 8(a).

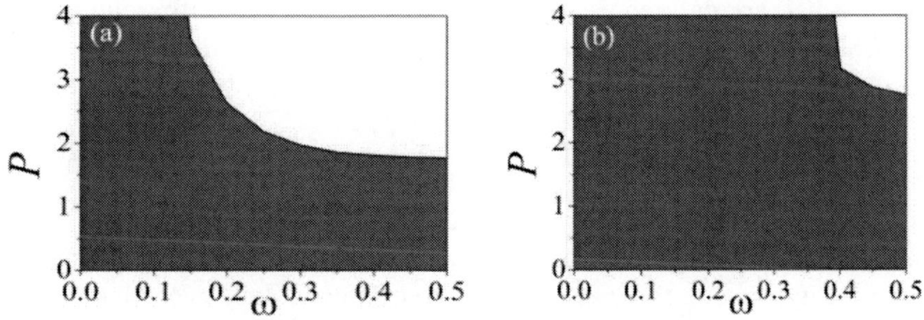

Figure 7. (Color online) In the case of the SF nonlinearity, the AnS (antisymmetric) mode exists and is stable in the red area, as shown here for (a) $A = 0.2$ and (b) $A = 0.4$.

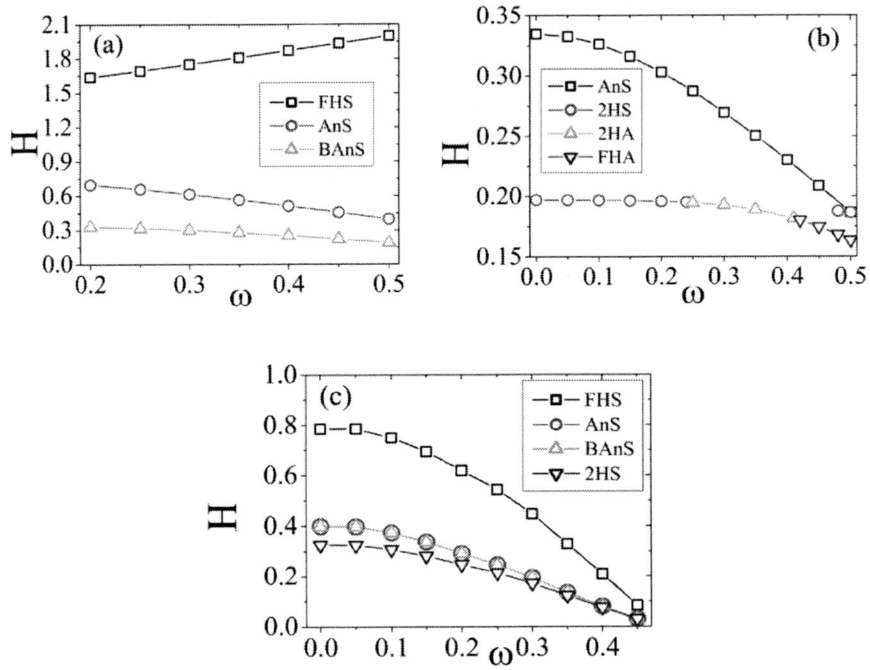

Figure 8. (Color online) (a) Energies of the FHS, FHA, and AnS modes, computed according to Equation (6) along the horizontal dotted line drawn in Figure 6(c), with $(P, A) = (1.25, 0.5)$, in the interval of $0.2 \leq \omega \leq 0.5$. (b) The energies of the 2HS, 2HA, and FHA modes, along with the coexisting AnS one, along the horizontal dotted line drawn in Figure 9(c), for $(P, A) = (0.5, 0.5)$. Panels (a) and (b) pertain to the system with the SF nonlinearity, $\sigma = \pm 1$. (c) The energies of the AnS, BAnS (brokenantisymmetry), and 2HS modes, along the boundary separating the existence regions of the AnS and BAnS solutions in the case of the SDF nonlinearity $\sigma = -1$, see Figure 12(b).

The multistability between the three species of the modes based on the fundamental harmonic—symmetric, asymmetric, and antisymmetric ones—makes it necessary to compare their energies, which are defined as per Equation (6). The comparison along the horizontal dotted line, which is drawn in Figure 6(c), is presented in Figure 8(a). The figure clearly shows the following relation between the energies:

$$H_{FHA} < H_{AnS} < H_{FHS} \, . \tag{18}$$

For the 2HS and 2HA patterns, which are based on the second angular harmonic, Figure 9 displays the stability areas in the (P, ω) plane, at different values of the lattice's strength, A [cf. Figure 6], together with the stability area for the FHA mode, which may stably coexist with the 2HS pattern. Figure 9 also demonstrates that, as mentioned above, the 2HA mode, which exists and is stable in the blue (intermediate) area in Figure 9, does not exist at $A = 0$, emerging and expanding with the increase of A. Note that, on the contrary to the situation for the patterns dominated by the fundamental angular harmonic (FHS and FHA) described above, there is no overlap between the stability areas of the symmetric and asymmetric second-harmonic-based modes, 2HS and 2HA, hence (as confirmed by an additional consideration of the numerical results) the symmetry-breaking 2HS→2HA transition, following the increase of P at $A \neq 0$, amounts to a supercritical bifurcation, which features no bistability [2-8, 40].

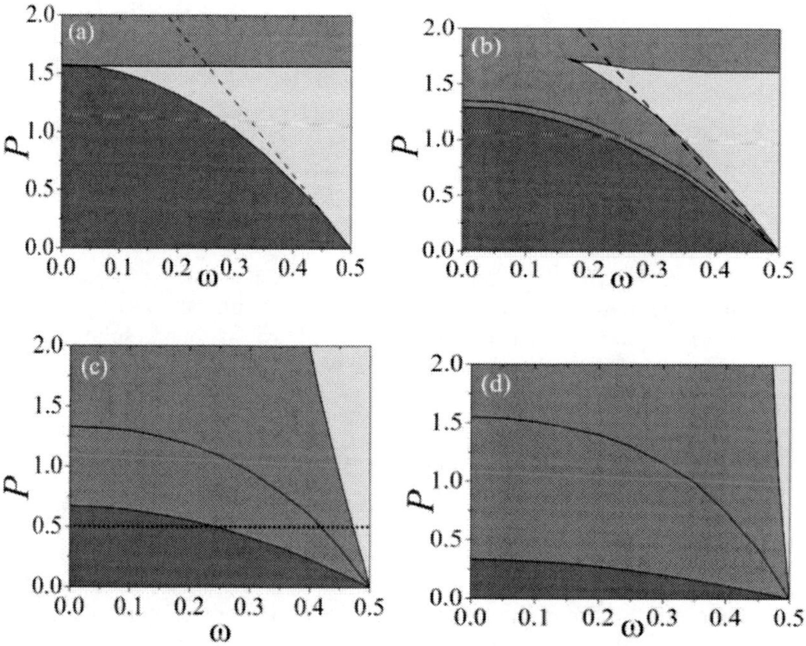

Figure 9. (Color online) The same as in Figure 6, but for the following types of the solutions. The green (top) area: the stable FHA mode; the red (bottom) area: the stable 2HS mode; the blue (intermediate) area: the stable 2HA mode; the yellow (rightmost) area: the bistability of the FHA and 2HS modes.

Another peculiarity of the symmetric mode based on the second harmonic (2HS) is that it is stable in two disjoint stability areas—the red and yellow ones in Figure 9 (the bottom and rightmost regions), in the latter one the 2HS featuring the bistability with the FHA mode, although they are not linked by any bifurcation. On the other hand, there is no overlap between the stability areas of the asymmetric modes based on the different angular harmonics, FHA and 2HA, therefore these modes may be generated by the same input waveform ($1 + sin\theta'$ in Table I), in different parameter regions.

The effective bistability occurring in Figure 9 makes it necessary to compare the energy of the coexisting stable patterns, which is done in Figure 8(b), along the horizontal dotted line drawn in Figure 9(c). The energy of the coexisting stable AnS mode is included too. In particular, the short segment corresponding to the 2HS mode appears, above the one pertaining to the FHA mode, when the horizontal dotted line in Figure 9(c) enters the yellow (rightmost) region of the 2HS-FHA bistability. Thus, adding the results from Figure 8(b), we extend energy relations (18) as follows:

$$H_{FHA} \ or \ H_{2HA} < H_{2HS} < H_{AnS} < H_{FHS} . \tag{19}$$

The conclusion is that the asymmetric modes, either FHA or 2HA, represent the ground state of the rotating ring carrying the harmonic potential and SF nonlinearity.

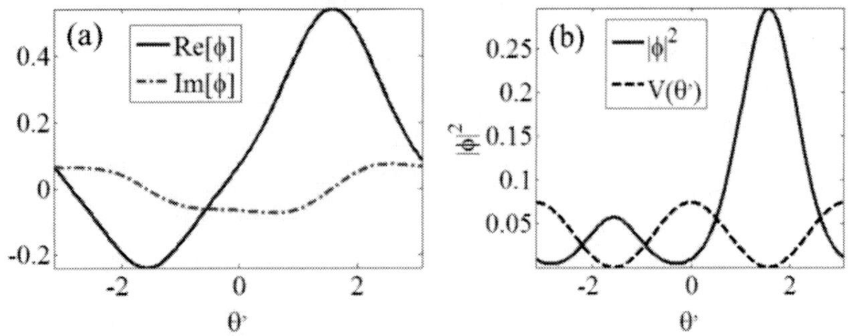

Figure 10. (Color online) A typical example of a stable BAnS mode (the one with the broken antisymmetry) in the system with the SDF (self-defocusing) nonlinearity, $\sigma = -1$, and parameters $(\omega, A, P) = (0.2, 1, 0.5)$: (a) the real and imaginary parts of the stationary solution; (b) its intensity profile, which features unequal peaks in the two potential wells.

4. Numerical Results for the Self-Defocusing Nonlinearity ($\sigma = -1$)

In the case of the SDF nonlinearity, symmetric modes of the FHS and 2HS types have been found too, although, on the contrary to the system with the SF sign, they do not undergo any bifurcations, hence the FHA and 2HA species do not exist in this case. On the other hand, the AnS mode features an antisymmetry-breaking bifurcation, which gives rise to a broken-antisymmetry (BAnS) mode (which does not exist under the SF nonlinearity). A typical example of the stable BAnS mode is displayed in Figure 10. In fact, antisymmetry-breaking bifurcations are characteristic to the double-well systems with the SDF nonlinearity [23-24]. The stationary solutions of the BAnS type can be numerically generated by the initial guess of the form of $(b + sin\theta')$, with $0 < b \le 1$, see Table I.

Proceeding to the systematic description of the modes supported by the SDF nonlinearity acting in the combination with the rotating harmonic potential, we note, first, that the numerical analysis demonstrates the existence and stability of the mode of the 2HS type at all values of (ω, P, A), therefore this solution is not included in the stability diagrams, which are displayed for the FHS mode in Figure 12, and for the AnS and newly found BAnS ones in Figure 11.

In the blank area of Figure 12, the FHS solutions exist too, featuring an instability accounted for by a quartet of complex eigenvalues [see Equation (17)]. In direct simulations, this instability converts the stationary mode into an oscillatory one (not shown here). Further, Figure 12 shows that the antisymmetry-breaking transition, which follows the increase of the total power, P, is of a supercritical type. Actually, the supercritical character of the antisymmetry breaking bifurcation is a characteristic feature of systems with the SDF sign of the nonlinearity [23-24].

Finally, the energy comparison for the model with the SDF nonlinearity is presented in Figure 8(c), along the boundary separating the existence (and stability) regions of the AnS and BAnS modes in Figure 12(b).

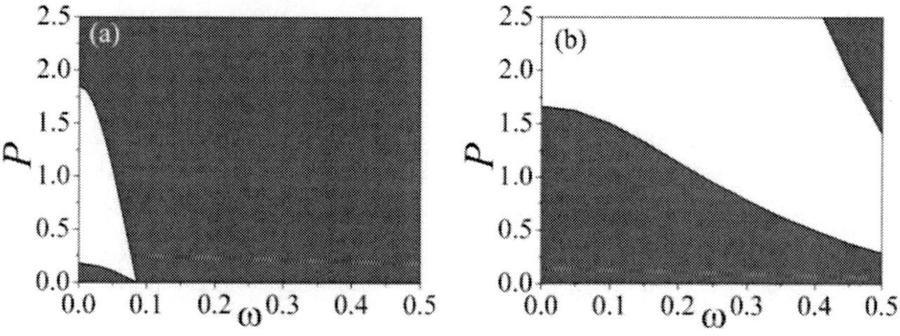

Figure 11. (Color online) The stability diagram for the FHS mode, in the case of the SDF nonlinearity ($\sigma = -1$), in the (ω, P) plane, for $A = 1.2$ (a) and $A = 1.75$ (b). The FHS modes are stable and unstable in the red and blank areas, respectively.

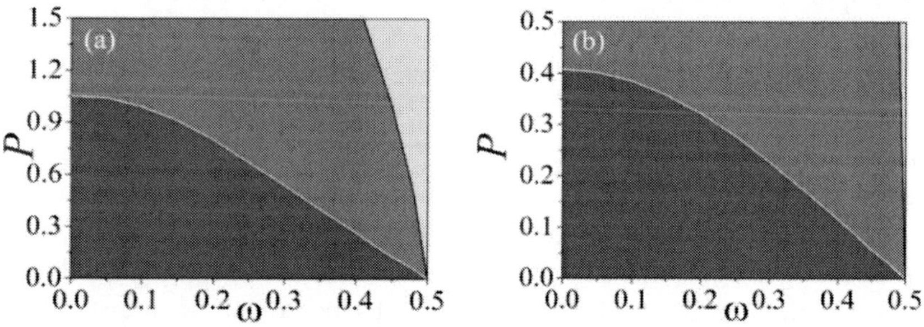

Figure 12. (Color online) The stability diagrams for the AnS (antisymmetric) and BAnS (broken-antisymmetry) modes in the case of the SDF nonlinearity for A=0.5 (a) and $A = 1$ (b). The red (bottom) and green (top) areas refer to the monostability of the AnS and BAnS modes, respectively. The bistability between these species occurs in the yellow (rightmost) region.

The figure shows the energies of these modes, which coincide along the boundary, along with the results for the FHS and 2HS symmetric modes. From here, it is concluded that the energies are related as follows:

$$H_{2HS} < H_{BAnS} = H_{AnS} < H_{FHS}, \tag{20}$$

i.e., the 2HS mode represents the ground state in the case of the SDF nonlinearity, cf. Equations (18) and (19).

Conclusion and Outlook

This chapter introduces the trapped modes of the symmetric, asymmetric, and antisymmetric types, and the symmetry- or antisymmetry-breaking transitions between them, in the rotating ring carrying the DWP (double-well potential) and the cubic nonlinearity with the SF or SDF (self-focusing or defocusing) sign. The analysis has been performed in the first rotational Brillouin zone. In the SF case, five types of different modes, and their stability, have been identified, three dominated by the fundamental angular harmonic (FHS, FHA, and

AnS), and two others based on the second harmonic, 2HS and 2HA modes. The SSB (spontaneous symmetry-breaking) transition between the FHS and FHA modes is of the subcritical type, featuring a conspicuous bistability area, while the transition between the 2HS and 2HA states is supercritical. There is no overlap between the existence regions of the asymmetric modes of the FHA and 2HA types, one of which realizes the ground state of the SF system. The SDF model supports four distinct species of the trapped modes, viz., FHS, 2HS, AnS, and, in addition, the BAnS (broken-antisymmetry) mode. The latter one appears from the AnS state as a result of supercritical antisymmetry-breaking transition. The 2HS mode represents the ground state of the SDF system.

The present chapter can be naturally extended in other directions. First, it is possible to consider the rotating potential in the form of $Acos(n\theta')$ with $n \geq 3$, while the present analysis corresponds to the DWP with $n = 2$ [note that Equation (1) suggests a possibility to create the helical photonic lattice with $n = 2S$ and $S > 1$]. This generalization will give rise to many new trapped modes; in particular, it may be possible to build one with a small width, $\Delta\theta' \ll 2\pi$, hence it may be considered as a soliton trapped in the lattice and rotating along with it, cf. Ref. [68-69]. It is also relevant to consider the effects of the two-dimensionality (the entire rotating plane, rather than a narrow ring). Our preliminary analysis, based on simulations of the 2D counterpart of Equation (3), demonstrates essentially the same types of trapped modes in rotating annuli of a finite radial size, as reported above in the one-dimensional limit. Another interesting extension is to replace the linear rotating potential by its nonlinear counterpart, generated by modulation of the local nonlinearity along the angular coordinate (as was done in Refs. [126-128] in the absence of the rotation). Results obtained for one-dimensional modes trapped in the rotating nonlinear potential will be reported elsewhere.

ACKNOWLEDGMENTS

We appreciate help in the use of numerical methods provided by Nir Dror and Shenhe Fu. This work was supported by Chinese agencies NKBRSF (grant No. G2010CB923204) and CNNSF(grant No. 11104083,10934011), by the German-Israel Foundation through grant No. I-1024-2.7/2009, and by the Tel Aviv University in the framework of the “matching” scheme.

REFERENCES

[1] Eilbeck, JC; Lomdahl, PS; Scott, AC. “The discrete self-trapping equation”, *Physica D*, 1985, 16, 318-338.

[2] Snyder, AW; Mitchell, DJ; Poladian, L; Rowland, DR; Chen, Y. “Physics of nonlinear fiber couplers”, *J. Opt. Soc. Am. B*, 1991, 8, 2102-2118;

[3] Paré, C; Fłrjańczyk, M. “Approximate model of soliton dynamics in all-optical couplers”, *Phys. Rev. A*, 1990, 41, 6287-6295.

[4] Maimistov, AI. “Propagation of a light pulse in nonlinear tunnel-coupled optical waveguides”, *Kvantovaya Elektron.*, 1991, 18, 758.

[5] Maimistov, AI. "Propagation of a light pulse in nonlinear tunnel-coupled optical waveguides", *Sov. J. Quantum Electron.*, 1991, 21, 687-690.

[6] Akhmediev, N; Ankiewicz, A. "Novel soliton states and bifurcation phenomena in nonlinear fiber couplers", *Phys. Rev. Lett.*, 1993, 70, 2395-2398.

[7] Chu, PL; Malomed, BA; Peng, GD. "Soliton switching and propagation in nonlinear fiber couplers: analytical results", *J. Opt. Soc. Am. B*, 1993, 10, 1379-1385.

[8] Malomed, BA. in *Progress optics*, edited by E. Wolf, Vol. 43 (North Holland, Amsterdam, 2002), p. 71.

[9] Mak, WCK; Malomed, BA; Chu, PL. "Solitary waves in coupled nonlinear waveguides with Bragg gratings", *J. Opt. Soc. Am. B*, 1998, 15, 1685-1692.

[10] Fleischer, JW; Segev, M; Efremidis, NK; Christodoulides, DN. "Observation of two-dimensional discrete solitons in optically induced nonlinear photonic lattices", *Nature*, 2003, 422, 147-150.

[11] Wang, X; Chen, Z; Kevrekidis, PG. "Observation of discrete solitons and soliton rotation in optically induced periodic ring lattices", *Phys. Rev. Lett.*, 2006, 96, 083904.

[12] Jia S; Fleischer, JW. "Nonlinear light propagation in rotating waveguide arrays", *Phys. Rev. A*, 2009, 79, 041804(R).

[13] Kevrekidis, PG; Chen, Z; Malomed, BA; Frantzeskakis, DJ; Weinstein, MI. "Spontaneous symmetry breaking in photonic lattices: theory and experiment", *Phys. Lett. A*, 2005, 340, 275-280.

[14] Milburn, GJ; Corney, J; Wright, EM; Walls, DF. "Quantum dynamics of an atomic Bose-Einstein condensate in a double-well potential", *Phys. Rev. A*, 1997, 55, 4318-4324.

[15] Mahmud, KW; Perry, H; Reinhardt, WP. "Quantum phase-space picture of Bose-Einstein condensates in a double well", *Phys. Rev. A*, 2005, 71, 023615.

[16] Schumm, T; Hofferberth, S; Andersson, LM; Wildermuth, S; Groth, S; Bar-Joseph, I; Schmiedmayer, J; Krüger, P. "Matter-wave interferometry in a double well on an atom chip", *Nature Phys.*, 2005, 1, 57-62.

[17] Dounas-Frazer, DR; Hermundstad, AM; Carr, LD. "Ultracold Bosons in a tilted multilevel double-well potential", *Phys. Rev. Lett.*, 2007, 99, 200402.

[18] Huang YP; Moore, MG. "Optimized double-well quantum interferometry with Gaussian squeezed states", *Phys. Rev. Lett.*, 2008, 100, 250406.

[19] He, QY; Reid, MD; Vaughan, TG; Gross, C; Oberthaler, M; Drummond, PD. "Einstein-Podolsky-Rosen entanglement strategies in two-well Bose-Einstein condensates", *Phys. Rev. Lett.*, 2011, 106, 120405.

[20] Zapata, I; Sols, F; Leggett, AJ. "Josephson effect between trapped Bose-Einstein condensates", *Phys. Rev. A*, 1998, 57, R28-R31.

[21] Radzihovsky, L; Gurarie, V. "Relation between ac Josephson effect and double-well Bose-Einstein-condensate oscillations", *Phys. Rev. A*, 2010, 81, 063609.

[22] Qi, R; Yu, XL; Li, ZB; Liu, WM. "Non-abelian Josephson effect between two F=2 spinor Bose-Einstein condensates in double optical traps", *Phys. Rev. Lett.*, 2009, 102, 185301.

[23] Matuszewski, M; Malomed, BA; Trippenbach, M. "Spontaneous symmetry breaking of solitons trapped in a double-channel potential", *Phys. Rev. A*, 2007, 75, 063621.

[24] Trippenbach, M; Infeld, E; Gocalek, J; Matuszewski, M; Oberthaler, M; Malomed, BA. "Spontaneous symmetry breaking of gap solitons and phase transitions in double-well traps", *Phys. Rev. A*, 2008, 78, 013603.

[25] Mazzarella, G; Malomed, B; Salasnich, L; Salerno, M; Toigo, F. "Rabi–Josephson oscillations and self-trapped dynamics in atomic junctions with two bosonic species", *J. Phys. B: At. Mol. Opt. Phys.*, 2011, 44, 035301.

[26] Ng HT; Leung, PT. "Two-mode entanglement in two-component Bose-Einstein condensates", *Phys. Rev. A*, 2005, 71, 013601.

[27] Satija, II; Balakrishnan, R; Naudus, P; Heward, J; Edwards, M; Clark, CW. "Symmetry-breaking and symmetry-restoring dynamics of a mixture of Bose-Einstein condensates in a double well", *Phys. Rev. A*, 2009, 79, 033616.

[28] Wang, C; Kevrekidis, PG; Whitaker, N; Malomed, BA. "Two-component nonlinear Schrödinger models with a double-well potential", *Physica D*, 2008, 237, 2922-2932.

[29] Albiez, M; Gati, R; Fölling, J; Hunsmann, S; Cristiani, M; Oberthaler, MK. "Direct observation of tunneling and nonlinear self-trapping in a single Bosonic Josephson junction", *Phys. Rev. Lett.*, 2005, 95, 010402.

[30] Gati, R; Albiez, M; Fölling, J; Hemmerling, B; Oberthaler, MK. "Realization of a single Josephson junction for Bose Einstein condensates", *Appl. Phys. B*, 2006, 82, 207-210.

[31] Gubeskys, A; Malomed, BA. "Symmetric and asymmetric solitons in linearly coupled Bose-Einstein condensates trapped in optical lattices", *Phys. Rev. A*, 2007, 75, 063602.

[32] Gubeskys, A; Malomed, BA. "Spontaneous soliton symmetry breaking in two-dimensional coupled Bose-Einstein condensates supported by optical lattices", *Phys. Rev. A,* 2007, 76, 043623.

[33] Salasnich, L; Malomed, BA; Toigo, F. "Competition between the symmetry breaking and onset of collapse in weakly coupled atomic condensates", *Phys. Rev. A,* 2010, 81, 045603.

[34] Adhikari, SK; Malomed, BA; Salasnich, L; Toigo, F. "Spontaneous symmetry breaking of Bose-Fermi mixtures in double-well potentials", *Phys. Rev. A*, 2010, 81, 053630.

[35] Ostrovskaya, EA; Kivshar, YS; Lisak, M; Hall, B; Cattani, F; Anderson, D. "Coupled-mode theory for Bose-Einstein condensates", *Phys. Rev. A*, 2000, 61, 031601.

[36] D'Agosta, R; Malomed, BA; Presilla, C. "Stationary solutions of the Gross–Pitaevskii equation with linear counterpart", *Phys. Lett. A*, 2000, 275, 424-434.

[37] Jackson RK; Weinstein, MI. "Geometric analysis of bifurcation and symmetry breaking in a Gross—Pitaevskii equation", *J. Stat. Phys.*, 2004, 116, 881-905.

[38] Ananikian, D; Bergeman, T. "Gross-Pitaevskii equation for Bose particles in a double-well potential: two-mode models and beyond", *Phys. Rev. A*, 2006, 73, 013604.

[39] Kirr, EW; Kevrekidis, PG; Shlizerman, E; Weinstein, MI. "Symmetry-breaking bifurcation in nonlinear Schrödinger/Gross–Pitaevskii equations", *SIAM J. Math. Anal.*, 2008, 40, 566-604.

[40] Iooss, G; Joseph, DD. *Elementary Stability and Bifurcation Theory*, (Springer: New York, 1980).

[41] Kivshar, YS; Agrawal, GP. *Optical Solitons: From Fibers to Photonic Crystals,* (Academic Press: San Diego).

[42] Fetter, AE. "Rotating trapped Bose-Einstein condensates", *Rev. Mod. Phys.*, 2009, 81, 647-691.

[43] Engels, P; Coddington, I; Haljan, PC; Schweikhard, V; Cornell, EA. "Observation of long-lived vortex aggregates in rapidly rotating Bose-Einstein condensates", *Phys. Rev. Lett.*, 2003, 90, 170405.
[44] Fetter, AL. "Rotating vortex lattice in a Bose-Einstein condensate trapped in combined quadratic and quartic radial potentials", *Phys. Rev. A*, 2001, 64, 063608.
[45] Lundh, E. "Multiply quantized vortices in trapped Bose-Einstein condensates", *Phys. Rev. A*, 2002, 65, 043604.
[46] Kasamatsu, K; Tsubota, M; Ueda, M. "Giant hole and circular superflow in a fast rotating Bose-Einstein condensate", *Phys. Rev. A*, 2002, 66, 053606.
[47] Kavoulakis, GM; Baym, G. "Rapidly rotating Bose-Einstein condensates in anharmonic potentials", *New J. Phys.*, 2003, 5, 51.
[48] Aftalion, A; Danaila, I. "Giant vortices in combined harmonic and quartic traps", *Phys. Rev. A*, 2004, 69, 033608.
[49] Fischer, UR; Baym, G. "Vortex states of rapidly rotating dilute Bose-Einstein condensates", *Phys. Rev. Lett.*, 2003, 90, 140402.
[50] Jackson, AD; Kavoulakis, GM. "Vortices and hysteresis in a rotating Bose-Einstein condensate with anharmonic confinement", *Phys. Rev. A*, 2004, 70, 023601.
[51] Ghosh, TK. "Vortex formation in a fast rotating Bose-Einstein condensate", *Phys. Rev. A*, 2004, 69, 043606.
[52] Wilkin, NK; Gunn, JMF; Smith, RA. "Do attractive Bosons condense?", *Phys. Rev. Lett.*, 1998, 80, 2265-2268.
[53] Mottelson, B. "Yrast spectra of weakly interacting Bose-Einstein condensates", *Phys. Rev. Lett.*, 1999, 83, 2695-2698.
[54] Pethick, CJ; Pitaevskii, LP. "Criterion for Bose-Einstein condensation for particles in traps", *Phys. Rev. A*, 2000, 62, 033609.
[55] Lundh, E; Collin, A; Suominen, KA. "Rotational states of bose gases with attractive interactions in anharmonic traps", *Phys. Rev. Lett.*, 2004, 92, 070401.
[56] Kavoulakis, GM; Jackson, AD; Baym, G. "Rotating Bose-Einstein condensates with attractive interactions", *Phys. Rev. A*, 2004, 70, 043603.
[57] Collin, A. "Collective modes and the broken symmetry of a rotating attractive Bose gas in an anharmonic trap", *Phys. Rev. A*, 2006, 73, 013611.
[58] Collin, A; Lundh, E; Suominen, KA. "Center-of-mass rotation and vortices in an attractive Bose gas", *Phys. Rev. A*, 2005, 71, 023613.
[59] Bargi, S; Kavoulakis, GM; Reimann, SM. "Rotating Bose-Einstein condensates confined in an anharmonic potential", *Phys. Rev. A*, 2006, 73, 033613.
[60] Sakaguchi, H; Malomed, BA. "Localized matter-wave patterns with attractive interaction in rotating potentials", *Phys. Rev. A*, 2008, 78, 063606.
[61] Schweikhard, V; Coddington, I; Engels, P; Tung, S; Cornell, EA. "Vortex-lattice dynamics in rotating spinor Bose-Einstein condensates", *Phys. Rev. Lett.*, 2004, 93, 210403.
[62] Tung, S; Schweikhard, V; Cornell, EA. "Observation of vortex pinning in Bose-Einstein condensates", *Phys. Rev. Lett.*, 2006, 97, 240402.
[63] Kakarantzas, G; Ortigosa-Blanch, A; Birks, TA; Russell, PStJ; Farr, L; Couny, F; Mangan, BJ. "Structural rocking filters in highly birefringent photonic crystal fiber", *Opt. Lett.*, 2003, 28, 158-160.

[64] Reijnders JW; Duine, RA. "Pinning and collective modes of a vortex lattice in a Bose-Einstein condensate", *Phys. Rev. A*, 2005, 71, 063607.

[65] Pu, H; Baksmaty, LO; Yi, S; Bigelow, NP. "Structural phase transitions of vortex matter in an optical lattice", *Phys. Rev. Lett.*, 2005, 94, 190401.

[66] Bhat, R; Carr, LD; Holland, MJ. "Bose-Einstein condensates in rotating lattices", *Phys. Rev. Lett.,* 2006, 96, 060405.

[67] Kasamatsu, K; Tsubota, M. "Dynamical vortex phases in a Bose-Einstein condensate diven by a rotating optical lattice", *Phys. Rev. Lett.,* 2006, 97, 240404.

[68] Sakaguchi, H; Malomed, BA. "Two-dimensional matter-wave solitons in rotating optical lattices", *Phys. Rev. A*, 2007, 75, 013609.

[69] Sakaguchi, H; Malomed, BA. "Solitary vortices and gap solitons in rotating optical lattices", *Phys. Rev. A*, 2009, 79, 043606.

[70] Kartashov, YV; Malomed, BA; Torner, L. "Guiding-center solitons in rotating potentials", *Phys. Rev. A*, 2007, 75, 061602(R).

[71] Brazhnyi, VA; Konotop, VV. "*Theory of nonlinear matter waves in optical lattices*", *Mod. Phys. Lett. B*, 2004, 18, 627-651.

[72] Saito, H; Ueda, M. "Emergence of Bloch bands in a rotating Bose-Einstein condensate", *Phys. Rev. Lett.*, 2004, 93, 220402.

[73] Schwartz, S; Cozzini, M; Menotti, C; Carusotto, I; Bouyer, P; Stringari, S. "One-dimensional description of a Bose–Einstein condensate in a rotating closed-loop waveguide", *New J. Phys.*, 2006, 8, 162.

[74] Carretero-González, R; Frantzeskakis, DJ; Kevrekidis, PG. "Nonlinear waves in Bose–Einstein condensates: physical relevance and mathematical techniques", *Nonlinearity*, 2008, 21, R139-R202.

[75] Wen, LH; Luo, XB. "Formation and structure of vortex lattices in a rotating double-well Bose-Einstein condensate", *Laser Phys. Lett.*, 2012, 9, 618-624.

[76] Wen, L; Xiong, H; Wu, B. "Hidden vortices in a Bose-Einstein condensate in a rotating double-well potential", *Phys. Rev. A*, 2010, 82, 053627.

[77] Efremidis, NK; Sears, S; Christodoulides, DN; Fleischer, JW; Segev, M. "Discrete solitons in photorefractive optically induced photonic lattices", *Phys. Rev. E*, 2002, 66, 046602.

[78] Fleischer, JW; Carmon, T; Segev, M; Efremidis, NK; Christodoulides, DN. "Observation of discrete solitons in optically induced real time waveguide arrays", *Phys. Rev. Lett.*, 2003, 90, 023902.

[79] Fleischer, JW; Segev, M; Efremidis, NK; Christodoulides, DN. "Observation of two-dimensional discrete solitons in optically-induced nonlinear photonic lattices", *Nature,* 2003, 422, 147-150.

[80] Martin, H; Eugenieva, ED; Chen, ZG; Christodoulides, DN. "Discrete solitons and soliton-induced dislocations in partially coherent photonic lattices", *Phys. Rev. Lett.*, 2004, 92, 123902.

[81] Neshev, DN; Alexander, TJ; Ostrovskaya, EA; Kivshar, YS; Martin, H; Makasyuk, I; Chen, ZG. "Observation of discrete vortex solitons in optically induced photonic lattices", *Phys. Rev. Lett.*, 2004, 92, 123903.

[82] Fleischer, JW; Bartal, G; Cohen, O; Manela, O; Segev, M; Hudock, J; Christodoulides, DN. "Observation of vortex-ring "discrete" solitons in 2D photonic lattices", *Phys. Rev. Lett.*, 2004, 92, 123904.

[83] Freedman, B; Bartal, G; Segev, M; Lifshitz, R; Christodoulides, DN; Fleischer, JW. "Wave and defect dynamics in nonlinear photonic quasicrystals", *Nature*, 2006, 440, 1166-1169.

[84] Trompeter, H; Królikowski, W; Neshev, DN; Desyatnikov, AS; Sukhorukov, AA; Kivshar, YS; Pertsch, T; Peschel, U; Lederer, F. "Bloch oscillations and Zener tunneling in two-dimensional photonic lattices", *Phys. Rev. Lett.*, 2006, 96, 053903.

[85] Fischer, R; Trager, D; Neshev, DN; Sukhorukov, AA; Królikowski, W; Denz, C; Kivshar, YS. "Reduced-symmetry two-dimensional solitons in photonic lattices", *Phys. Rev. Lett.*, 2006, 96, 023905.

[86] Schwartz, T; Bartal, G; Fishman, S; Segev, M. "Transport and Anderson localization in disordered two-dimensional photonic lattices", *Nature*, 2007, 446, 52-55.

[87] Fleischer, JW; Bartal, G; Cohen, O; Schwartz, T; Manela, O; Freedman, B; Segev, M; Buljan, H; Efremidis, NK. "Spatial photonics in nonlinear waveguide arrays", *Opt. Express*, 2005, 13, 1780-1796.

[88] Lederer, F; Stegeman, GI; Christodoulides, DN; Assanto, G; Segev, M; Silberberg, Y. "Discrete solitons in optics", *Phys. Rep.*, 2008, 463, 1-126.

[89] Fleischhauer, M; Imamoğlu, A; Marangos, JP. "Electromagnetically induced transparency: Optics in coherent media", *Rev. Mod. Phys.*, 2005, 77, 633-673.

[90] Schmidt, H; Imamoğlu, A. "Giant Kerr nonlinearities obtained by electromagnetically induced transparency", *Opt. Lett.*, 1996, 21, 1936-1938.

[91] Wang, H; Goorskey, D; Xiao, M. "Enhanced Kerr nonlinearity via atomic coherence in a three-level atomic system", *Phys. Rev. Lett.*, 2001, 87, 073601.

[92] Li, Y; Malomed, BA; Feng, M; Zhou, J. "Arrayed and checkerboard optical waveguides controlled by the electromagnetically induced transparency", *Phys. Rev. A*, 2010, 82, 063813.

[93] Pang, W; Wu, J; Yuan, Z; Liu Y; Chen, G. "Lattice solitons in optical lattice controlled by electromagnetically induced transparency", *J. Phys. Soc. Jpn.*, 2011, 80, 113401.

[94] Wu, J; Feng, M; Pang, W; Fu, S; Li, Y. "The transmission of quasi-discrete solitons in resonant waveguide arrays activated by the electromagnetically induced transparency", *J. Nonlin. Opt. Phys. Mat.*, 2011, 20, 193-203.

[95] Li, Y; Pang, W; Fu, S; Malomed, BA. "Two-component solitons with a spatially modulated linear coupling: Inverted photonic crystals and fused couplers", *Phys. Rev. A*, 2012, 85, 053821.

[96] Kuznetsova, E; Kocharovskaya, O; Hemmer, PR; Scully, MO. "Two-component solitons with a spatially modulated linear coupling: inverted photonic crystals and fused couplers", *Phys. Rev. A*, 2002, 66, 063802.

[97] Turukhin, AV; Sudarshanam, VS; Shahriar, MS; Musser, JA; Ham, BS; Hemmer, PR. "Observation of ultraslow and stored light pulses in a solid", *Phys. Rev. Lett.*, 2002, 88, 023602.

[98] Gupta, S; Murch, KW; Moore, KL; Purdy, TP; Stamper-Kurn, DM. "Bose-Einstein condensation in a circular waveguide", *Phys. Rev. Lett.*, 2005, 95, 143201.

[99] Arnold, AS; Garvie, CS; Riis, E. "Large magnetic storage ring for Bose-Einstein condensates", *Phys. Rev. A,* 2006, 73, 041606(R).

[100] Lesanovsky, I; von Klitzing, W. "Time-averaged adiabatic potentials: versatile matter-wave guides and atom traps", *Phys. Rev. Lett.*, 2007, 99, 083001.

[101] Ryu, C; Andersen, MF; Cladé, P; Natarajan, V; Helmerson, K; Phillips, WD. "Observation of persistent flow of a Bose-Einstein condensate in a toroidal trap", *Phys. Rev. Lett.*, 2007, 99, 260401.

[102] Ramanathan, A; Wright, KC; Muniz, SR; Zelan, M; Hill III, WT; Lobb, CJ; Helmerson, K; Phillips, WD; Campbell, GK. "Superflow in a Toroidal Bose-Einstein Condensate: An Atom Circuit with a Tunable Weak Link", *Phys. Rev. Lett.,* 2011, 106, 130401.

[103] Salasnich, L; Parola, A; Reatto, L. "Thermodynamics of solitonic matter waves in a toroidal trap", *Phys. Rev. A*, 2006, 74, 031603(R).

[104] Kanamoto, R; Saito, H; Ueda, M. "Critical fluctuations in a soliton formation of attractive Bose-Einstein condensates", *Phys. Rev. A*, 2006, 73, 033611.

[105] Modugno, M; Tozzo, C; Dalfovo, F. "Detecting phonons and persistent currents in toroidal Bose-Einstein condensates by means of pattern formation", *Phys. Rev. A*, 2006, 74, 061601 (R).

[106] Bhat, R; Holland, MJ; Carr, LD. "Bose-Einstein condensates in rotating lattices", *Phys. Rev. Lett.*, 2006, 96, 060405.

[107] Carpentier AV; Michinel, H. "A ring accelerator for matter-wave solitons", *Eur. Phys. Lett.*, 2007, 78, 10002-.

[108] Mason, P; Berloff, NG. "Dynamics of quantum vortices in a toroidal trap", *Phys. Rev. A*, 2009, 79, 043620.

[109] Brand, J; Haigh, TJ; Zülicke, U. "Rotational fluxons of Bose-Einstein condensates in coplanar double-ring traps", *Phys. Rev. A*, 2009, 80, 011602 (R).

[110] Capuzzi, P; Jezek, DM. "Stationary arrays of vortices in Bose–Einstein condensates confined by a toroidal trap", *J. Phys. B: At. Mol. Opt. Phys.*, 2009, 42, 145301.

[111] Aftalion, A; Maso, P. "Rotation of a Bose-Einstein condensate held under a toroidal trap", *Phys. Rev. A*, 2010, 81, 023607.

[112] Smyrnakis, J; Magiropoulos, M; Kavoulakis, GM; Jackson, AD. "Solitary waves in mixtures of Bose gases confined in annular traps", *Phys. Rev. A*, 2010, 81, 063601.

[113] Zöllner, S; Bruun, GM; Pethick, CJ; Reimann, SM. "Bosonic and Fermionic dipoles on a ring", *Phys. Rev. Lett.*, 2011, 107, 035301.

[114] Zhou, ZW; Zhang, SL; Zhou, XF; Guo, GC; Zhou, X; Pu, H. "Quantum phase transition of Bose-Einstein condensates on a nonlinear ring lattice", *Phys. Rev. A*, 2011, 83, 043626.

[115] Abad, M; Guilleumas, M; Mayol, R; Pi, M; Jezek, DM. "Phase slippage and self-trapping in a self-induced bosonic Josephson junction", *Phys. Rev. A*, 2011, 84, 035601.

[116] Zhou, X; Zhang, S; Zhou, Z; Malomed, BA; Pu, H. "Bose-Einstein condensates in a ring-shaped trap with a nonlinear double-well potential", *Phys. Rev. A*, 2012, 85, 023603.

[117] Adhikari, SK. "Dipolar Bose-Einstein condensate in a ring or in a shell", *Phys. Rev. A*, 2012, 85, 053631.

[118] Smyrnakis, J; Bargi, S; Kavoulakis, GM; Magiropoulos, M; Kärkkäinen, K; Reimann, SM. "Mixtures of Bose gases confined in a ring potential", *Phys. Rev. Lett.*, 2009, 103, 100404.

[119] Kittel, C. *Introduction to Solid State Physics*, (Wiley: New York, 1995).

[120] Vakhitov, M; Kolokolov, A; "Stationary solutions of the wave equation in a medium with nonlinearity saturation", *Izvestiya VUZov Radiofizika*, 1973, 161020 [in Russian;

English translation: "Stationary solutions of the wave equation in a medium with nonlinearity saturation", *Radiophys. Quantum. Electron.*, 1973, 16, 783-789].

[121] Bergé, L. "Wave collapse in physics: principles and applications to light and plasma waves", *Phys. Rep.*, 1998, 303, 259-370.

[122] Kuznetsov, EA; Dias, F. "Bifurcations of solitons and their stability", *Phys. Rep.*, 2011, 507, 43-105.

[123] Sakaguchi, H; Malomed, BA. "Solitons in combined linear and nonlinear lattice potentials", *Phys. Rev. A*, 2010, 81, 013624.

[124] Yang, J; Lakoba, TI. "Universally-convergent squared-operator iteration methods for solitary waves in general nonlinear wave equations", *Stud. Appl. Math.*, 2007, 118, 153-197.

[125] Yang, J; Lakoba, TI. "Accelerated imaginary-time evolution methods for the computation of solitary waves", *Stud. Appl. Math.*, 2008, 120, 265-292.

[126] Qian, LC; Wall, ML; Zhang, S; Zhou, Z; Pu, H. "Bose-Einstein condensates on a ring with periodic scattering length: Spontaneous symmetry breaking and entanglement", *Phys. Rev. A*, 2008, 77, 013611.

[127] Zhou, ZW; Zhang, SL; Zhou, XF; Guo, GC; Zhou, X; Pu, H. "Quantum phase transition of Bose-Einstein condensates on a nonlinear ring lattice", *Phys. Rev. A*, 2011, 83, 043626.

[128] Zhou, XF; Zhang, SL; Zhou, ZW; Malomed, BA; Pu, H. "Bose-Einstein condensates in a ring-shaped trap with a nonlinear double-well potential", *Phys. Rev. A*, 2012, 85, 023603.

In: Nonlinear Performance and Characterization Methods … ISBN: 978-1-62808-093-3
Editor: Jingsong Wei

Chapter 9

SHARP CRITERION OF GLOBAL EXISTENCE FOR A CLASS OF NONLINEAR SCHRÖDINGER EQUATION

Chuang Xu[1] and Runzhang Xu[2,*]
[1]Department of Mathematics, Harbin Institute of Technology, Harbin, Heilongjiang, China
[2]College of Science, Harbin Engineering University, Harbin, China

ABSTRACT

This chapter deals with two problems. First, we discuss the nonhomogeneous nonlinear Shrödinger equation with critical exponents. Second, we discuss a class of nonlinear Shrödinger equations with combined power-type nonlinearities and harmonic potential. By constructing a constrained variational problem and so-called invariant manifolds of the evolution flow, we derive a sharp criterion for blowup and global existence of the solutions.

Keywords: Shrödinger equation, constrained variational problem, critical exponent, invariant manifold, global existence, combined power-type nonlinearities, harmonic potential, blowup

1. INTRODUCTION

In this chapter, we first study the nonhomogeneous nonlinear Shrödinger equation with critical exponents:

* Corresponding author e-mail: xurunzh@yahoo.com.cn.

$$i\varphi_t + \Delta\varphi - k_1(x)|\varphi|^{\frac{4}{N}}\varphi - k_2(x)|\varphi|^{p_2}\varphi = 0, \quad t>0, \ x\in\mathbb{R}^N, \varphi(0,x)=\varphi_0(x), \tag{1}$$

where $\varphi=\varphi(x,t):\mathbb{R}^N\times[0,T)\to\mathbb{C}$ is a complex valued wave function, and $0<T\leqslant+\infty$ is the maximal existence time. $N\geqslant 3$ is the space dimension, $i=\sqrt{-1}$, Δ is the Laplace operator on $\mathbb{R}^N$ and the nonlinear power exponent p_2 satisfies $1+\frac{4}{N}<p_2\leqslant 2$.

We assume that for $i=1,2$, k_i is a given C^1 function such that there are $k_{i,1}>0$, $k_{i,2}>0$ and $c_0>0$ such that:

$(H1)$ $\forall x\in\mathbb{R}^N, k_{i,1}\leqslant k_i(x)\leqslant k_{i,2}$,

$(H2)$ $\forall x\in\mathbb{R}^N, |\nabla k_i(x)|+|x\cdot\nabla k_i(x)|\leqslant c_0$,

$(H3)$ there is $x_{i,0}$, $k_i(x_{i,0})=k_{i,2}$.

Second, the following initial value problem for the nonlinear Schödinger equation with combined power-type nonlinearities and harmonic potential:

$$i\varphi_t + \Delta\varphi - |x|^2\varphi - \lambda_1|\varphi|^{q_1}\varphi - \lambda_2|\varphi|^{q_2}\varphi = 0, \ t>0, \ x\in\mathbb{R}^N, \ \varphi(0,x)=\varphi_0(x). \tag{2}$$

In mathematical physics, the Schödinger equation is the most fundamental equation in non-relativistic quantum mechanics, playing the same role as Hamilton's laws of motion. In pure mathematics, the Schödinger equation and its variants are one of the basic equations studied in the field of partial differential equations, and has applications to geometry, spectral and scattering theory, and to integrable systems [1].

When $\lambda_1=-1$, $\lambda_2=0$ and $p_1=3$, Equation (2) describes attractive Bose-Einstein condensation. The slowing of atoms by use of a cooling apparatus produces a singular quantum state known as a Bose condensate or Bose-Einstein condensate. Its the state of matter of a dilute gas of weakly interacting bosons, confined in an external potential and cooled to temperatures very near to absolute zero. Under such conditions, a large fraction of the bosons occupy the lowest quantum state of the external potential, and all wave functions overlap each other, at which point quantum effects become apparent on a macroscopic scale.

Many scientists studied Equation (1) for the special cases. When $k_i(x)=\lambda_i$ (constant), Tao, Visan and Zhang [2] established the global existence, blowup and most importantly the scattering in different energy spaces. When the nonlinear term becomes single pattern, many mathematicians studied the problem. Ginibre and Velo [3], and Kato [4] established the local and global existence in $H^1\mathbb{R}^N$; Glassey [5], and Ogawa and Tsutsumi [6, 7] studied the blowup properties for some initial data. Berestycki and Cazenave [8], Weinstein [9] and Zhang [10, 11] showed some sharp criteria for blowup and global existence by the variational method. When $k_1(x)=0$, $k_2(x)\neq k_0$ (constant), Merle [12] proved the existence of blowup solutions and found minimal blowup solutions in L^2. When $k_1(x) \neq k_0$ (constant), $k_2(x) = 0$, Shu and Zhang [13] demonstrated sharp criterion of blowup and global existence for a class of nonlinear Shrödinger equation with critical exponent.

It is also well-known that

$$i\varphi_t + \Delta\varphi + |\varphi|^{p-1}\varphi = 0, t > 0, x \in \mathbb{R}^N, \ \varphi(0,x) = \varphi_0(x) \tag{3}$$

is one of the basic evolution models for nonlinear waves in various branches of physics, such as the quantum mechanics and the field theory.

Many scientists have studied Equation (3). In [11], Zhang studied the global existence of Equation (3) and the relationship between the Schrödinger equation and its ground state. It was found that for sufficiently small initial data, the solutions of the Cauchy problem for Equation (3) globally exists on $t \in [0,\infty)$. (cfs. [14, 15, 16, 17, 18]). For the problem with harmonic potential, that is:

$$i\varphi_t + \Delta\varphi - |x|^2\varphi + |\varphi|^p\varphi = 0, t > 0, x \in \mathbb{R}^N, \varphi(0,x) = \varphi_0(x), \tag{4}$$

or

$$i\varphi_t + \Delta\varphi - V(x)\varphi + |\varphi|^p\varphi = 0, t > 0, x \in \mathbb{R}^N, \varphi(0,x) = \varphi_0(x), \tag{5}$$

There are also many mathematicians who addressed these problems. In [19], Chen and Zhang derived a global existence condition for the supercritical case for Equation (4), which coincides with the critical case. For Equation (5), Fujiwara [20] proved that the smoothness of Schrödinger kernel shows potential for quadratic growth, and Yajima [21] showed that for super-quadratic potentials, the Schrödinger kernel is nowhere C^1. From Oh [22], it is known that quadratic potentials are the highest order potentials for local well-posedness of the equation. When $1 < p < 1 + 4/N$, Zhang [23] proved that global solutions of the Cauchy problem of Equation (4) exist for any initial data in the energy space. When $p = 1 + 4/N$, Zhang [24] showed that there exists a sharp condition of the global existence. When $p > 1 + 4/N$, Cazenave [25], Carles [26, 27] and Tsurumi and Wadati [28] showed that the solutions of the Cauchy problem of Equation (4) blow up in finite time for some initial data, especially for a class of sufficiently large initial data. But the solutions of the Cauchy problem of Equation (4) globally exist for other initial data, especially for a class of sufficiently small initial data, see [26, 27, 28]. Moreover, Shu and Zhang [29] also studied Equation (4) for its global existence and blowup. For nonlinear Schrödinger equations with combined power-type nonlinearities

$$i\varphi_t + \Delta\varphi = \lambda_1|\varphi|^{q_1}\varphi + \lambda_2|\varphi|^{q_2}\varphi, t > 0, x \in \mathbb{R}^N, \varphi(0,x) = \varphi_0(x), \tag{6}$$

In [30], Shu and Zhang described the existence of standing wave associated with the ground states of Equation (6) with $\lambda_1 = -1$ and $\lambda_2 = 1$. Furthermore they obtained a sharp condition for blowup and global existence to the solutions of the Cauchy problem and showed for small initial data, that the solutions globally exist and proved the instability of the standing wave. Xu, Zhang and Wu [31] attained the vacuum isolating of solutions and the threshold result of global existence and nonexistence of solutions of Equation (7), i.e., the reaction-diffusion equation

$$\begin{aligned} &u_t - \Delta u = a|u|^{p-1}u - b|u|^{q-1}u, t>0, x\in\Omega, u(x,0)=u_0(x);\\ &u_t(x,0)=u_1(x), x\in\Omega, u(x,t)=0, x\in\partial\Omega, t\geqslant 0. \end{aligned} \tag{7}$$

From the work mentioned above, we are apt to consider the Schrödinger equation with combined power-type nonlinearities and harmonic potential. We generalize the results in [29].

In this chapter, we study the sharp criteria for global existence of Equations (1) and (2). We also strive for some difference in conditions that lead to global existence and blowup between the combined power-type and single power-type nonlinearities.

The main path is to establish a variational problem and derive some invariant manifolds. Then we can derive the sharp criterion for global existence for the solutions by applying the potential well theory Payne and Sattinger [32] put up with, and the concavity method by Levine [33].

This chapter is, to a large extent, inspired by Shu and Zhang [13]. Moreover, it is also inspired by Berestycki and Cazenave [8], Merle [12], Weinstein [9] and Zhang [34].

This chapter is organized as follows. In Section 2, we state some concerned preliminaries. In Section 3, we establish the variational problem and invariant manifolds for problem in Equation (1). In Section 4, we derive the sharp criterion for global existence for problem in Equation (1). In Section 5, we establish the variational problem and invariant manifolds for problem in Equation (1). In Section 6, we derive the sharp criterion for global existence for problem in Equation (1). In Section 7, we give some remarks on the results we derive and put up some unsolved problems.

2. PRELIMINARIES

For problems in Equations (1) and (2), we define the energy space in the course of nature by

$$H := \{\psi \in H^1(\mathbb{R}^N) : \int |x|^2 |\psi|^2\, dx < \infty\}. \tag{8)'$$

Here and hereafter, for simplicity, we denote $\int_{\mathbb{R}^N} dx$ by $\int dx$, H becomes a Hilbert space, continuously embedded into $H^1\mathbb{R}^N$, when endowed with the inner product as follows:

$$<\psi,\phi>:=\int\nabla\psi\nabla\bar{\phi}+\psi\bar{\phi}+|x|^2\psi\bar{\phi}dx,$$

whose associated norm we denote by $\|\cdot\|_H$. From Ginibre and Velo [3], and Kato [4], we can get the local well-posedness for the Cauchy problems in Equations (1) and (2) in energy space H. For more details, see Cazenave [25] and Merle [12].

First we have the following lemmas [25, 12].

Lemma 2.1: Let $\varphi_0\in H$. Then there exists a unique solution φ of the Cauchy problem in Equations (1) and (2), respectively) in $C([0,T];H)$ for some $T\in(0,\infty]$ (maximal existence time), and either $T=\infty$ (global existence) or else $T<\infty$ and

$\lim_{t\to T}\|\varphi\|_H=\infty$ (blow-up).

Lemma 2.2: Let $\varphi_0\in H$ and φ be a solution of the Cauchy problem in Equation (1) in $C([0,T];H)$, $\forall x_0\in\mathbb{R}^N, |x-x_0|\varphi_0\in L^2(\mathbb{R}^N)$. Set $J(t)=\int|x-x_0|^2|\varphi|^2\,dx$. Then one has

$$J''(t)=16E(\varphi_0)+\frac{4N}{N+2}\int(x-x_0)\cdot\nabla k_1(x)|\varphi|^{2+\frac{4}{N}}\,dx+\frac{8}{p_2+2}\int(x-x_0)\cdot\nabla k_2(x)|\varphi|^{2+p_2}\,dx. \tag{9}$$

Lemma 2.3: Let $\varphi_0\in H$ and φ be a solution of the Cauchy problem in Equation (2) in $C([0,T];H)$, $\forall x_0\in\mathbb{R}^N, |x-x_0|\varphi_0\in L^2(\mathbb{R}^N)$. Set $V(t)=\int|x-x_0|^2|\varphi|^2\,dx$. Then one has

$$V''(t)=8\int\left(|\nabla\varphi|^2-|x|^2|\varphi|^2+\frac{\lambda_1 Nq_1}{2(q_1+2)}|\varphi|^{q_1+2}+\frac{\lambda_2 Nq_2}{2(q_2+2)}|\varphi|^{q_2+2}\right)dx.$$

Next usinging similar arguments in [25, 5, 35], we have the following lemmas.

Lemma 2.4: Let $\varphi_0\in H$ and φ be a solution of the Cauchy problem in Equation (1) in $C([0,T];H)$. Set the energy functional

$$E(\varphi)=\int\frac{1}{2}|\nabla\varphi|^2-\frac{N}{2(N+2)}k_1(x)|\varphi|^{2+\frac{4}{N}}-\frac{1}{p_2+2}k_2(x)|\varphi|^{p_2+2}\,dx$$

Then one has

$$\int|\varphi|^2\,dx=\int|\varphi_0|^2\,dx, \tag{10}$$

$$E(\varphi) \equiv E(\varphi_0). \tag{11}$$

Lemma 2.5: Let $\varphi_0 \in H$ and φ be a solution of the Cauchy problem in Equation (2) in $C([0,T];H)$. Set the energy functional

$$E(\varphi) = \int \frac{1}{2}|\nabla\varphi|^2 + \frac{1}{2}|x|^2|\varphi|^2 + \frac{\lambda_1}{q_1+2}|\varphi|^{q_1+2} + \frac{\lambda_2}{q_2+2}|\varphi|^{q_2+2}\,dx,$$

Then one has

$$\int|\varphi|^2\,dx = \int|\varphi_0|^2\,dx, \tag{12}$$

$$E(\varphi) \equiv E(\varphi_0). \tag{13}$$

3. THE CROSS-CONSTRAINED VARIATIONAL PROBLEM AND INVARIANT MANIFOLDS FOR PROBLEM IN EQUATION (1)

For $u \in H$, we define the following functions for problem in Equation (1).

$$I(u) = \int \frac{1}{2}|\nabla u|^2 + \frac{1}{2}|u|^2 + \frac{1}{2}|x|^2|u|^2 - \frac{N}{2(N+2)}k_1(x)|u|^{2+\frac{4}{N}} + \frac{1}{p_2+2}k_2(x)|u|^{p_2+2}\,dx, \tag{14}$$

$$S(u) = \frac{1}{2}\int|\nabla u|^2\,dx, \tag{15}$$

$$Q(u) = \int \frac{1}{2}|\nabla u|^2 + \frac{1}{2}|x|^2|u|^2 - \frac{N}{2(N+2)}k_1(x)|u|^{2+\frac{4}{N}} + \frac{1}{p_2+2}k_2(x)|u|^{p_2+2}\,dx \tag{16}$$

and define the manifold

$$M := \{u \in H\setminus\{0\} : Q(u) = 0\}.$$

Now we consider the following constrained variational problem:

$$d = \inf_{u\in M} S(u). \tag{17}$$

First, from Equation (17) we have as follows.

Lemma 3.1: $d > 0$.

Proof: By Equation (15), it follows that $d \geqslant 0$. Now we claim that

$d > 0.$

By Sobolev embedding inequality, we have:

$$\int |u|^{2+\frac{4}{N}} dx \leqslant c_1 \left(\int |u|^2 dx \right)^{1+\frac{2}{N}}$$

and

$$\int |u|^{p_2+2} dx \leqslant c_2 \left(\int |u|^2 dx \right)^{\frac{p_2+2}{2}}.$$

Here and hereafter c_i and c denote various positive constants. From $Q(u) = 0$, it follows that

$$\begin{aligned}
\int |u|^2 &= \frac{N}{(N+2)} \int k_1(x) |u|^{2+\frac{4}{N}} + \frac{2}{p_2+2} \int k_2(x) |u|^{p_2+2} dx \\
&\leqslant c_1 \left(\int |u|^2 dx \right)^{1+\frac{2}{N}} + c_2 \left(\int |u|^2 dx \right)^{\frac{p_2+2}{2}} \\
&\leqslant \max \left\{ c_1 \left(\int |u|^2 dx \right)^{1+\frac{2}{N}}, c_2 \left(\int |u|^2 dx \right)^{\frac{p_2+2}{2}} \right\}.
\end{aligned}$$

Therefore, $\int u \|^2 dx \geqslant c$. By the Poincaré inequality, we have $\int |\nabla u|^2 dx \geqslant c$. Thus $d = \inf_{u \in M} S(u) \geqslant c > 0$.

Theorem 3.2: Define

$$K_+ := \{\phi \in H \setminus \{0\}, I(\phi) < d, Q(\phi) > 0\},$$

and

$$K_- := \{\phi \in H \setminus \{0\}, I(\phi) < d, Q(\phi) < 0\}.$$

Then K_+ and K_- are both the invariant manifolds under the flow generated by Equation (1), in other words, if $\varphi_0 \in K_+$ or $\varphi_0 \in K_-$, then the solution $\varphi(x,t)$ of the Cauchy problem in Equation (1) also satisfies $\varphi(x,t) \in K_+$ or $\varphi(x,t) \in K_-$ for any $t \in [0,T)$, respectively.

Proof: Let $\varphi_0 \in K_+$. By Lemma 2.1, there exists a unique $\varphi(x,t) \in C([0,T);H)$ with $T \leqslant \infty$ such that $\varphi(x,t)$ is a solution of the Cauchy problem in Equation (1).

From Equations (9) and (10), we have

$$I(\varphi) = I(\varphi_0),\ t \in [0,T).$$

Thus $I(\varphi_0) < d$ implies that $I(\varphi) < d$ for any $t \in [0,T)$.

Now we show $Q(\varphi) > 0$ for $t \in [0,T)$. If otherwise, from the continuity, there were a $t_1 \in [0,T)$ such that $Q(\varphi(x,t_1)) = 0$. Apparently, $\varphi(x,t_1) \neq 0$. Otherwise, by Equation (9), we have the contradiction that $Q(\varphi_0) > 0$. Hence $\varphi(x,t_1) \in M$. From the definition of d, we have that $S(\varphi(x,t_1)) \geqslant d$. Due to $I(\varphi) = S(\varphi) + Q(\varphi)$, we know that $I(\varphi(t_1) \geqslant d$, which contradicts $I(\varphi) < d$ for any $t \in [0,T)$. Hence we show that $Q(\varphi) > 0$ for $t \in [0,T)$. So K_+ is invariant under the flow generated by Equation (1).

By the same argument as the above, we can prove that K_- is also invariant under the flow generated by Equation (1). So far we complete the proof of the theorem.

4. Sharp Conditions for Global Existence of Problem in Equation (1)

Theorem 4.1: If $\varphi_0 \in K_+$, then the solution $\varphi(x,t)$ of the Cauchy problem in Equation (1) globally exists on $t \in [0,\infty)$.

Proof: First, let $\varphi_0 \in K_+$. Thus Theorem 3.2 implies that the solution $\varphi(x,t)$ of the Cauchy problem in Equation (1) satisfies that $\varphi(x,t) \in K_+$ for $t \in [0,T)$. For fixed $t \in [0,T)$, denote $\varphi(x,t) = \varphi$. Thus we have $I(\varphi) < d$, $Q(\varphi) > 0$. Since $I(\varphi) = S(\varphi) + Q(\varphi)$, we know that $\int |\nabla \varphi|^2\, dx < 2d$. By the Sobolev inequality and $I(\varphi) < d$, we have $\int |x|^2 |u|^2\, dx < \infty$, which indicates $\| u \|_H < \infty$. Hence φ globally exists on $t \in [0,\infty)$ in H.

Theorem 4.2: Assume there is an $x_0 \in \mathbb{R}^N$ such that

$$\forall x \in \mathbb{R}^N, (x - x_0) \cdot \nabla k_i(x) \leqslant 0, i = 1,2,$$

i.e., x_0 is a global maximum of both $k_1(x)$ and $k_2(x)$.

If $\varphi_0 \in K_-$ and $|x - x_0| \varphi_0 \in L^2(\mathbb{R}^N)$, then the solution $\varphi(x,t)$ of the Cauchy problem in Equation (1) blows up in finite time.

Proof: From $\varphi_0 \in K_-$, Theorem3.2 implies that the solution $\varphi(x,t)$ of the Cauchy problem in Equation (1) satisfies that $\varphi(x,t) \in K_-$ for $t \in [0,T)$. For $J(t) = \int |x|^2 |\varphi|^2 \, dx$, Equations (11) and (16) imply that

$$J''(t) \leqslant 16E(\varphi_0), \; t \in [0,T).$$

For fixed $t \in [0,T)$, there exists a $0 < \lambda < 1$ such that $Q(\lambda\varphi_0) = 0$, that is

$$\int |\varphi_0|^2 \, dx - \frac{N}{N+2} \lambda^{\frac{4}{N}} \int k_1(x) |\varphi_0|^{2+\frac{4}{N}} \, dx - \frac{2}{p_2+2} \lambda^{p_2} \int k_2(x) |\varphi_0|^{2+p_2} \, dx = 0. \tag{18}$$

It is clear that $\varphi_0 \neq 0$. By the definition of d, we have $S(\lambda\varphi_0) \geqslant d$.
Thus $I(\varphi_0) < S(\lambda\varphi_0)$, that is

$$\int |\nabla\varphi_0|^2 + |\varphi_0|^2 \, dx - \frac{N}{N+2} \lambda^{\frac{4}{N}} \int k_1(x) |\varphi_0|^{2+\frac{4}{N}} \, dx - \frac{2}{p_2+2} \lambda^{p_2} \int k_2(x) |\varphi_0|^{2+p_2} \, dx < \lambda^2 \int |\nabla\varphi_0|^2 \, dx. \tag{19}$$

By Equations (18) and (19), we have

$$(\lambda^2 - 1) \int |\nabla\varphi_0|^2 \, dx - (\lambda^{\frac{4}{N}} - 1) \int \frac{N}{N+2} k_1(x) |\varphi_0|^{2+\frac{4}{N}} - (\lambda^{p_2} - 1) \int \frac{2}{p_2+2} k_2(x) |\varphi_0|^{2+p_2}.$$

Notice that $0 < \lambda < 1$, $N \geqslant 2$ and $p_2 \leqslant 2$, it follows that

$$\int |\nabla\varphi_0|^2 \, dx - \int \frac{N}{N+2} k_1(x) |\varphi_0|^{2+\frac{4}{N}} - \int \frac{2}{p_2+2} k_2(x) |\varphi_0|^{2+p_2}.$$

Therefore, $E(\varphi_0) < 0$, which contributes to $J''(t) < 0$. By the same argument as that in [35] and due to Lemma 2.1, it must be the case that $T < \infty$, which implies that

$$\lim_{t \to T} \| \varphi \|_H = \infty,$$

i.e., the solution $\varphi(x,t)$ of the Cauchy problem in Equation (1) blows up in finite time in H.

5. The Variational Problem and Invariant Manifolds for Problem in Equation (2)

For $\varphi \in H$, we define the following functions for problem in Equation (2).

$$P(\varphi)=\int\frac{1}{2}|\nabla\varphi|^2+\frac{1}{2}|\varphi|^2+\frac{1}{2}|x|^2|\varphi|^2+\frac{\lambda_1}{q_1+2}|\varphi|^{q_1+2}+\frac{\lambda_2}{q_2+2}|\varphi|^{q_2+2}\,dx,$$

and

$$I_0(\varphi)=\int|\nabla\varphi|^2+|\varphi|^2+|x|^2|\varphi|^2+\frac{\lambda_1 Nq_1}{2(q_1+2)}|\varphi|^{q_1+2}+\frac{\lambda_2 Nq_2}{2(q_2+2)}|\varphi|^{q_2+2}\,dx.$$

Now we define the manifold

$$M:=\{\psi\in H\setminus\{0\}:I_0(\psi)=0\}.$$

We consider the following variational problem,

$$d_0=\inf_{\{\psi\in H\setminus\{0\}:I(\psi)=0\}}P(\psi). \tag{20}$$

One can prove the following lemma easily by the similar argument to that in [29].

Lemma 5.1: $d_0>0$.

Theorem 5.2: Define

$$W:=\{\psi\in H,P(\psi)<d_0,I_0(\psi)>0\}.$$

Then W is an invariant manifold of Equation (2), that is, if $\varphi_0\in W$, then the solution $\varphi(x,t)$ of the Cauchy problem in Equation (2) also satisfies $\varphi(x,t)\in W$ for any $t\in[0,T)$.

Proof: Let $\varphi_0\in W$. So we have $P(\varphi_0)<d_0$. By Lemma 2.1, there exists a unique $\varphi(x,t)\in C([0,T);H)$ with $T\leqslant\infty$ such that $\varphi(x,t)$ is a solution of the Cauchy problem in Equation (2). By Lemma 2.5, we have

$$P(\varphi)<d_0,\text{ for all, } t\in[0,T). \tag{21}$$

Now we show $I_0(\varphi)>0$ for $t\in[0,T)$. First note that $I_0(\varphi_0)>0$ due to $\varphi_0\in W$. Arguing by contradiction, from the continuity of $I_0(\varphi)$, there is a $t_2\in[0,T)$ such that $I_0(\varphi(x,t_2))=0$. If $\varphi(x,t_2)=0$, then by Lemma 2.5, we have $0=\int|\varphi(x,t_2)|^2\,dx=\int|\varphi_0|^2\,dx$, which indicates $\varphi_0=0$. So we have $I_0(\varphi_0)=0$, which contradicts $I_0(\varphi_0)>0$. If $\varphi(x,t_2)\neq0$, by Equation (20), we have $P(\varphi(x,t_2))\geqslant d_0$, which contradicts Equation (21). Therefore $I_0(\varphi)>0$ for all $t\in[0,T)$. From the above, we proved that $\varphi(x,t)\in W$ for any $t\in[0,T)$. This completes the proof of the theorem.

By the same argument as Theorem 5.2, we can get the following result.

Theorem 5.3: Define

$$V := \{\psi \in H, P(\psi) < d_0, I_0(\psi) < 0\}.$$

Then, V is an invariant manifold of Equation (2).

6. Sharp Conditions for Global Existence of Problem in Equation (2)

Theorem 6.1: If $\varphi_0 \in W$, then the solution $\varphi(x,t)$ of the Cauchy problem in Equation (2) globally exists on $t \in [0,\infty)$.

Proof: First, we let $\varphi_0 \in W$. Thus Theorem 5.2 implies that the solution $\varphi(x,t)$ of the Cauchy problem in Equation (2) satisfies that $\varphi(x,t) \in W$ for $t \in [0,T)$. For fixed $t \in [0,T)$, denote $\varphi(x,t) = \varphi$. Thus we have $P(\varphi) < d,\ I(\varphi) > 0$. From $I(\varphi) > 0$, we have

$$\begin{aligned}
&\int \frac{\lambda_1 N q_1}{2(q_1+2)} |\varphi|^{q_1+2} + \frac{\lambda_2 N q_1}{2(q_2+2)} |\varphi|^{q_2+2}\, dx \\
&> \int \frac{\lambda_1 N q_1}{2(q_1+2)} |\varphi|^{q_1+2} + \frac{\lambda_2 N q_2}{2(q_2+2)} |\varphi|^{q_2+2}\, dx \\
&> -\int |\nabla \varphi|^2 + |\varphi|^2 + |x|^2 |\varphi|^2 .
\end{aligned}$$

Hence

$$\begin{aligned}
d_0 &> P(\varphi) \\
&= \int \frac{1}{2}|\nabla \varphi|^2 + \frac{1}{2}|\varphi|^2 + \frac{1}{2}|x|^2|\varphi|^2 + \frac{\lambda_1}{q_1+2}|\varphi|^{q_1+2} + \frac{\lambda_2}{q_2+2}|\varphi|^{q_2+2}\, dx \\
&> \int \frac{1}{2}|\nabla \varphi|^2 + \frac{1}{2}|\varphi|^2 + |x|^2|\varphi|^2 - \frac{2}{Nq_1}\int |\nabla \varphi|^2 + |\varphi|^2 + |x|^2|\varphi|^2\, dx \\
&= \left(\frac{1}{2} - \frac{2}{Nq_1}\right) \int |\nabla \psi|^2 + |\varphi|^2 + |x|^2|\varphi|^2\, dx,
\end{aligned}$$

which implies

$$\int |\nabla\varphi|^2 + |\varphi|^2 + |x|^2|\varphi|^2\, dx < \frac{2Nq_1 d_0}{Nq_1 - 4}.$$

Therefore the local well-posedness implies that φ globally exists on $t \in [0,\infty)$.
This completes the proof of this theorem.

Theorem 6.2: If $\varphi_0 \in V$, then the solution $\varphi(x,t)$ of the Cauchy problem in Equation (2) blows up in finite time.

Proof: From $\varphi_0 \in V$, Theorem 5.2 implies that the solution $\varphi(x,t)$ of the Cauchy problem in Equation (2) satisfies that $\varphi(x,t) \in V$ for $t \in [0,T)$. For $V(t) = \int |x|^2 |\varphi|^2\, dx$, Lemma 2.3 implies that

$$\begin{aligned} &V''(t) \\ &= 8\int \left(|\nabla\varphi|^2 - |x|^2|\varphi|^2 + \frac{\lambda_1 N q_1}{2(q_1+2)}|\varphi|^{p_1+2} + \frac{\lambda_2 N q_2}{2(q_2+2)}|\varphi|^{q_2+2} \right) dx \\ &< 8\int \left(|\nabla\varphi|^2 + |\varphi|^2 + |x|^2|\varphi|^2 + \frac{\lambda_1 N q_1}{2(q_1+2)}|\varphi|^{q_1+2} + \frac{\lambda_2 N q_2}{2(q_1+2)}|\varphi|^{q_1+2} \right) dx - 16\int |x|^2|\varphi|^2\, dx \\ &< -16\int |x|^2|\varphi|^2\, dx = -16V(t). \end{aligned}$$

Now we show that there exists a $T_1 \in (0,\infty)$ such that $V(t) > 0$ for $t \in [0,T_1]$ and $V(T_1) = 0$. Arguing by contradiction, suppose $\forall t \in [0,\infty)$, $V(t) > 0$. Set

$$g(t) = \frac{V'(t)}{V(t)}.$$

It is easy to show that

$$g'(t) = \frac{V''(t)}{V(t)} - \left(\frac{V'(t)}{V(t)}\right)^2 < -16 - g^2(t). \tag{22}$$

Next we like to show $g(t) \neq 0$ for any $t \in [0,\infty)$. Once again arguing by contradiction, suppose that there is a t_0 such that $g(t_0) = 0$, i.e., $V'(t_0) = 0$. By $V''(t) < 0$, we have $V'(t) < 0$ for $t \in (t_0,\infty)$. Hence we have $g(t) < 0$ for $t \in (t_0,\infty)$.

For any fixed $t_1 > t_0$, dividing Equation (22) by $g^2(t)$, we have

$$\frac{g'(t)}{g^2(t)} < -\frac{16}{g^2(t)} - 1 < -1.$$

Further we derive

$$\int_{t_1}^{t} \frac{g'(\tau)}{g^2(\tau)} d\tau < \int_{t_1}^{t} -1 d\tau,$$

namely,

$$\frac{1}{g(t)} > \frac{1}{g(t_1)} + (t - t_1), \tag{23}$$

which indicates there exists a $t_2 > t_1$ such that

$$g(t) > 0, t \in (t_2, \infty). \tag{24}$$

This contradicts $g(t) < 0$ for $t \in (t_0, \infty)$. Hence we have $g(t) \neq 0$ for any $t \in [0, \infty)$. By Equations (23) and (24), for $t \in (0, \infty)$, we have

$$\frac{1}{g(t)} > \frac{1}{g(0)} + t.$$

Hence, $V'(t) > 0$ for $t \in [|\frac{1}{g(0)}|, \infty)$. Therefore $V(t)$ is increasing on $(|\frac{1}{g(0)}|, \infty)$ and

$$V''(t) < -16V(t) < 0.$$

Further we have $\int_0^t V''(\tau) d\tau < -16V(0)t$, i.e.,

$$V'(t) - V'(0) < -16V(0)t,$$

that is,

$$V'(t) < V'(0) - 16V(0)t$$

Again, we have

$$\int_0^t V'(\tau) d\tau < V'(0)t - 8V(0)t^2.$$

Namely

$$V(t)-V(0)<V'(0)t-8V(0)t^2.$$

Therefore we have

$$V(t)<V(0)+V'(0)t-8V(0)t^2,$$

which contradicts $V(t)>0$ for $t\in[0,\infty)$. So far we have shown that there exists a $T_1\in(0,\infty)$ such that $V(t)>0$ for $t\in[0,T_1]$ and $V(T_1)=0$.

Again by the inequality [35],

$$\|\varphi\|^2\leqslant\frac{2}{N}\|\nabla\varphi\|\cdot\|x\varphi\|,$$

we get

$$\lim_{t\to T_1}\|\nabla\varphi\|=\infty,$$

which indicates

$$\lim_{t\to T_1}\|\varphi\|_H=\infty,$$

i.e., the solution of the Cauchy problem in Equation (2) blows up in finite time. This completes the proof of the theorem.

7. Some Remarks and an Open Problem

Remark 7.1: It is clear that

$$\{\phi\in H\setminus\{0\},I(\phi)<d\}=K_+\cup K_-.$$

Thus Theorem 4.2 shows that Theorem 4.1 is sharp.

Remark 7.2: It is easy to know that the prerequisite $p_2\leqslant 2$ can be relaxed to $p_2<\dfrac{4}{N-2}$, that is the condition can be removed for Theorem 4.1.

By Remark 7.2, we immediately have the corollary as follows.

Corollary 7.3: Let $\varphi_0 \in H\setminus\{0\}$ and satisfy $\|\varphi_0\|_H^2 < 2d$, then the solution φ of the Cauchy problem in Equation (1) globally exists on $t\in[0,\infty)$.

Proof: From $\|\varphi_0\|_H^2 < 2d$, we have $I(\varphi_0) < d$. Moreover, we claim that

$$Q(\varphi_0) > 0.$$

Otherwise, there exists a $0<\mu\leqslant 1$ such that $Q(\mu\varphi_0)=0$. Thus $S(\mu\varphi_0)\geqslant d$. Notice that $I(\varphi_0)=S(\varphi_0)+Q(\varphi_0)$. It follows that $I(\mu\varphi_0)\geqslant d$. On the other hand,

$$\|\mu\varphi_0\|_H^2 = \mu^2\|\varphi_0\|_H^2 < 2\mu^2 d \leqslant 2d.$$

It follows that $I(\mu\varphi_0)<d$. This is a contradiction. Therefore we have $\varphi_0\in K_+$. Thus Theorem 4.1 implies this corollary.

Remark 7.4: It is easy to show that $\varphi_0=0$ leads to a trivial solution $\varphi=0$ of the Cauchy problem in Equation (1).

Remark 7.5: When $k_2(x)\equiv 0$, that is the combined-power type reduces to a single term, all the main results are applicable to the equation discussed in [13], which means that this chapter generalizes the results in [13].

Remark 7.6:
It is clear that

$$\{\psi\in H, P(\psi)<d\} = W\cup V\cup\{0\}.$$

Thus Theorem 6.1 and Theorem 6.2 are both sharp.

By Theorem 6.1, we get another condition for global existence of the solution of Equation (2).

Corollary 7.7:
If φ_0 satisfies $\|\varphi_0\|_H^2 < 2d_0$, then the solution φ of the Cauchy problem in Equation (2) globally exists on $t\in[0,\infty)$.

Proof: If $\varphi_0 = 0$, then $\varphi = 0$. Hence it is obvious that the solution φ of the Cauchy problem in Equation (2) globally exists on $t \in [0, \infty)$. If $\varphi_0 \neq 0$, from $\| \varphi_0 \|_H^2 < 2d_0$, we have $P(\varphi_0) < d_0$. Moreover, we claim that

$$I_0(\varphi_0) > 0.$$

If otherwise, there is a $0 < \mu \leqslant 1$ such that $I_0(\mu\varphi_0) = 0$. Thus $P(\mu\varphi_0) \geqslant d$. On the other hand,

$$\| \mu\varphi_0 \|_H^2 = \mu^2 \| \varphi_0 \|_H^2 < 2\mu^2 d_0 < 2d_0.$$

It follows that $P(\mu\varphi_0) < d$. This is a contradiction. Therefore we have $\varphi_0 \in W$. Thus Theorem 6.1 implies this corollary.

Remark 7.8: It is easy to show that by setting $\lambda_1 = 0$, $\lambda_2 = -1$ (or $\lambda_2 = 0, \lambda_1 = -1$), results in [29] can be derived.

Remark 7.9: Section 4 and Section 5 are subtracted from [1].

An Open Problem: Can the prerequisite be removed for Theorem 4.2? In other words, is there another way to find a sharp criterion for blowup without the assumption in Theorem 4.2?

ACKNOWLEDGMENTS

This work was supported by the National Natural Science Foundation of China (No.11101102, No.11101104, No.11031002), Ph.D. Programs Foundation of Ministry of Education of China (20102304120022), the Support Plan for the Young College Academic Backbone of Heilongjiang Province (1252G020), the Natural Science Foundation of Heilongjiang Province (A201014), Science and Technology Research Project of Department of Education of Heilongjiang Province (12521401), Foundational Science Foundation of Harbin Engineering University and Fundamental Research Funds for the Central Universities (HEUCF20131101). The first author was supported by China Scholarship Council.

REFERENCES

[1] Xu, RZ; Xu, C. "Nonlinear Schrödinger equation with combined power-type nonlinearities and harmonic potential", *Appl. Math. Mech.*, 2010, 31, 521-528.

[2] Tao, T; Visan, M; Zhang, X. "The nonlinear Schrödinger equation with combined power-type nonlinearities", *Commun in Partial Differential Equations*, 2007, 32, 1281-1343.

[3] Ginibre, J; Velo, G. "On a class of nonlinear Schrödinger equations", *J. Funct. Anal.*, 1979, 32, 1-71.

[4] Kato, T. "On nonlinear Shrödinger equations", *Annales de I'institut Henri Poincaré (A) Physique théorique*, 1987, 49, 113-129.

[5] Glassey, RT. "On the blowup of nonlinear Schrödinger equations", *J. Math. Phys.*, 1977, 18, 1794-1797.

[6] Ogawa, T ; Tsutsumi, Y. "Blow-up of H^1 solution for the nonlinear Shrödinger equation", *J. Differential Equations*, 1991, 92, 317-330.

[7] Ogawa, T ; Tsutsumi, Y. "Blow-up of H^1 solution for the nonlinear Shrödinger equation with critical power nonlinearity", *Proc. Am. Math. Soc.*, 1991, 111, 487-496.

[8] Berestycki, H; Cazenave, T. "Instabilité des états stationnaires dans les équations de Schrödinger et de Klein-Gordon non linéarires", *C. R. Acad. Sci., Ser. I: Math.*, 1981, 293, 489-492.

[9] Weinstein, MI. "Nonlinear Schrödinger equations and sharp interpolations estimates", *Commun. Math. Phys.*, 1983, 87, 567-576.

[10] Zhang, J. "Cross-constrained variational problem and nonlinear Schrödinger equations", *Foundation of Computational Mathematics-Proceedings of the Smalefest*, 2001, 457-469.

[11] Zhang, J. "Sharp conditions of global existence for nonlinear Schrödinger and Klein-Gordon equations", *Nonlinear Anal. Theory, Methods Appl.*, 2002, 48, 191-207.

[12] Merle, F. "Nonexistence of minimal blow-up solutions of equation $iu_t = -\Delta u - k(x)|u|^{\frac{4}{N}u}$ in $\mathbb{R}^N$", *Ann. Inst. Henri Poincaré, Physique Théorique*, 1996, 64, 33-85.

[13] Shu, J ; Zhang, J. "Sharp criterion of global existence for a class of nonlinear Schrödinger equation with critical exponent", *Appl. Math. Comput.*, 2006, 182, 1482-1487.

[14] Ginibre, J; Velo, G. "The Cauchy problem for the nonlinear Schrödinger equation revisited", *Ann. Inst. Henri. Poincaré, Analysé non linéaire*, 1985, 2, 309–327.

[15] Ginibre, J; Ozawa, T. "Long range scattering for nonlinear Schr¨odinger and Hartree equations in space dimensions $n \geqslant 2$", *Commun. Math. Phys.*, 1993, 151, 619–645.

[16] Hayashi, N; Nakamitsu, K; Tsutsumi, M. "On solutions of the initial value problem for the nonlinear Schr¨odinger equations", *J. Funct. Anal.*, 1987, 71, 218–245

[17] Hayashi, N; Tsutsumi Y. "Scattering theory for Hartree type equations", *Annales de I'institut Henri Poincaré (A) Physique théorique*, 1987, 46, 187–213.

[18] Kenig, C; Ponce, G; Vega, L. "Small solution to nonlinear Schrödinger equations", *Annales de. I'Institut Henri Poincaré Nonlinear Analysis*, 1993, 10, 255–288.

[19] Chen, GG; Zhang, J. "Remarks on global existence for the supercritical nonlinear Schrödinger equation with a harmonic potential", *J. Math. Anal. Appl.*, 2006, 320, 591–598.

[20] Fujiwara, D. "Remarks on convergence of the Feynman path integrals", *Duke Math. J.*, 1980, 47, 559–600.

[21] Yajima, K. "On fundamental solution of time dependent Schrödinger equations", *Contemp. Math.*, 1998, 217, 49–68.

[22] Oh, YG. "Cauchy problem and Ehrenfest's law of nonlinear Schrödinger equations with potentials", *J. Differential Equations*, 1989, 81, 255–274.

[23] Zhang, J. "Stability of standing waves for nonlinear Schrödinger equations with unbounded potentials", *Zeitschrift für Angewandte Mathematik und Physik*, 2000, 51, 498–503.

[24] Zhang, J. "Stability of attractive Bose-Einstein condensates", *J. Stat. Phys.*, 2000, 101, 731–746.

[25] Cazenave, T. *An introduction to nonlinear Schrödinger equations*. Textos de Metodos Matematicos, Rio de Janeiro, 1993.

[26] Carles, R. "Critical nonlinear Schrödinger equation with and without harmonic potential", *Math. Mod. Meth. Appl. S.*, 2002, 12, 1513–1523.

[27] Carles, R. "Remarks on the nonlinear Schrödinger equation with harmonic potential", *Annales Henri Poincaré.*, 2002, 3, 757–772.

[28] Tsurumi, T; Wadati, M. "Collapses of wave functions in multidimensional nonlinear Schrödinger equations under harmonic potential", *J. Phys. Soc. Jpn.*, 1997, 66, 3031–3034.

[29] Shu, J ; Zhang, J. "Nonlinear Shrödinger equation with harmonic potential", *J. Math. Phys.*, 2006, 47, 063503.

[30] Shu, J; Zhang, J. "Instability of standing waves for a class of nonlinear Schrödinger equations", *J. Math. Anal. Appl.*, 2007, 327, 878–890.

[31] Xu, RZ; Zhang, WY; Wu, WN. *Nonlinear Analysis Research Trends,* Nova Science Publishers, Inc., 2008, 259–281.

[32] Payne, LE ; Sattinger, DH. "Saddle points and instability of nonlinear hyperbolic equations", *Israel J. Math.*, 1975, 22, 273-303.

[33] Levine, HA. "Instability and non-existence of global solutions to nonlinear wave equations of the form $Pu_t t = -Au + F(u)$", *Trans. Am. Math. Soc.*, 1974, 192, 1-21.

[34] Zhang, J. "Sharp threshold of blowup and global existence in nonlinear Schrödinger equations under a harmonic potential", *Commun. Partial Differential Equations*, 2005, 30, 1429-1443.

[35] Tsutsumi, Y ; Zhang, J. "Instability of optical solutions for two-wave interaction model in cubic nonlinear media", *Adv. Math. Sci. Appl.*, 1998, 8, 691-713.

In: Nonlinear Performance and Characterization Methods … ISBN: 978-1-62808-093-3
Editor: Jingsong Wei

Chapter 10

QUENCHING PHENOMENA FOR SECOND AND FOURTH ORDER NONLINEAR PARABOLIC EQUATIONS WITH SPECIAL ABSORPTION TERM

***Junsheng Zhao*[1*], *Xiaotong Qiu*[2] *and Senyang Wu*[3]**
[1]College of Mathematical Science, Heilongjiang University, China
[2]College of Science, Harbin Engineering University, China
[3]School of Accounting, Harbin Commercial University, China

ABSTRACT

In this chapter, we introduced the quenching phenomena of the Cauchy problem for the second-order nonlinear parabolic equation on unbounded domain and fourth-order nonlinear parabolic equations with special absorption term. By some assumptions on the exponents and initial data for a class of equations with the special absorption term, we did not only obtain the quenching phenomena in finite time but also estimated the quenching time. Our main tools were comparison and maximum principles. We extended the result to the case of a more general nonlinear absorption.

Keywords: Quenching phenomena, unbounded domain, special absorption term

1. INTRODUCTION

In this chapter, we consider the initial boundary value problem of a nonlinear parabolic equation with nonlocal nonlinear source

* Corresponding author e-mail:zhaojunsh@yeah.net.

$$\begin{cases} \dfrac{\partial u}{\partial t} - \Delta u = \dfrac{\alpha\left(1+|x|\right)^{m_1}}{\left(1-u\right)^n}, & (x,t) \in \mathbb{R}^N \times (0,T) \ , \\ u(x,0) = 0, & x \in \mathbb{R}^N \ , \\ u(x,t) \to 0, & |x| \to \infty \ , \end{cases} \tag{1}$$

where $m_1, n \in (-\infty, +\infty)$ are real numbers and $\alpha > 0$, and the fourth-order nonlinear parabolic equation

$$\begin{cases} \dfrac{\partial u}{\partial t} - \Delta^2 u = \dfrac{\alpha\left(1+|x|\right)^{m_2}}{1-u}, & (x,t) \in \mathbb{R}^N \times (0,T) \\ u(x,0) = 0, & x \in \mathbb{R}^N \ , \\ u(x,t) \to 0, & |x| \to \infty \ , \end{cases} \tag{2}$$

where $m_2 \geq 0$, $\alpha > 0$, $\Delta = \sum_{i=1}^{N} \frac{\partial^2}{\partial x^2}$ and $\Delta^2 u = \Delta(\Delta u)$.

For convenience, we first present the following definition of the quenching phenomenon.

Definition 1

Let $u(x,t)$ be a classical solution of problem in Equation (1) or Equation (2). We say that $u(x,t)$ quenches in finite time if there exists a real number $T \in (0,\infty)$ such that

$$\lim_{t \to T} \sup_{x \in \Omega} u(x,t) = b.$$

The quenching problem has a 36 year history. In 1975, Kawarada first introduced the concept of quenching when considering a famous initial boundary problem for the parabolic equation $u_t = u_{xx} + 1/(1-u)$ and derived many interesting results (see [1-6] and references therein). In [5], Acker and Walter had considerably sharpened and generalized Kawardada's results. Levine [6] thoroughly considered the following problem which greatly generalized the results of [1] and [5]

$$\begin{cases} u_t = u_{xx} + \varepsilon (1-u)^{-\beta}, & 0<x<1,\ 0<t<T, \\ u(0,t) = u(1,t) = 0, & 0 \leq t \leq T, \\ u(x,0) = u_0(x), & u_0 < 1,\ 0 \leq x \leq 1, \end{cases} \tag{3}$$

then gave the criteria for the quenching, nonquenching and beyond quenching of the solution. There are many authors who extended the results of Equation (3) in [7-11]. In [7], Levine determined quenching sets and derived quenching rate estimates. Furthermore, many scientists [12-14] presented recent progress in the analysis of the quenching phenomena on a bounded domain. It's clear that these results were obtained in a bounded domain.

When compared to the case of bounded domain, it seems that very few results concerning the quenching phenomena were obtained on unbounded domain (see [15, 16]). In [17], Dai investigated the quenching phenomena of problem in Equation (1) when $n = 1$ and obtained some results that the solution quenches in finite time. In addition, it was shown in [1, 10, 17] that the absorbing term $(1-u)^{-1}$ played an essential role in quenching phenomena. In the case that the absorbing term is $(1-u)^{-n}$, will the solution quench in finite time? How about is the higher order? The questions above are still open-ended up to now. We will restrict our attention to more complex problems and the cases for $(1-u)^{-n}$ and the fourth-order.

2. Preliminaries about Quenching Phenomena

In this section, we give some definitions and recall some preliminary lemmas.

For problem of Equations (1) and (2), we assume that there are constants $c_1 > 0$ and $c_2 \in \mathbb{R}$, we can definition four conditions for $g(u):[0,b) \mapsto (0,\infty)$.

(G1): $g(u)$ is a locally Lipschitz on $[0,b)$ and $g(0) > 0$,

(G2): $\lim_{u\to b^-} g(u) = +\infty$,

(G3): $g(u) \ge c_1 + c_2 u, \quad u \in [0,b)$,

(G4): $c_2 + \dfrac{c_1}{b} > 0$.

Let $\Omega \subset \mathbb{R}^N$ be a bounded domain, $\lambda_1(\Omega)$ and $\Psi_1(x)$ denote the first eigenvalue and the first eigenfunction of the following eigenvalue problem, respectively.

$$\begin{cases} \Delta\Psi + \lambda\Psi = 0, & x \in \Omega, \\ \Psi = 0, & x \in \partial\Omega. \end{cases}$$

For convenience, we choose $\Psi_1(x)$ such that

$$\Psi_1(x) > 0, \quad x \in \Omega$$

and

$$\int_\Omega \Psi_1(x)dx = 1.$$

One can take $p(t,x,y)=(4\pi t)^{-N/2}\exp\left(-\frac{|x-y|^2}{4t}\right)$, denote the fundamental solution of the heat operator and set

$$I_m(x,t)=\int_{\mathbb{R}^N} p(t,x,y)(1+|y|)^m dy.$$

Then, we have

Lemma 2.1 ([9], Lemma 1)

If $m\geq 0$ and $I_m(x,t)\leq 1$ for $t\geq 0$.

Lemma2.2: ([10], Lemma 6)

If $m\geq 0$ and $t\geq 0$, then the function $I_m(x,t)$ attains its maximum at $x=0$.

Lemma 2.3: ([9], Lemma 3)

Assume that $m\in\mathbb{R}$ and $t\geq 1$. Then we have

- If $m>-N$, then there exist two positive constants C_1, C_2 such that $C_1 t^{m/2}\leq I_m(0,t)\leq C_2 t^{m/2}$,
- If $m=-N$, then there exist two positive constants C_1, C_2 such that $C_1 t^{-N/2}\ln(1+t)\leq I_m(0,t)\leq C_2 t^{-N/2}\ln(1+t)$.
- If $m<-N$, then there exist two positive constants C_1, C_2 such that $C_1 t^{-N/2}\leq I_m(0,t)\leq C_2 t^{-N/2}$.

3. SECOND-ORDER NONLINEAR PARABOLIC EQUATION

In this section, we consider the initial boundary value problem (IBVP) of a nonlinear parabolic equation with Equation (1). For the problem, we can obtain the following results.

Theorem 3.1

If $m=0$, $n>0$, then the solution of problem in Equation (1) always quenches in finite time for any $\alpha>0$.

Proof: if $m=0$, $n>0$, then the absorption of problem in Equation (1) is $\alpha/(1-u)^n$, which can be defined by

$$g(u)=\frac{\alpha}{(1-u)^n}.$$

By taking $b = 1$, $c_1=\alpha$ and $c_2=n\alpha$, we have $g(u):[0,b)\mapsto(0,\infty)$ which satisfies

(G1): $g(u)$ is a locally Lipschitz on $[0,b)$ and $g(0)>0$,

(G2): $\lim_{u\to b^-} g(u)=+\infty$,

(G3): $g(u)\geq c_1+c_2u, \quad u\in[0,b)$,

(G4): $c_2+\frac{c_1}{b}>0$.

In view of [11], we can know that the solution of problem in Equation (1) always quenches in finite time for any $\alpha>0$. In fact, we take $u(x,t)\to 0$ as $|x|\to\infty$, then we can get a real number $M>0$ so that $u(x,t)<\varepsilon$ as $x>M$. Hence

$$u_x \to 0 \quad as \quad |x|\to\infty.$$

We assume by contradiction that there exist two points $x_1,x_2>M$ such that for all $x\in[x_1,x_2]$

$$u_x>\delta>0 \quad or \quad u_x<\delta<0.$$

Without loss of generality, we may assume that $u_x>\delta>0$. Then, we can obtain

$$u>u_x(x_2-x_1)>\delta(x_2-x_1)>0.$$

By taking eventually $\varepsilon=\delta(x_2-x_1)$ we conclude $u(x,t)>\varepsilon$, which is impossible. Moreover, similar calculations lead to

$$u_{xx}\to 0 \quad as \quad |x|\to\infty \tag{4}$$

To prove that $\frac{\partial u}{\partial t}\to 0$ as $|x|\to\infty$, we assume by contradiction that there exists $\varepsilon>0$ such that

$$\frac{\partial u}{\partial t}>\varepsilon.$$

We integrate partially over $[0,t]$ to obtain

$$u>\varepsilon t+C,$$

which contradicts the fact that $u(x,t)<\varepsilon$. Thus, we prove that $\frac{\partial u}{\partial t}\to 0$ as $|x|\to\infty$. Therefore, combining this with Equation (4), we get

$$\frac{\partial u}{\partial t}\text{-}\Delta u\to 0 \quad as \quad |x|\to\infty,$$

which is contradictory to problem in Equation (1). So we imply that the solution of problem in Equation (1) does not exist globally if $m=0$, n=0 .

If $m=0$, n<0 , then problem in Equation (1) becomes the following problem

$$\begin{cases} \frac{\partial u}{\partial t}-\Delta u=\alpha(1\text{-}u)^{-n}, & (x,t)\in\mathbb{R}^N\times(0,T) \\ u(x,0)=0, & x\in\mathbb{R}^N, \\ u(x,t)\to 0, & |x|\to\infty. \end{cases}$$

We can get

$$(1\text{-}u)^{-n}\to 1 \text{ as } |x|\to\infty,$$

thus

$$\frac{\partial u}{\partial t}\text{-}\Delta u\to\alpha\neq 0 \text{ as } |x|\to\infty.$$

This contradicts in Equation (1) and Theorem 3.1 is proved.

Theorem 3.2

Assume that $N\geq 3$, $n\in\mathbb{R}$. Then we have

1) If m<-2, then the solution of problem in Equation (1) exists globally for α small enough and quenches in finite time for α large enough.
2) If m<-2, then the solution of problem in Equation (1) always quenches in finite time for any $\alpha > 0$.

Proof: the mild solution of problem in Equation (1) is

$$u(x,t)=\alpha\int_0^t\int_{\mathbb{R}^N}p(t\text{-}s,x,y)\frac{(1+|y|)^m}{(1\text{-}u)^n}dyds$$

(I) In order to obtain global solution for α small enough, which suffices to prove by the priori assumption that

$$u(x,t)\le\frac{1}{2}\quad for\ all\ (x,t)\in\mathbb{R}^N\times(0,T) \tag{5}$$

Indeed, it is valid for α small enough, since the solution may be found by successive substitution in right hand side of the last equation.

By comparison principle, we obtain

$$u(x,t)\ge 0\quad for\ (x,t)\in\mathbb{R}^N\times(0,\infty) \tag{6}$$

Combining Equations (5) and (6) gives

$$\left(\frac{1}{2}\right)^{-n}\le(1\text{-}u)^{-n}\le 1.$$

Then we have

$$\begin{aligned}u(x,t)&=\alpha\int_0^t\int_{\mathbb{R}^N}p(t\text{-}s,x,y)\frac{(1+|y|)^m}{(1\text{-}u)}dyds\\&\le\alpha\int_0^t\int_{\mathbb{R}^N}p(t\text{-}s,x,y)(1+|y|)^m dyds\\&=\alpha\int_0^t I_m(x,t\text{-}s)ds,\end{aligned}$$

where $I_m(x,t\text{-}s)=\int_{\mathbb{R}^N}p(t\text{-}s,x,y)(1+|y|)^m dy$. We distinguish the following two cases as follows.

Case 1

If $t\le 1$ occurs, by Lemma 2.1 we conclude

$$\begin{aligned}u(x,t)&\le\alpha\int_0^t I_m(x,t\text{-}s)ds\\&\le\alpha\int_0^t ds\le\alpha.\end{aligned}$$

Case 2

We now suppose that $t>1$ occurs, then

$$u(x,t)\leq\alpha\int_0^t I_m(x,t\text{-}s)ds$$
$$=\alpha\int_0^{t-1} I_m(x,t\text{-}s)ds+\alpha\int_{t-1}^t I_m(x,t\text{-}s)ds.$$

By Lemma 2.1, Lemma 2.2 and Lemma 2.3, we have

$$u(x,t)\leq\alpha+\alpha\int_0^{t-1} I_m(0,t\text{-}s)ds$$
$$=\alpha\left(1+\int_0^{t-1} I_m(0,\theta)\right)d\theta.$$

Hence, there exists sufficiently small α such that

$$u(x,t)<\frac{1}{2},\quad (x,t)\in\mathbb{R}^N\times(0,\infty),$$

so we find that problem in Equation (1) has a global solution for $\alpha>0$ small enough.

Next, we are going to prove that the solution of problem in Equation (1) quenches in finite time for α large enough. By taking $t>1$ we have

$$u(x,t)=\alpha\int_0^t\int_{\mathbb{R}^N} p(t\text{-}s,x,y)\frac{(1+|y|)^m}{(1\text{-}u)^n}dyds$$
$$\geq\alpha\int_0^{t-1} I_m(x,t\text{-}s)\left(\frac{1}{2}\right)^{-n}ds. \tag{7}$$

Moreover, by Equation (7) we obtain

$$u(0,t)\geq\alpha\left(\frac{1}{2}\right)^{-n}\int_0^{t-1} I_m(0,t\text{-}s)ds$$
$$=\alpha\left(\frac{1}{2}\right)^{-n}\int_1^t I_m(0,\theta)d\theta. \tag{8}$$

By Lemma 2.3, this implies

$$0<\int_1^\infty I_m(0,\theta)d\theta=J_m<+\infty\quad for\ N\geq 3,m<\text{-}2.$$

Combining Equation (8) and $\alpha > \frac{(1/2)^n}{J_m}$, we get

$$u(0,t) > 1,$$

which is impossible. This implies that $u(x,t)$ quenches in finite time for α large enough. The proof is finished.

(II) We now prove (2) of this theorem. Arguing by contradiction, we assume that

$$u(x,t) < 1 \quad for\ (x,t) \in \mathbb{R}^N \times (0,\infty).$$

By applying Equation (8) and Lemma 2.3, we obtain

$$\lim_{x\to\infty} \int_1^t I_m(0,\theta)\,d\theta = +\infty \quad for\ any\ \alpha > 0, m \geq -2, N \geq 3.$$

This is a contradiction, which implies that $u(x,t)$ quenches in finite time for $\alpha > 0$.

Theorem 3.3

If $N \leq -2$, $n \in \mathbb{R}$, then for any $m \in (-\infty, +\infty)$ and $\alpha \in (0,\infty)$, the solution of problem in Equation (1) always quenches in finite time.

Proof: by supposing $1 \leq N \leq 2$ and $n < 0$, we obtain

$$\int_1^t I_m(0,\theta)\,d\theta \to +\infty \quad as\ t \to \infty \quad for\ any\ m \in (-\infty, +\infty).$$

A similar argument as the proof of (2) of this theorem immediately deduces Theorem 3.3. If we let $n = 0$, then

$$u(x,t) = \alpha \int_0^t \mathrm{I}_m(x,t-s)\,ds.$$

We have Theorem 3.3 by the same proof. If $n > 0$, then by a priori assume

$$u(x,t) \leq \frac{1}{2} \quad for\ (x,t) \in \mathbb{R}^N \times (0,\infty),$$

we have

$$1 \leq \frac{1}{(1-u)^n} \leq 2^n.$$

Similarly, for $n>0$, we prove Theorem 3.3.

4. FOURTH-ORDER NONLINEAR PARABOLIC EQUATION

In this section, we generalize quenching phenomena for fourth-order nonlinear parabolic equations with special absorption term

$$\begin{cases} \dfrac{\partial u}{\partial t} - \Delta^2 u = \dfrac{\alpha\left(1+|x|\right)^{m_2}}{1-u}, & (x,t) \in \mathbb{R}^N \times (0,T) \\ u(x,0) = 0, & x \in \mathbb{R}^N, \\ u(x,t) \to 0, & |x| \to \infty. \end{cases}$$

The fourth order parabolic equations had been researched for many years and many results had been obtained (see [18-22]). The extremum principle of the solutions of fourth order elliptic equations had been studied since 1972 (see [23]). Then others studied the extremum principle of the solutions of fourth order parabolic equations. But the research about the quenching phenomena of fourth-order nonlinear parabolic equations does not begin. So we introduce the quenching phenomena of fourth order nonlinear parabolic equations with special absorption term and obtain the following result.

Theorem 4.1

Let $u(x,t)$ be the classical solution of problem in Equation (1), then $u(x,t)$ must quench in a finite time $T_{\max}$ and for $T_{\max}$ we have the estimate

$$0 < T_{\max} \le \frac{1}{\alpha + \lambda_1^2} \ln \frac{2\alpha + \lambda_1^2}{\alpha}.$$

Proof: the absorption of problem in Equation (2) is $\dfrac{\alpha(1+|x|)^m}{1-u}$, which can be defined by

$g(u) = \dfrac{\alpha\left(1+|x|\right)^m}{1\text{-}u}$. If $u(x,t) \to 0$ as $|x| \to \infty$, then we can find $\Delta u \to 0$ as $|x| \to \infty$ and $\dfrac{\partial u}{\partial t} \to 0$ as $|x| \to \infty$.

In fact, we have

$$\frac{\partial u}{\partial t} = \lim_{\Delta t \to 0} \frac{u(x,t+\Delta t) - u(x,t)}{\Delta t}.$$

If we assume by contradiction that $\frac{\partial u}{\partial t} \neq 0$, there must be a real number $\varepsilon > 0$ so that $\left|\frac{\partial u}{\partial t}\right| > \varepsilon$. Then we have

$$|u(x,t+\Delta t) - u(x,t)| > \varepsilon \Delta t > 0.$$

Hence, we have $u(x,t) \neq 0$. This is impossible. So we have $\Delta u \to 0$ and $\frac{\partial u}{\partial t} \to 0$ as $|x| \to \infty$.

If $m \geq 0$, we get

$$g(u) = \frac{\alpha(1+|x|)^m}{1-u} \geq \frac{\alpha}{1-u}.$$

Then we can have

$$\lim_{u \to 1^-} g(u) \to \infty.$$

So we know $g(u)$ satisfies (G2).

If we take $h(u) = \frac{\alpha}{1-u}$, $c_1 = \alpha$, $c_2 = \alpha$, $b = 1$, then we may infer that

$$h(0) = \frac{\alpha}{1-0},$$

$$h'(0) = \frac{\alpha}{(1-0)^2},$$

$$h''(\eta) = \frac{2\alpha}{(1-\eta)^3} > 0, \eta \in (0,u).$$

By Taylor's expansion theorem, there is a real number $\varepsilon \in (0\ ,\ u)$ such that

$$h(u) = h(0) + h'(0)u + \frac{1}{2}h''(\varepsilon)u^2 > \alpha + \alpha u, u \in [0,1).$$

Since

$$g(u) = \frac{\alpha(1+|x|)^m}{1-u} \geq \frac{\alpha}{1-u},$$

we have

$$g(u) \geq \alpha + \alpha = 2\alpha > 0.$$

Then we can know $g(u)$ satisfies

(G3): $g(u) \geq c_1 + c_2 u$, $u \in [0,b)$,

(G4): $c_2 + \left(\frac{c_1}{b}\right) > 0$.

So $g(u)$ satisfies (G1)-(G4).

Because $u(x,t) \to 0$ as $|x| \to \infty$, we can get a large enough number $L > 0$ so that

$$u = 0 \text{ as } t = 0$$

and

$$u(x,t) \to 0 \text{ as } x \to L.$$

So we can get a large enough domain Ω which contains L.

Let $(0, T_{\max})$ be the maximum time interval in which the classical solution $u(x,t)$ of problem in Equation (1) exists. By (G1) and the comparison principle we have

$$0 < u(x,t), \ (x,t) \in \Omega \times (0, T_{\max}).$$

Since b is a singular point of $g(s)$ and $T_{\max}$ is maximal, we can conclude that if $T_{\max} < +\infty$, then

$$\lim_{t \to T_{\max}^-} \sup_{x \in \Omega} u(x,t) = b.$$

Otherwise, $u(x,t)$ can be extended beyond. This is impossible. Now in order to prove Theorem 4.1, it is sufficient to prove that T_{max} is finite and

$$0 < T_{\max} \le \frac{1}{\alpha + \lambda_1^2} \ln \frac{2\alpha + \lambda_1^2}{\alpha}.$$

Through the research of problem in Equation (1), we can know

$$\int_\Omega \Psi_1 \Delta^2 u dx = \lambda_1^2 \int_\Omega \Psi_1 u dx.$$

In fact, as $u = 0$,$for\ (x,t) \in \partial\Omega \times (0,T)$, we have

$$\begin{aligned}
\int_\Omega \Psi_1 \Delta u dx &= \Psi_1 \frac{\partial u}{\partial x}\Big|_{\partial\Omega} - \int_\Omega \frac{\partial \Psi_1}{\partial x} \frac{\partial u}{\partial x} dx \\
&= -\int_\Omega \frac{\partial \Psi_1}{\partial x} \frac{\partial u}{\partial x} dx \\
&= -\frac{\partial \Psi_1}{\partial x} u\Big|_{\partial\Omega} + \int_\Omega \Delta \Psi_1 u dx \\
&= -\lambda_1 \int_\Omega \Psi_1 u dx
\end{aligned}$$

If we take $\Delta u = v$, then we have

$$v = 0, (x,t) \in \partial\Omega \times (0,T).$$

Hence, we conclude that

$$\begin{aligned}
\int_\Omega \Psi_1 \Delta^2 u dx &= -\lambda_1 \int_\Omega \Psi_1 v dx \\
&= \lambda_1^2 \int_\Omega \Psi_1 u dx.
\end{aligned}$$

To estimate the quenching time, multiplying the differential equation in Equation (2) by $\Psi_1(x)$ and integration on Ω with respect to x , we can include

$$\frac{d}{dt} \int_\Omega u \Psi_1 dx - \lambda_1^2 \int_\Omega u \Psi_1 dx = \int_\Omega g(u) \Psi_1 dx. \tag{9}$$

Since $u(x,t)$ is the classical solution of problem in Equation (1), we have

$$0 < u(x,t) < b \ , \ x \in \Omega \ , \ t \in (0, T_{\max}).$$

Hence, by (G3) we have

$$g(u) \ge c_1 + c_2 u \ , \ (x,t) \in \Omega \times (0, T_{\max}) \tag{10}$$

Substituting Equation (10) into Equation (9), we can obtain that

$$\frac{d}{dt}\int_{\Omega} u\Psi_1 dx - \lambda_1^2 \int_{\Omega} u\Psi_1 dx \geq c_1 + c_2 \int_{\Omega} u\Psi_1 dx . \tag{11}$$

Taking $y(t) = \int_{\Omega} u\Psi_1 dx$, we can get that

$$\frac{dy}{dt} \geq c_1 + \left(c_2 + \lambda_1^2\right)y \quad , t \in \left(0, T_{\max}\right). \tag{12}$$

Since

$$0 < y(t) = \int_{\Omega} u\Psi_1 dx < b, t \in (0, T_{\max}),$$

from the condition

$$0 < c_2 + (c_1 / b), \text{ i.e., } c_1 + c_2 b > 0,$$

we infer that

$$c_1 + (c_2 + \lambda_1^2)y > 0 , \quad t \in (0, T_{\max}) . \tag{13}$$

Taking into account Equations (12) and (13), we have

$$\frac{dy}{c_1 + (c_2 + \lambda_1^2)y} \geq dt .$$

And which implies that

$$t \leq \frac{1}{c_2 + \lambda_1^2} \ln \frac{c_1 + (c_2 + \lambda_1^2)y(t)}{c_1 + (c_2 + \lambda_1^2)y(0)} . \tag{14}$$

Let $t \to T_{\max}$, from Equation (14) it follows that

$$T_{\max} \leq \frac{1}{c_2 + \lambda_1^2} \ln \frac{c_1 + (c_2 + \lambda_1^2)y(T_{\max})}{c_1 + (c_2 + \lambda_1^2)y(0)} . \tag{15}$$

Since

$$0 < y(T_{\max}) = \int_{\Omega} u(x, T_{\max})\Psi_1 dx \le b,$$

we can obtain that

$$T_{\max} \le \frac{1}{c_2 + \lambda_1^2} \ln \frac{c_1 + (c_2 + \lambda_1^2)b}{c_1 + (c_2 + \lambda_1^2)m} \tag{16}$$

where $m = y(0) = \int_{\Omega} \phi(x)\Psi_1(x)dx = 0$.

It is obvious that

$$\begin{aligned} T_{\max} &\le \frac{1}{c_2 + \lambda_1^2} \ln \frac{c_1 + c_2 + \lambda_1^2}{c_1} \\ &= \frac{1}{\alpha + \lambda_1^2} \ln \frac{2\alpha + \lambda_1^2}{\alpha} \\ &< +\infty. \end{aligned}$$

To obtain a lower bound of $T_{\max}$ let us consider the initial value problem

$$\begin{cases} \dfrac{d\eta(t)}{dt} = g(\eta), \\ \eta(0) = M, \end{cases} \tag{17}$$

where $M = \sup_{x \in \Omega} \varphi(x) < b$. Since $g(s) > 0$, problem in Equation (17) implies that

$$\int_M^{\eta(T)} \frac{ds}{g(s)} = t. \tag{18}$$

Let t^* be the time for which $\lim_{t \to t^*} \eta(t) = b$, and from Equation (18) we have

$$t^* = \int_M^b \frac{ds}{g(s)}. \tag{19}$$

Obviously, $\eta(t)$ is a super-function of $u(x,t)$, thus

$$T_{\max} \ge t^* = \int_M^b \frac{ds}{g(s)}. \tag{20}$$

Then we can know

$$\frac{\alpha}{1-u} \le g(u) = \frac{\alpha(1+|x|)^m}{1-u} \le \frac{\alpha(1+L)^m}{1-u}$$

and

$$\frac{1-u}{\alpha(1+L)^m} \le \frac{1}{g(u)} \le \frac{1-u}{\alpha}.$$

We can obtain that

$$\int_0^1 \frac{1-s}{\alpha(1+L)^m} ds \le \int_0^1 \frac{ds}{g(s)} \le \int_0^1 \frac{1-s}{\alpha} ds. \tag{21}$$

Through calculation, we have

$$\int_0^1 \frac{ds}{g(s)} \ge \frac{\int_0^1 (1-s)ds}{\alpha(1+L)^m} = \frac{1}{2\alpha(1+L)^m} > 0$$

and

$$\int_0^1 \frac{ds}{g(s)} \le \frac{\int_0^1 (1-s)ds}{\alpha} \le \frac{1}{2\alpha},$$

Then

$$\begin{aligned} T_{\max} &\ge \int_0^1 \frac{ds}{g(s)} \\ &\ge \frac{1}{2\alpha(1+L)^m} \\ &> 0. \end{aligned}$$

Hence $u(x,t)$ quenches in finite time and Theorem 4.1 is proved.

CONCLUSION

In this chapter, we give the quenching phenomena of second-order and fourth-order nonlinear parabolic equations. Through analyzing the characteristics of the absorption term, we do not only find that the absorption satisfies (G1)-(G4) and the solution of problem in

Equations (1) and (2) quenches in finite time but also estimate the quenching time. Further, we extend the result to the case of more generally nonlinear absorption.

ACKNOWLEDGMENTS

This work was supported by the National Natural Science Foundation of China (11101102), Ph.D. Programs Foundation of Ministry of Education of China (20102304120022), the Support Plan for the Young College Academic Backbone of Heilongjiang Province (1252G020), the Natural Science Foundation of Heilongjiang Province (A201014), Science and Technology Research Project of Department of Education of Heilongjiang Province (12521401), Foundational Science Foundation of Harbin Engineering University and Fundamental Research Funds for the Central Universities (HEUCF20131101).

REFERENCES

[1] Kawarada, H. "On solutions of initial-boundary problem for $u_t = u_{xx} + 1/(1-u)$ ", *Publ. RIMS, Kyoto Univ.*, 1975, 10, 729-736.

[2] Levine, HA; Montgomery, JT. "The quenching of solutions of some nonlinear parabolic equations", *SIAM J. Math. Anal.*, 1980, 11, 842-847.

[3] Chiy, YC ; Kwong, MK. "Quenching phenomena for singular nonlinear parabolic equations", *Nonlinear Anal., Theory, Methods & Appl.*, 1988, 12, 1377-1383.

[4] Chiu, YC ; Hans, K. "Quenching for semilinear singular parabolic problems", *SIAM J. Math. Anal.*, 1989, 20, 558-566.

[5] Pinsky, RG. "Existence and Nonexistence of Global Solutions for $u_t = \Delta u + a(x)\, u^p \text{in} R^d$", *J. Differential Equations*, 1997, 33, 152-177.

[6] Acker, A ; Walter, W. *The quenching problem for nonlinear parabolic differential equations,* Lecture Notes in Mathematics Springer-Verlag, New York (1976) p.1-12.

[7] Levin, HA. "Quenching, nonquenching, and beyond quenching for solution of some parabolic equations", *Annali di Matematica Puraed Applicata*, 1989, 155, 243-260.

[8] Zhi YH; Mu, CL. "The quenching behavior of a nonlinear parabolic equation with nonlinear boundary outflux", *Appl. Math. Comput.*, 2007, 184, 624-630.

[9] Marek, F; Kawohl, B ; Levine, HA. "Quenching for quasilinear equations", *Commun. in Partial Differential Equations*, 1992, 17, 593-614.

[10] Boni, TK. "On quenching of solutions for some semilinear parabolic equations of second order", *Bull. Belg. Math. Soc. Simon Stevin*, 2000, 7, 73-95.

[11] Deng, K; Levine, HA. "On the blow up of u_t at quenching", Proc. Amer. Math. Soc., 1989, 106, 1049-1056.

[12] Olmstead, WE; Roberts, CA. "Quenching for the heat equation with a nonlocal nonlinearity", in *Nonlinear Problems in Applied Mathematics*, Cook, A; Olmstead K. (Eds.), SIAM, Philadelphia, 1995, pp.199-205.

[13] Chan, CY. "New results in quenching", Proc. *1st World Congress Nonlinear Anal.*, 1996, 7, 427-434.

[14] Levin, HA. "Advances in quenching", *Progress in Nonlinear Differential Equations and Their Applications,* 1992, 7, 319-346.

[15] Levine, HA. "The phenomenon of quenching: A survey", *North-Holland Mathematics Studies*, 1985, 110, 275-286.

[16] Dai, Q; Gu, Y. "A short note on quenching phenomena for semilinear parabolic equations", *J. Differential Equations*, 1997, 137, 240-250.

[17] Dai, Q; Zeng, X. "The quenching phenomena for the Cauchy problem of semilinear parabolic equations", *J. Differential Equations*, 2001, 175, 163-174.

[18] Kim, TY; Dolbow, JE. "An edge-bubble stabilized finite element method for fourth-order parabolic problems", *Finite Elements in Analysis and Design*, 2009, 45, 485-494.

[19] Yu, H. "Null controllability for a fourth order parabolic equation", *Science in China Series F : Information Sciences*, 2009, 52, 2127-2132.

[20] Liu, YN; Jian, HY. "A curve flow evolved by a fourth order parabolic equation", *Science in China. Series A : Math.*, 2009, 52, 2177-2184.

[21] Byun, SS; Wang, L. "Fourth-order parabolic equations with weak BMO coefficients in Reifenberg domains", *J. Differential Equations*, 2008, 245, 3217-3252.

[22] Jiang, YM; Shi, KH; Wang, YJ. "Large deviation principle for the fourth-order stochastic heat equations with fractional noises", *Acta Math. Sinica*, 2010, 26, 89-106.

[23] Goyal, VB; Schaefer, PW. "On a subharmonic functional in fourth order nonlinear elliptic problems", *J. Math. Anal. Appl.*, 1981, 83, 20-25.

INDEX

#

4*f*, 8, 9, 11, 13, 16, 17, 19, 20, 22, 25, 27, 28, 30, 39, 86
4*f* coherent imaging, 1, 10, 12, 14, 15, 21, 23, 24, 29, 38, 40, 92

A

Abraham, 20
absorption coefficients, 58, 67, 69, 140, 143, 148, 149
accelerator, 192
acquisitions, 81
advancement(s), 2, 6, 35
agencies, 186
amplitude, 3, 6, 7, 8, 9, 11, 13, 16, 21, 22, 25, 28, 31, 79, 80, 102, 103, 136, 150, 156, 158, 159, 160, 161, 162, 163, 164, 165, 173, 175, 179
angular semi-aperture, 97, 98, 99, 104, 105, 109
antisymmetry-breaking, 171, 184, 185
aperture, 9, 11, 12, 13, 14, 15, 16, 18, 21, 24, 25, 26, 27, 28, 31, 32, 50, 51, 61, 64, 67, 68, 69, 70, 71, 72, 75, 78, 80, 82, 88, 93, 95, 96, 98, 99, 102, 105, 107, 109, 110, 114, 119, 120, 131, 132, 134, 135, 136, 137, 138, 139, 140, 142, 148, 149, 152, 153, 154, 155, 156, 157, 158, 159, 161, 162, 165, 168
ASL, 37
assessment, 2
astigmatic, 41, 43, 53, 56, 57, 58, 60, 61, 66, 86
astigmatism, 43, 71
asymmetric, 123, 172, 173, 175, 176, 177, 178, 179, 181, 182, 183, 185, 188
asymmetry, 53, 54, 55, 176
atoms, 196

B

band gap, 40, 81, 85
bandwidth, 20
beam dimension(s), vii, 41, 42, 49, 56, 58, 61, 62
beam distortion, 14, 148
beam profile, 13, 14, 15, 27, 31, 41, 42, 49, 50, 58, 64, 74, 75, 76, 86, 96, 148
beam profiler, 41, 49, 50
beam radius, 22, 41, 42, 43, 49, 50, 51, 61, 63, 64, 73, 74, 75, 78, 120, 132, 134, 149
beam radius measurement, 74
beam shape, 63, 74, 75, 81, 82, 83, 84, 86, 87, 88
beam waist, 18, 33, 46, 53, 64, 65, 69, 82, 86, 96, 97, 98, 133, 135, 150, 151
beams, 11, 19, 21, 26, 35, 37, 41, 44, 46, 48, 53, 54, 58, 59, 61, 62, 69, 71, 78, 86, 96, 112, 114, 123, 132, 148, 168
BEC, 171, 172, 173, 174
Beijing, 95
Bessel function, 66, 127
birefringence, 26
blowup, 195, 196, 197, 198, 210, 211, 212
Bose-Einstein condensate, 171, 172, 187, 188, 189, 190, 191, 192, 193, 196, 212
bosons, 196
boundary value problem, 213, 216
bright ring, 6, 73
Brillouin zone, 171, 175, 185

C

cancer, 2
carbon, 79, 80, 87, 90, 91, 93
Cauchy problem, vii, 197, 198, 199, 200, 201, 202, 203, 204, 205, 206, 208, 209, 210, 211, 212, 213, 230
causality, 5
CCD camera, 5, 7, 11, 12, 13, 14, 15, 17, 21, 22, 24, 27, 28, 41, 49, 81, 82, 83, 86, 87

challenges, 35
characterization, vii, 1, 9, 20, 34, 37, 53, 92, 124, 147, 148, 167
chemical, 145, 174
China, viii, 1, 95, 123, 125, 144, 147, 167, 171, 195, 210, 229, 230
circular Gaussian beam, 41, 42, 47, 50, 53, 60, 61, 76
cleavage, 111, 112, 113
closed aperture, 80, 88, 89, 90, 120
coherence, 124, 191
color, 20, 37, 38
combined power-type nonlinearities, vii, 195, 196, 197, 198, 210, 211
communication, 2
competition, 33
complexity, 15
complications, 24
compounds, 146
computation, 167, 177, 193
computer, 15, 24, 77, 132, 153
computing, vii, 2, 69, 138
condensation, 189, 191, 196
conduction, 126
conductivity, 127, 143
conductor, 124
configuration, 8, 11, 15, 17, 22, 24, 26, 27, 28, 31, 50, 70, 74, 104, 149, 172
Congress, 230
constrained variational problem, 195, 200, 211
contradiction, 202, 217, 223
convergence, 69, 71, 212
cooling, 196
correlation function, 38
cost, 15, 17, 27
critical exponent, vii, 195, 197, 211
critical value, 172
crystalline, 39, 82, 118, 121, 140, 142
crystallization, 121, 124
crystals, 172, 191
CTA, 36
cubic nonlinearity, 65, 171, 185
culture, 7
curvature radius, 64, 76, 77, 132
cylindrical lens, 79, 80, 81, 82, 86, 87, 91

D

damage, 18, 27, 80, 82, 85, 95, 96, 121, 122
data analysis, 135, 138, 144
data processing, 17, 24, 95, 96, 107, 116, 148
database, 10
decay, 20, 23, 56
decomposition, 42, 43, 44, 48, 49, 53, 55, 61, 63, 67, 69, 71, 73, 74, 76, 77, 78, 176
defects, 91
deformation, 76
degenerate, 20, 34, 36, 40, 148, 172
demonstrations, 35
Department of Education, 210, 229
deposition, 145
depth, 152
detection, 34, 37, 40, 74, 117, 118, 119, 120, 136, 137
detection system, 157
deviation, 159, 230
differential equations, 229
diffraction, vii, 5, 16, 22, 25, 31, 42, 63, 64, 65, 66, 67, 68, 72, 73, 74, 77, 78, 82, 88, 97, 111, 118, 119, 123, 132, 133, 136, 145, 150, 152, 159, 164, 168
diffusion, 26, 198
diffusivity, 127
dilute gas, 196
dimensionality, 186
diodes, 111, 124
dipoles, 192
direct measure, vii, 41, 58, 61
disability, 63, 77
dispersion, 5, 34, 36, 145
displacement, 3, 81
distribution, 13, 14, 16, 47, 49, 63, 66, 68, 70, 72, 81, 82, 87, 88, 91, 119, 136, 148, 156, 157
divergence, 69
double-well potential, 171, 172, 185, 187, 188, 190, 192, 193
drawing, 81
dyes, 39

E

effective incident light intensity, 149
EIT, 173
elaboration, 61
electric field, 3, 4, 7, 11, 12, 25, 42, 43, 44, 45, 46, 47, 51, 58, 59, 60, 63, 64, 65, 66, 67, 68, 69, 76, 77, 96, 97, 100, 103, 118, 119, 132, 133, 134, 136, 147, 149, 150, 155, 156
electromagnetic, 3, 69
electron(s), 19, 20, 23, 63
ellipsometry measurement, 143, 144
elliptic Gaussian beam, 47, 66, 72
e-mail, 1, 41, 63, 79, 95, 125, 171, 195, 213
emission, 112, 145

energy, 2, 11, 13, 14, 15, 24, 32, 34, 35, 41, 42, 49, 50, 52, 58, 71, 85, 90, 111, 126, 130, 134, 143, 174, 178, 183, 184, 196, 197, 198, 199, 200
environment, 98, 114, 120, 140
European Community, 91
evidence, 121
evolution, 5, 58, 178, 193, 195, 197
excitation, 2, 13, 31
excited state refraction, 35
experimental condition, 49
expertise, 2
exploitation, 10
external self-action, 133, 150
extinction, 82
extracts, 86

F

fabrication, vii
far field, vii, 6, 15, 18, 24, 25, 26, 31, 32, 41, 42, 47, 49, 58, 61, 64, 67, 69, 71, 72, 74, 75, 77, 79, 80, 86, 87, 148
feedback light, 110, 111, 112, 113, 115, 116, 117, 118
femtosecond, 36, 38, 49, 83, 85, 87, 145
ferromagnetic, 36
fiber(s), 5, 172, 186, 187, 189
field theory, 197
film thickness, 140, 151
films, 1, 10, 118, 121, 124, 140, 142, 168
filters, 13, 91, 189
finite element method, 230
fitting, 13, 15, 17, 23, 32, 34, 49, 51, 52, 53, 56, 70, 71, 72, 74, 82, 85, 89, 91, 96, 104, 105, 107, 108, 109, 119, 121, 138, 140, 141, 142, 143, 148, 153, 154
fluctuations, 41, 42, 49, 58, 192
fluorescence, 35
focal plane, 5, 6, 8, 11, 15, 16, 17, 24, 27, 28, 33, 79, 80, 81, 82, 86, 87, 89, 90, 91, 142
force, 172
formation, 172, 189, 192
formula, 9, 42, 45, 47, 51, 77, 100, 147, 167
Fourier transform, 11, 12, 16, 22, 28, 102
fourth-order nonlinear parabolic equations, vii, 213, 222, 228
four-wave mixing, vii, 36, 148
free-carrier absorption, 34, 130
free-carrier refraction, 34
Fresnel diffraction integrals, 25
Fresnel reflection, vii, 82, 119, 133, 147, 148, 149, 150, 151, 155, 167
Fresnel reflection loss, vii, 119, 133, 147, 148, 149, 151, 155, 167
Fresnel–Kirchhoff diffraction integrals, vii

G

Gaussian beam, 18, 33, 41, 42, 44, 46, 48, 52, 53, 54, 56, 58, 59, 60, 61, 62, 64, 65, 66, 68, 69, 71, 73, 74, 75, 76, 77, 78, 86, 96, 99, 123, 124, 148, 150, 159, 164, 168
Gaussian decomposition, 42, 43, 44, 48, 49, 53, 55, 61, 63, 67, 69, 71, 73, 74, 76, 77, 78
geometry, 18, 138, 172, 196
glasses, 36
global existence, 195, 196, 197, 198, 199, 209, 211, 212
graduate students, viii
gratings, 187
Greece, 41
ground state, 178, 183, 185, 186, 197, 198
growth, 55, 82, 197
Guangdong, 171
Guangzhou, 171

H

halogen, 6
Hamiltonian, 174, 177
harmonic potential, vii, 177, 179, 183, 184, 195, 196, 197, 198, 210, 211, 212
heat capacity, 127, 142
heat conduction, 126
Hilbert space, 198
history, 1, 214
homogeneity, 81
hybrid, 39, 40
hysteresis, 189

I

ideal, 9, 14
illumination, 6, 9, 14, 18, 20, 21, 27, 33
image(s), 5, 6, 7, 8, 9, 11, 13, 14, 15, 17, 24, 29, 30, 31, 38, 81, 86, 87, 88, 92
improvements, 148
incidence, 29, 37, 82, 85, 92, 110, 123, 155
incident angle, 21, 95, 96, 97, 98, 99, 101, 102, 103, 105, 110, 114
induction, 172
inequality, 201, 208
information processing, 10
integration, 13, 17, 24, 74, 77, 102, 225

interaction process, 2, 20, 21, 30, 35
interface, 100, 123, 147, 151, 155, 167, 168
interference, 6, 8, 31, 105, 110, 147, 151, 155, 156, 157, 158, 159, 161, 162, 164, 167
invariant manifold, 195, 198, 201, 204, 205
inversion, 4
ions, 173
IR spectroscopy, 38
irradiation, 8, 10, 26, 82, 85, 87, 88, 89, 90, 95, 116, 121, 126, 130, 131, 142, 148, 154, 155
isotropic media, 38
iteration, 193

J

Josephson junction, 188, 192

K

kinetics, 20, 21, 24, 25

L

laser(s), vii, 2, 3, 8, 9, 10, 11, 12, 13, 14, 15, 16, 18, 19, 20, 21, 23, 26, 27, 28, 29, 32, 33, 36, 38, 41, 43, 49, 52, 53, 60, 61, 63, 66, 69, 71, 78, 79, 80, 81, 84, 85, 86, 87, 88, 89, 90, 91, 92, 97, 99, 102, 103, 105, 107, 108, 109, 110, 111, 112, 113, 114, 115, 116, 117, 120, 121, 122, 124, 125, 126, 127, 128, 129, 130, 131, 132, 133, 134, 136, 137, 140, 141, 142, 143, 144, 145, 148, 149, 150, 151, 152, 153, 154, 155, 164, 168, 172
laser ablation, 118
laser device instability, 95, 96
laser pulse width tunable, vii, 125, 126, 129, 131, 132, 140, 141, 142, 143, 144
laser radiation, 79, 80, 85, 86, 87
lattices, 172, 187, 190, 191, 192
laws, 196
LBA, 49
lead, 32, 77, 80, 110, 118, 147, 198, 217
lens, 6, 8, 11, 15, 16, 17, 24, 26, 27, 33, 43, 44, 49, 69, 79, 80, 81, 82, 86, 87, 88, 91, 92, 95, 96, 97, 98, 99, 104, 105, 108, 109, 110, 114, 118, 132, 142, 145, 148, 149, 153, 164, 166
lens-like, 155, 156, 157, 158, 159, 160, 161, 162, 163
light, vii, 1, 2, 3, 4, 5, 6, 7, 8, 10, 11, 12, 16, 17, 18, 19, 20, 21, 23, 28, 30, 32, 33, 34, 37, 38, 86, 95, 96, 97, 100, 101, 102, 103, 105, 109, 110, 111, 112, 113, 114, 115, 116, 117, 118, 121, 123, 132, 133, 134, 136, 137, 141, 142, 143, 149, 152, 153, 157, 159, 171, 186, 187, 191, 193
light beam, 97, 105, 114, 133, 134, 149, 157
light transmittance, 102
light-emitting diodes, 20
limited transparency, 80
linear transmittance, 13, 22, 51, 69
liquids, 1, 10, 20, 26, 35, 36, 37, 123, 145
localization, 191
Luo, 39, 190
lying, 7

M

magnetostriction, 147
magnitude, 5, 6, 9, 13, 19, 23, 66, 76, 77, 141, 143, 159, 161
manifolds, 195, 198, 200, 201, 203
manipulation, 36
manufacturing, 85
mass, 172, 189
materials, vii, 2, 8, 9, 10, 16, 18, 20, 21, 26, 27, 28, 30, 31, 33, 34, 35, 36, 37, 39, 40, 63, 70, 78, 79, 80, 86, 87, 91, 92, 93, 95, 96, 110, 118, 120, 125, 140, 142, 144, 145, 146, 147, 149
mathematics, 196
matter, 1, 2, 3, 5, 10, 20, 21, 23, 30, 32, 35, 172, 189, 190, 191, 192, 196
Maxwell equations, 3
measurement(s), vii, 1, 2, 3, 5, 8, 9, 10, 14, 15, 16, 17, 18, 19, 20, 21, 24, 25, 26, 27, 30, 34, 35, 36, 37, 38, 39, 40,41, 42, 49, 50, 51, 52, 58, 61, 62, 64, 69, 70, 71, 72, 73, 74, 75, 77, 78, 79, 80, 81, 85, 87, 88, 89, 90, 91, 92, 95, 96, 97, 98, 102, 103, 104, 105, 107, 108, 110, 114, 115, 116, 117, 118, 119, 120, 121, 122,123, 124, 125, 126, 128, 129, 131, 132, 133, 134, 138, 140, 141, 142, 143, 144, 145, 146, 147, 148, 149, 150, 152, 153, 154, 155, 164, 167, 168
media, 3, 4, 37, 80, 86, 87, 145, 146, 172, 175, 191, 212
melting, 121
melting temperature, 145
metal ions, 169
microcavity, 40
microscope, 5, 7, 8
microscopy, 3, 5, 6, 7, 8, 9, 37, 38, 168
Ministry of Education, 210, 229
mixing, 2, 3, 18, 19, 25, 27, 37, 38
models, 71, 172, 188, 197
modifications, 10, 79, 80, 85, 86
molecular dynamics, 38
molecular liquids, 40

molecular reorientation, 87, 90
molecular structure, 38
molecules, 7
morphology, 95, 96, 142
multidimensional, 38, 212
multi-interference, 147, 151, 155, 156, 157, 158, 159, 161, 162, 164, 167
multi-photon absorption, vii, 148
multiple shots, 15, 24, 85
multiple-shot, 11, 18, 27
multiplication, 55

N

nanocrystals, 39
nanofilm, 147, 155, 167
nanomaterials, 1
nanoparticles, 71, 72, 78, 92
nanosecond, 131, 142, 145
nanostructures, 36, 37
nanotechnology, 147
neutral, 13
Nobel Prize, 5
non-degenerate, 20, 34, 40
nonlinear absorption, 9, 10, 13, 16, 18, 19, 20, 21, 23, 25, 26, 29, 34, 37, 38, 41, 53, 55, 56, 58, 62, 65, 67, 69, 70, 71, 82, 87, 88, 90, 91, 92, 93, 96, 98, 103, 104, 105, 107, 108, 110, 116, 119, 121, 124, 130, 131, 132, 134, 135, 138, 139, 140, 142, 143, 144, 145, 146, 147, 148, 149, 152, 154, 156, 157, 159, 164, 165, 167, 168, 213, 229
nonlinear deviation, 63
nonlinear dynamics, 3, 34, 40
nonlinear imaging, 9, 11, 12, 15, 20, 22, 24, 38
nonlinear index characterization method, vii
non-linear lens, 69
nonlinear measurement, 9, 10, 18, 25, 35, 38
nonlinear optical response, 10, 15, 16, 17, 18, 22, 23, 30, 34, 39, 80
nonlinear optics, vii, viii, 1, 2, 3, 10, 19, 35, 36, 38, 39, 172
nonlinear polarization, 2, 3, 4, 5, 8, 20, 26
nonlinear refraction, 8, 9, 10, 14, 17, 18, 21, 23, 24, 25, 26, 29, 30, 32, 33, 34, 37, 38, 39, 40, 41, 53, 54, 61, 63, 65, 67, 69, 70, 71, 76, 78, 79, 80, 82, 85, 86, 88, 91, 92, 93, 119, 123, 129, 130, 131, 132, 134, 135, 138, 140, 141, 144, 147, 148, 149, 150, 151, 152, 153, 154, 155, 157, 167, 168
nonlinear sample, 8, 63, 68, 71, 148, 150
nonlinear Schrödinger equation(s), vii, 174, 178, 197, 211, 212
nonlinear systems, vii, 171, 172
normalized reflectance, 80, 82, 85
normalized transmittance, 13, 14, 16, 17, 18, 22, 23, 25, 31, 32, 42, 69, 79, 88, 89, 90, 103, 105, 107, 110, 120, 137, 138
numerical analysis, 179, 184
numerical aperture, 12, 95, 96, 98, 99, 105, 107, 109, 110, 132, 142

O

one dimension, 172
opaque material, 27, 28, 79, 80
open aperture, 16, 25, 34, 41, 70, 71, 72, 80, 82, 86, 107, 109, 116, 119, 120, 138, 140, 152, 154
optical fiber, 36
optical kerr effect, 25, 37
optical lattice, 172, 188, 190, 191
optical limiters, vii, 10
optical microscopy, 5
optical nonlinearity characterization, vii
optical parameters, 50, 80, 85, 86, 87, 89, 92
optical phase conjugation, vii, 36
optical properties, 2, 4, 10, 14, 18, 27, 29, 34, 37, 39, 92, 148
optimization, 9
optoelectronics, 36
opto-isolator, 95, 114, 115, 116, 117, 118
organic polymers, 10
oscillation, 35, 36, 164
overlap, 182, 183, 186, 196
oxidation, 141

P

parallel, 100
partial differential equations, 196
passing beam, 64, 69, 72, 74
path integrals, 212
peak-valley, 18, 31, 50, 55, 56, 57, 70, 138
permittivity, 42
phase change, 12, 26, 63, 118, 124, 145, 146, 152, 159, 164
phase conjugation, vii, 2, 36
phase contrast, 3, 5, 6, 7, 8, 9, 10, 13, 30, 31, 38
phase object(s), vii, 1, 2, 5, 10, 11, 12, 13, 14, 15, 16, 17, 18, 19, 20, 21, 22, 24, 25, 27, 28, 30, 31, 33, 34, 35, 38, 39, 40, 86, 92
phase shift(s), 6, 7, 13, 28, 42, 43, 44, 45, 50, 51, 54, 55, 56, 57, 58, 59, 63, 64, 65, 66, 67, 68, 70, 72, 73, 74, 77, 86, 118, 123, 133, 135, 138, 156, 157, 158, 159, 160, 161, 162, 163, 164, 165, 166
phase transitions, 188, 190

phase variation, 6, 7, 21, 86, 87, 91, 132, 138, 139, 151, 155, 157
phenomenology, vii, 172, 173
phonons, 192
photo-excitation, 35
photonics, 36, 38, 191
photons, 34
physical mechanisms, 19
physical properties, 2, 35
physics, 1, 5, 39, 41, 193, 196, 197
picosecond, 34
pitch, 172
platform, 107, 110
playing, 196
Poincaré, 201, 211, 212
polar, 7, 13, 173
polarization, 2, 3, 4, 5, 8, 20, 26, 27, 38, 95, 96, 99, 100, 101, 102, 103, 104, 105, 106, 107, 108, 109, 110, 114, 143, 147, 173
polarization azimuth angle, 96, 99, 100, 101, 102, 103, 105, 107, 108, 109, 110
polarization-resolved, 26
polymer(s), 10, 39
polymer solutions, 10
p-polarization, 99, 100, 101, 104, 105, 107, 110, 114
principles, 2, 3, 10, 17, 25, 35, 80, 193, 213
probe, 1, 20, 21, 22, 23, 24, 25, 26, 27, 33, 34, 36, 39, 40, 86, 87, 90, 91, 126, 173
probe beam, 21, 22, 23, 24, 25, 26, 86
propagation, 12, 16, 18, 22, 23, 25, 28, 32, 44, 61, 65, 76, 78, 79, 80, 86, 87, 90, 99, 126, 132, 146, 152, 171, 173, 174, 176, 187
proportionality, 56
proposition, 9
protection, 10
pseudo-nonlinearity, vii, 95, 96, 116, 117, 118, 122
pulse width, 125, 126, 128, 129, 130, 131, 140, 141, 142, 143, 144
pump beam, 21, 22, 23, 25, 26
pump-probe, 1, 20, 21, 22, 23, 24, 25, 26, 27, 33, 34, 39, 87, 126
pure line, 96

Q

quantum mechanics, 196, 197
quantum state, 196
quartz, 145
quenching phenomena, vii, 213, 215, 222, 228, 230

R

radiation, 5, 79, 80, 81, 82, 86, 87, 90, 91
radio, 140
radius, 7, 11, 12, 13, 15, 16, 22, 25, 41, 42, 43, 49, 50, 51, 52, 55, 56, 58, 61, 63, 64, 69, 73, 74, 75, 76, 77, 78, 82, 96, 97, 98, 120, 132, 134, 137, 149, 150, 157, 174
Rayleigh length, 12, 44, 47, 57, 59, 67, 107, 110, 148
readership, 1
real numbers, 214
real time, 190
real-time imaging, 95, 122
reasoning, 98
recall, 175, 180, 215
reference frame, 171, 174
reflected beam, 27, 28, 29, 81, 82, 83
reflected z-scan, 152, 153, 154
reflection z-scan, 27, 37, 40, 80, 92, 96, 123, 148
reflectivity, 113, 118
refraction index, 32, 63, 79, 129
refractive index, 6, 7, 8, 9, 10, 12, 13, 16, 18, 21, 23, 29, 31, 32, 35, 36, 37, 38, 41, 42, 44, 50, 51, 55, 56, 57, 58, 62, 72, 74, 81, 82, 85, 86, 87, 90, 92, 100, 103, 118, 119, 125, 126, 128, 130, 140, 141, 148, 149, 150, 154, 156, 161, 168, 169, 172
refractive index variation, 129
refractive indices, 58, 81, 85, 91
relaxation, 49
relevance, 36, 190
researchers, viii, 27
resolution, 5, 20, 24, 145, 168
resonator, 52, 111, 112, 113, 124
resources, 2
response, 4, 5, 10, 16, 20, 21, 23, 26, 27, 34, 124
response time, 10, 34
restrictions, 27
retardation, 6, 11
reverse saturation absorption, 29
ring pattern, 64, 68, 72, 73, 74
room temperature, 120, 140, 141
rotating ring, vii, 171, 172, 173, 183, 185
RZ-scan, 27

S

sample damage, 95, 118, 119, 120, 122
sapphire, 81, 86, 90
saturation, 29, 192
saturation absorption, 30
scattering, vii, 2, 35, 40, 193, 196, 211

Schrödinger equation, vii, 174, 178, 197, 198, 210, 211, 212
science, 2, 3
scope, 17, 19, 20
second harmonic generation, 2
second-order nonlinear parabolic equation, vii, 213
selectivity, 38
self-defocusing, vii, 69, 72, 74, 75, 77, 171, 173, 184
self-focusing, vii, 69, 168, 171, 173, 185
self-phase modulation, vii
self-refraction, 133, 150
semiconductor(s), 10, 20, 23, 33, 34, 36, 39, 85, 92, 110, 111, 114, 118, 124, 142
sensitivity, 2, 6, 9, 15, 26, 27, 30, 36, 41, 42, 43, 49, 51, 52, 58, 61, 78, 85, 96, 123, 146, 148, 167, 168
shape, 18, 32, 33, 63, 66, 74, 75, 80, 81, 82, 86, 87, 88, 91, 96, 121, 159, 161, 178
showing, 82, 103, 156
Shrödinger equation, vii, 195, 197, 211, 212
signals, 132
signal-to-noise ratio, 10, 15, 26, 27, 43
signs, 39, 175
silica, 36, 87, 169
silicon, 21, 28, 29, 38, 40, 85, 92, 168
silver, 92
simulation(s), 14, 15, 16, 29, 30, 51, 53, 55, 56, 63, 64, 68, 74, 78, 96, 103, 147, 157, 160, 161, 163, 164, 167, 179, 184, 186
single-shot, 1, 3, 15, 19, 20, 24, 27, 30, 79, 80, 86, 87, 91
single–shot technique, vii
smoothness, 197
Sobolev inequality, 202
software, 50
solar cells, 2
solid state, 110
solitons, 172, 173, 180, 187, 188, 190, 191, 192, 193
solution, 25, 32, 33, 38, 70, 174, 175, 176, 177, 178, 179, 184, 199, 200, 201, 202, 203, 204, 205, 206, 208, 209, 210, 211, 212, 214, 215, 216, 217, 218, 219, 220, 221, 222, 224, 225, 228, 229
spatial frequency, 102
special absorption term., vii, 213
specialists, 1
species, 175, 176, 177, 182, 184, 185, 186, 188
spectroscopy, 20, 21, 26, 36, 37, 38, 39, 40
speed of light, 42
s-polarization, 100, 101, 104, 107, 110, 114
spontaneous symmetry breaking phenomenology, vii
stability, 110, 149, 153, 159, 167, 171, 172, 176, 177, 178, 180, 182, 183, 184, 185, 193
state(s), 2, 20, 26, 35, 50, 95, 96, 102, 110, 118, 121, 124, 130, 131, 140, 145, 148, 171, 172, 175, 178, 183, 185, 186, 187, 189, 196, 197, 198
stimulated Brillouin scattering, vii
storage, vii, 10, 118, 120, 125, 145, 147, 153, 191
storage ring, 191
stress, 35, 80, 91
structural transformation, 95, 96, 118, 121, 122, 145
structure, 1, 6, 7, 27, 35, 85, 145, 172, 176, 178, 190
substitution, 67, 175, 177, 219
substrate, 28, 103, 120, 140, 153, 155, 156, 161
superflow, 189
supervision, 49, 58
surface reflection, 149
susceptibility, 4, 18, 38, 168
symmetry, vii, 4, 171, 176, 180, 182, 185, 187, 188, 189, 191, 193
symmetry-breaking, 171, 180, 182, 186
synthesis, 39

T

tau, 171
techniques, vii, 1, 2, 3, 5, 10, 11, 18, 19, 20, 21, 24, 25, 26, 27, 30, 34, 35, 37, 61, 79, 80, 86, 126, 145, 148, 173, 190
technology, 2, 20, 39, 110
temperature, 32, 125, 126, 132, 140, 141, 142, 144
temperature dependence, 143
temperature rise, 127, 130, 131
temporal variation, 137
testing, 110, 120
textbook, 114, 151
thermal characteristic time, 128
thermal diffusivity, 127
thermal energy, 127
thermal lens, 49, 64, 92, 126, 145
thermal-lensing effect, 32, 39
thermally-induced, 126, 128, 129, 130, 131, 140
thermal-optic coefficient, vii
thermo-optic coefficient, 32, 125, 126, 140, 145
thin films, 19, 92, 118, 120, 121, 124, 125, 140, 141, 142, 144, 145, 154
thin sample, 44, 58, 63, 65, 67, 118, 126, 139, 151, 152, 155
third harmonic generation, vii
third-order optical nonlinearities, 1, 18, 31, 32, 35, 39
third-order optical nonlinearity measurement techniques, vii
tilted sample, 79, 80, 81, 85, 86, 87, 91
time-resolved, 1, 15, 20, 21, 22, 23, 24, 25, 26, 27, 34, 36, 37, 40, 62, 80, 96, 123, 148, 168

tissue, 5, 7
toluene, 32, 33
top-hat beam, 11, 13, 14, 30, 37, 123, 168
torus, 171
TPA, 10, 18
transformation, 95, 96, 105, 118, 121, 122, 127, 145, 174, 175
transient lens spectroscopy, 26
transient nonlinear dynamics, 3
translation, 193
transmission, 50, 80, 81, 85, 99, 101, 103, 155, 156, 158, 159, 160, 161, 162, 163, 164, 165, 166, 168, 191
transmittance z-scan, 50, 56, 61, 102, 118, 138
transmitted beam, 15, 63
transmitted z-scan, 152, 154
transparency, vii, 79, 80, 87, 173, 191
transparent, 5, 6, 7, 8, 9, 11, 24, 27, 28, 29, 79, 80, 81, 82, 86, 87, 91, 92, 124
trapped mode, vii, 171, 173, 185, 186
treatment, 40
tungsten, 6
tunneling, 188, 191
twisted pipe waveguide, 171
two-photon-absorption, 10, 34

U

UK, 79
unbounded domain, vii, 213, 215
uniform, 28, 168, 175, 178
unique features, 3, 10
USA, 36, 38

V

vacuum, 42, 103, 140, 143, 198
valley-peak, 50, 74
vapor, 145
variable-temperature, 125, 142, 144
variations, 6, 7, 23, 28, 42, 43, 51, 60, 81, 86, 87, 91, 105, 132, 157, 159, 164
varieties, 172
vector, 4, 7, 8, 100
velocity, 32, 172
versatility, 2
visualization, 5, 9

W

wave number, 43, 82, 150
wave vector, 4, 11, 26
wavelengths, 28, 34
wavelet, 137
wavenumber, 42
wealth, 2
wells, 178, 184
wide-bandgap semiconductor, 40

Z

Zernike phase contrast microscopy, 3, 5, 6, 7, 8, 9
z-scan, 2, 10, 11, 17, 18, 19, 25, 26, 27, 31, 32, 33, 37, 38, 39, 40, 42, 49, 50, 52, 53, 54, 55, 56, 57, 58, 61, 62, 69, 71, 74, 78, 79, 80, 81, 85, 86, 87, 88, 89, 90, 91, 92, 93, 95, 96, 97, 98, 99, 102, 103, 104, 105, 107, 108, 109, 110, 111, 114, 115, 116, 117, 118, 119, 120, 121, 123, 124, 125, 126, 127, 128, 129, 131, 132, 133, 134, 135, 136, 137, 138, 140, 141, 142, 143, 144, 145, 146, 147, 148, 149, 150, 151, 152, 153, 154, 155, 157, 159, 160, 161, 163, 164, 166, 167, 168
z–scan technique, vii, 41, 51